Springer Series in Statistics

Advisors:
P. Bickel, P. Diggle, S. Fienberg, K. Krickeberg,
I. Olkin, N. Wermuth, S. Zeger

Springer
New York
Berlin
Heidelberg
Barcelona
Hong Kong
London
Milan
Paris
Singapore
Tokyo

Springer Series in Statistics

(continued after index)

Michael J. Kolen
Robert L. Brennan

Test Equating

Methods and Practices

With 36 Illustrations

 Springer

Michael J. Kolen
University of Iowa
Iowa City, IA 52242
USA

Robert L. Brennan
University of Iowa
Iowa City, IA 52242
USA

Library of Congress Cataloging in Publication Data
Kolen, Michael J.
    Test equating : methods and practices / Michael J. Kolen,
Robert L. Brennan.
        p.   cm.—(Springer series in statistics)
    Includes bibliographical references.
    ISBN 0-387-94486-9
    1. Examinations—Scoring.   2. Examinations—Interpretation.
3. Examinations—Design and construction.   4. Psychological tests—
Standards.   5. Educational tests and measurements—standards.
I. Brennan, Robert L.   II. Title.   III. Series.
LB3060.77.K65   1995
371.2´71—dc20   95-5883

Printed on acid-free paper.

Production coordinated by Publishing Network and managed by Terry Kornak; manufactur-
ing supervised by Joe Quatela.
Typeset by Asco Trade Typesetting Ltd., Hong Kong.
Printed and bound by Edwards Brothers, Inc., Ann Arbor, MI.
Printed in the United States of America.

9 8 7 6 5 4 3 2 (Corrected second printing, 1999)

ISBN 0-387-94486-9 Springer-Verlag New York Berlin Heidelberg SPIN 10717594

To Amy, Rachel, and Daniel
—M.J.K.

To Cicely and Sean
—R.L.B.

Preface

Until recently, the subject of equating was ignored by most people in the measurement community except for psychometricians, who had responsibility for equating. In the early 1980s, the importance of equating began to be recognized by a broader spectrum of people associated with testing. For example, the AERA, APA, NCME (1985) *Standards for Educational and Psychological Testing* devoted a substantial portion of a chapter to equating, whereas the previous *Standards* did not even list equating in the index.

This increased attention to equating is attributable to at least three developments during the past 15 years. First, there has been an increase in the number and variety of testing programs that use multiple forms of tests, and the testing professionals responsible for such programs have recognized that scores on multiple forms should be equated. Second, test developers and publishers often have referenced the role of equating in arriving at reported scores to address a number of issues raised by testing critics. Third, the accountability movement in education and issues of fairness in testing have become much more visible. These developments have given equating an increased emphasis among measurement professionals and test users.

In addition to statistical procedures, successful equating involves many aspects of testing, including procedures to develop tests, to administer and score tests, and to interpret scores earned on tests. Of course, psychometricians who conduct equating need to become knowledgeable about all aspects of equating. The prominence of equating, along with its interdependence with so many aspects of the testing process, also suggests that test developers and all other testing professionals should be familiar with the concepts, statistical procedures, and practical issues associated with equating.

The need for a book on equating is evident to us from our experiences in equating hundreds of test forms in many testing programs during the past 15 years, in training psychometricians to conduct equating, in conducting seminars and courses on equating, and in publishing on equating and other areas of psychometrics. Our experience suggests that relatively few measurement professionals have sufficient knowledge to conduct equating. Also, many do not fully appreciate the practical consequences of various changes in testing procedures on equating, such as the consequences of many test-legislation initiatives, the use of performance assessments, and the introduction of computerized test administration. Consequently, we believe that measurement professionals need to be educated in equating methods and practices; this book is intended to help fulfill this need. Although several general published references on equating exist (e.g., Angoff, 1971; Harris and Crouse, 1993; Holland and Rubin, 1982; Petersen *et al.*, 1989), none of them provides the broad, integrated, in-depth, and up-to-date coverage in this book.

We anticipate that many of the readers of this book will be advanced graduate students, entry-level professionals, or persons preparing to conduct equating for the first time. Other readers likely will be experienced professionals in measurement and related fields who will want to use this book as a reference. To address these varied audiences, we make frequent use of examples and stress conceptual issues. This book is not a traditional statistics text. Instead, it is a book on equating for instructional use and a reference for practical use that is intended to address both statistical and applied issues. The most frequently used equating methodologies are treated, as well as many of the practical issues involved in equating. Although we are unable to cover all of the literature on equating, we provide many references so that the interested reader may pursue topics of particular interest.

The principal goals of this book are for the reader to understand the principles of equating, to be able to conduct equating, and to interpret the results of equating in reasonable ways. After studying this book, the reader should be able to:

- Understand the purposes of equating and the context in which it is conducted.
- Distinguish between equating and other related methodologies and procedures.
- Appreciate the importance to equating of test development and quality control procedures.
- Understand the distinctions among equating properties, equating designs, and equating methods.
- Understand the fundamental concepts of equating—including equating designs, equating methods, equating error, and statistical assumptions necessary for equating.

- Compute equating functions and choose among methods.
- Interpret results from equating analyses.
- Design reasonable and useful equating studies.
- Conduct equating in realistic testing situations.
- Identify appropriate and inappropriate uses and interpretations of equating results.

We have covered nearly all of the material in this book in a two-semester-hour graduate seminar at The University of Iowa. In our course, we supplemented the materials here with general references (Angoff, 1971; Holland and Rubin, 1982; Petersen *et al.* 1989) so that the students could become familiar with other perspectives and notational schemes. We have used much of the material in this book in training sessions at American College Testing (ACT) and at the annual meetings of the National Council on Measurement in Education and the American Educational Research Association.

A conceptual overview in Chapter 1 provides a general introduction. In this chapter, we define equating, describe its relationship to test development, and distinguish equating from other processes that might lead to comparable scores. We also present properties of equating and equating designs, and introduce the concept of equating error.

In Chapter 2, using the random groups design, we illustrate traditional equating methods, such as equipercentile and linear methods. We also discuss here many of the key concepts of equating, such as properties of converted scores and the influence of the resulting scale scores on the choice of an equating result.

In Chapter 3, we cover smoothing methods in equipercentile equating. We show that the purpose of smoothing is the reduction of error in estimating equating relationships in the population. We describe methods based on log-linear models, cubic splines, and strong true score models.

In Chapter 4, we treat linear equating with nonequivalent groups of examinees. We derive statistical methods and stress the need to disconfound examinee group and test form differences. Also, we distinguish observed score equating from true score equating. We continue our discussion of equating with nonequivalent groups in Chapter 5 with our presentation of equipercentile methods.

In Chapter 6, we describe item response theory (IRT) equating methods under various designs. This chapter covers issues that include scaling person and item parameters, IRT true and observed score equating methods, and equating using item pools.

Chapter 7 focuses on standard errors of equating; both bootstrap and analytic procedures are described. Then we illustrate the use of standard errors to choose sample sizes for equating and to compare the precision in estimating equating relationships for different designs and methods.

In Chapter 8, we describe many practical issues in equating and scaling to achieve comparability, including the importance of test development procedures, test standardization conditions, and quality control procedures. We stress conditions that are conducive to adequate equating. Also, we discuss comparability issues for performance assessments and computerized tests.

We use one random groups illustrative example and a second non-equivalent groups example throughout the book. Each chapter has a set of exercises that are intended to reinforce the concepts and procedures in the chapter. The answers to the exercises are in Appendix A. On request, we will make available the illustrative data sets in the book along with Macintosh computer programs that can be used to conduct many of the analyses. We describe the data sets and computer programs and how to obtain them in Appendix B.

We want to acknowledge the generous contributions that others made to this book. We have benefited from interactions with very knowledgeable psychometricians at ACT and elsewhere, and many of the ideas in this book have come from conversations and interactions with these people. Specifically, we thank Bradley Hanson for reviewing the entire manuscript and making valuable contributions, especially to the statistical presentations. He conducted the bootstrap analyses that are presented in Chapter 7 and, along with Lingjia Zeng, developed much of the computer software used in the examples. Deborah Harris reviewed the entire manuscript, and we thank her especially for her insights on practical issues in equating. Chapters 1 and 8 benefited considerably from her ideas and counsel. Lingjia Zeng also reviewed the entire manuscript and provided us with many ideas on statistical methodology, particularly in the areas of standard errors and IRT equating. Thanks to Dean Colton for his thorough reading of the entire manuscript, Xiaohong Gao for her review and for working through the exercises, and Ronald Cope and Tianqi Han for reading portions of the manuscript.

We are grateful to Nancy Petersen of Educational Testing Service for her in-depth review of an earlier draft, her insights, and her encouragement. Bruce Bloxom, formerly of the Defense Manpower Data Center, provided valuable feedback, as did Barbara Plake and her graduate class at the University of Nebraska–Lincoln. We thank an anonymous reviewer, and the reviewer's graduate student, for providing us with their valuable critique. We are indebted to all who have taken our equating courses and training sessions. Amy Kolen deserves thanks for her patience and superb editorial advice.

Also, we acknowledge Richard L. Ferguson—President, ACT—and Thomas Saterfiel—Vice President, Research, ACT—for their support.

Iowa City, Iowa   Michael J. Kolen
                                                           Robert L. Brennan

Contents

Notation*

Arabic Letters

* Chapter where first introduced shown in parentheses.

$d_Y(x)$ = expected value of a cubic spline estimator of $e_Y(x)$ (Chapter 3)
$\mathbf{E}$ = expected value   (Chapter 1)
E = number correct error score   (Chapter 4)
e = the equipercentile equating function, such as $e_Y(x)$ (Chapter 2)
eq = general equating function, such as $eq_Y(x)$ (Chapter 1)
$e_Y(x)$ = the Form Y equipercentile equivalent of a Form X
       score   (Chapter 1)
$e_X(y)$ = the Form X equipercentile equivalent of a Form Y
       score   (Chapter 2)
$\exp$ = exponential   (Chapter 6)
$F(x) = Pr(X \leq x)$ is the cumulative distribution for X (Chapter 1)
F^* = cumulative distribution function of $eq_X(y)$ (Chapter 2)
F^{-1} = inverse of function F (Chapter 2)
f = a general function   (Chapter 7)
f' = the first derivative of f (Chapter 7)
$f(x) = Pr(X = x)$ is the discrete density for X (Chapter 2)
$f(x,v) = Pr(X = x$ and $V = v)$ is the joint density of X and V (Chapter 5)
$f(x|v) = Pr(X = x$ given $V = v)$ is the conditional density of
      x given v (Chapter 5)
$func$ = function solved for in Newton-Raphson iterations   (Chapter 6)
$func'$ = first derivative of function solved for in Newton-
      Raphson iterations   (Chapter 6)
g = item subscript in IRT   (Chapter 6)
$G(y) = Pr(Y \leq y)$ is the cumulative distribution for Y (Chapter 1)
$g(y) = Pr(Y = y)$ is the discrete density for Y (Chapter 2)
$g(y,v) = Pr(Y = y$ and $V = v)$ is the joint density of Y and V (Chapter 5)
$g(y|v) = Pr(Y = y$ given $V = v)$ is the conditional density of
      y given v (Chapter 5)
G^* = the cumulative distribution function of e_Y (Chapter 1)
G^{-1} = inverse of function G (Chapter 2)
g_{adj} = density adjusted by adding 10^{-6} to each density and
      then standardizing   (Chapter 2)
$h(v) = Pr(V = v)$ is the discrete density for V (Chapter 5)
$Hcrit$ = criterion function for Haebara's method   (Chapter 6)
$Hdiff$ = difference function for Haebara's method   (Chapter 6)
I = IRT scale   (Chapter 6)
i and i' = individuals   (Chapter 6)
$intercept$ = intercept of an equating function   (Chapter 2)
irt = IRT true-score equating function   (Chapter 6)
J = IRT scale   (Chapter 6)
j and j' = items   (Chapter 6)
K = number of items   (Chapter 2)
k = Lord's k in the Beta4 method   (Chapter 3)
ku = kurtosis, such as $ku(X) = \mathbf{E}[X - \mu(X)]^4/\sigma^4(X)$ (Chapter 2)
$l_Y(x)$ = the Form Y linear equivalent of a Form X score   (Chapter 2)
$l_X(y)$ = the Form X linear equivalent of a Form Y score   (Chapter 2)

mse = mean squared error   (Chapter 3)
$m_Y(x)$ = the mean equating equivalent of a Form X score   (Chapter 2)
$m_X(y)$ = the mean equating equivalent of a Form Y score   (Chapter 2)
N = number of examinees   (Chapter 3)
p = probability of a correct response in IRT   (Chapter 6)
$P(x)$ = the percentile rank function for X (Chapter 2)
P^* = a given percentile rank   (Chapter 2)
P^{**} = P/100   (Chapter 7)
P^{-1} = the percentile function for X (Chapter 2)
p' = first derivative of p (Chapter 6)
$Q(y)$ = the percentile rank function for Y (Chapter 2)
Q^{-1} = the percentile function for Y (Chapter 2)
R = number of bootstrap replications   (Chapter 7)
r = index for calculating observed score distribution in IRT   (Chapter 6)
r = index for bootstrap replications   (Chapter 7)
$regression\ intercept$ = intercept constant in linear regression   (Chapter 5)
$regression\ slope$ = slope constant in linear regression   (Chapter 5)
S = smoothing parameter in postsmoothing   (Chapter 3)
s = synthetic population   (Chapter 4)
sc = scale score transformation, such as $sc(y)$ (Chapter 2)
sc_{int} = scale score rounded to an integer   (Chapter 2)
se = standard error, such as $se(x)$ is the standard error at
    score x (Chapter 3)
sem = standard error of measurement   (Chapter 7)
sk = skewness, such as $sk(X) = \mathbf{E}[X - \mu(X)]^3/\sigma^3(X)$ (Chapter 2)
$SLcrit$ = criterion function for Stocking-Lord method   (Chapter 6)
$SLdiff$ = difference function for Stocking-Lord method   (Chapter 6)
$slope$ = slope of equating function   (Chapter 2)
T = number correct true score   (Chapter 4)
t = realization of number correct true score   (Chapter 4)
$t_Y(x)$ = expected value of an alternate estimator of $e_Y(x)$ (Chapter 3)
U = uniform random variable   (Chapter 2)
u = standard deviation units   (Chapter 7)
V = the random variable indicating raw score on Form V   (Chapter 4)
v = a realization of V (Chapter 4)
var = sampling variance   (Chapter 3)
w = weight for synthetic group   (Chapter 4)
X = the random variable indicating raw score on Form X   (Chapter 1)
x = a realization of X (Chapter 1)
$X^* = X + U$ used in the continuization process   (Chapter 2)
x^* = integer closest to x such that $x^* - .5 \leq x < x^* + .5$ (Chapter 2)
x^* = Form X_2 score equated to the Form X_1 scale   (Chapter 7)
x_{high} = upper limit in spline calculations   (Chapter 3)
x_L^* = the largest integer score with a cumulative percent less
    than P^* (Chapter 2)
x_{low} = lower limit in spline calculations   (Chapter 3)

$x_U^* =$ the smallest integer score with a cumulative percent
    greater than P^* (Chapter 2)
$Y =$ the random variable indicating raw score on Form Y   (Chapter 1)
$y =$ a realization of Y (Chapter 1)
$y_i^* =$ largest tabled raw score less than or equal to $e_Y(x)$ in
    finding scale scores   (Chapter 2)
$y_L^* =$ the largest integer score with a cumulative percent less
    than Q^* (Chapter 2)
$y_U^* =$ the smallest integer score with a cumulative percent
    greater than Q^* (Chapter 2)
$Z =$ the random variable indicating raw score on Form Z   (Chapter 4)
$z =$ a realization of Z (Chapter 4)
$z =$ unit normal variable   (Chapter 7)

Greek Letters

$\alpha(X|V)$ and $\alpha(Y|V) =$ linear regression slopes   (Chapter 4)
$\beta(X|V)$ and $\beta(Y|V) =$ linear regression intercepts   (Chapter 4)
$\chi^2 =$ chi-square test statistic   (Chapter 3)
$\delta =$ location parameter in congeneric models   (Chapter 4)
$\phi =$ normal ordinate   (Chapter 7)
$\gamma =$ expansion factor in linear equating with the common-
    item nonequivalent groups design   (Chapter 4)
$\lambda =$ effective test length in congeneric models   (Chapter 4)
$\mu =$ mean as in $\mu(X)$ and $\mu(Y)$ (Chapter 2)
$\Theta =$ parameter used in developing the delta method   (Chapter 7)
$\theta =$ ability in IRT   (Chapter 6)
$\theta^+ =$ new value in Newton-Raphson iterations   (Chapter 6)
$\theta^- =$ initial value in Newton-Raphson iterations   (Chapter 6)
$\rho =$ correlation, such as $\rho(X, V)$ (Chapter 4)
$\rho(X, X') =$ reliability   (Chapter 4)
$\sigma(X, V)$ and $\sigma(Y, V) =$ covariance   (Chapter 4)
$\sigma^2 =$ variance such as $\sigma^2(X) = \mathbf{E}[X - \mu(X)]^2$ (Chapter 2)
$\tau =$ true score   (Chapter 1)
$\tau^* =$ true score outside of range of possible true scores   (Chapter 6)
$\upsilon =$ spline coefficient   (Chapter 3)
$\omega =$ weight in log-linear smoothing   (Chapter 3)
$\Psi =$ function that relates true scores   (Chapter 4)
$\psi =$ distribution of a latent variable   (Chapter 3)
$\partial =$ partial derivative   (Chapter 7)

CHAPTER 1

Introduction and Concepts[1]

This chapter provides a general overview of equating and briefly considers important concepts. The concept of equating is described, as is why it is needed, and how to distinguish it from other related processes. Equating properties and designs are considered in detail, because these concepts provide the organizing themes for addressing the statistical methods treated in subsequent chapters. Some issues in evaluating equating are also considered. The chapter concludes with a preview of subsequent chapters.

Equating and Related Concepts

Scores on tests often are used as one piece of information in making important decisions. Some of these decisions focus at the *individual level*, such as when a student decides which college to attend or the course in which to enroll. For other decisions the focus is more at an *institutional level*. For example, an agency or institution might need to decide what test score is required to certify individuals for a profession or to admit students into a college, university, or the military. Still other decisions are made at the *public policy level*, such as addressing what can be done to improve education in the United States and how changes in educational practice can be evaluated. Regardless of the type of decision that is to be made, it should be based on the most accurate information possible: All other things being equal, *the more accurate the information, the better the decision.*

Making decisions in many of these contexts requires that tests be ad-

---

[1] Some of the material in this chapter is based on Kolen (1988).

ministered on multiple occasions. For example, college admissions tests typically are administered on particular days, referred to as *test dates*, so examinees can have some flexibility in choosing when to be tested. Tests also are given over many years to track educational trends over time. If the same test questions were routinely administered on each test date, then examinees might inform others about the test questions. Or, an examinee who tested twice might be administered the same test questions on the two test dates. In these situations, a test might become more of a measure of exposure to the specific questions that are on the test than of the construct that the test is supposed to measure.

Test Forms and Test Specifications

These test security problems can be addressed by administering a different collection of test questions, referred to as a *test form*, to examinees who test on different test dates. A test form is a set of test questions that is built according to content and statistical *test specifications* (Millman and Greene, 1989). Test specifications provide guidelines for developing the test. Those responsible for constructing the test, the *test developers*, use these specifications to ensure that the test forms are as similar as possible to one another in content and statistical characteristics.

Equating

The use of different test forms on different test dates leads to another concern: the forms might differ somewhat in difficulty. *Equating* is a statistical process that is used to adjust scores on test forms so that scores on the forms can be used interchangeably. Equating adjusts for differences in difficulty among forms that are built to be similar in difficulty and content.

The following situation is intended to develop further the concept of equating. Suppose that a student takes a college admissions test for the second time and earns a higher reported score than on the first testing. One explanation of this difference is that the reported score on the second testing reflects a higher level of achievement than the reported score on the first testing. However, suppose that the student had been administered exactly the same test questions on both testings. Rather than indicating a higher level of achievement, the student's reported score on the second testing might be inflated because the student had already been exposed to the test items. Fortunately, a new test form is used each time a test is administered for most college admissions tests. Therefore, a student would not likely be administered the same test questions on any two test dates.

The use of different test forms on different test dates might cause another problem, as is illustrated by the following situation. Two students apply for the same college scholarship that is based partly on test scores. The two students take the test on different test dates, and Student 1 earns a higher reported score than Student 2. One possible explanation of this difference is that Student 1 is higher achieving than Student 2. However, if Student 1 took an easier test form than Student 2, then Student 1 would have an unfair advantage over Student 2. In this case, the difference in scores might be due to differences in the difficulty of the test forms rather than in the achievement levels of the students. To avoid this problem, equating is used with most college admissions tests. If the test forms are successfully equated, then the difference in equated scores for Student 1 and Student 2 is not attributable to Student 1 taking an easier form.

The process of equating is used in situations where such *alternate forms* of a test exist and scores earned on different forms are compared to each other. Even though test developers attempt to construct test forms that are as similar as possible to one another in content and statistical specifications, the forms typically differ somewhat in difficulty. Equating is intended to adjust for these difficulty differences, allowing the forms to be used interchangeably. *Equating adjusts for differences in difficulty, not for differences in content.* After successful equating, for example, examinees who earn an equated score of, say, 26 on one test form could be considered, on average, to be at the same achievement level as examinees who earn an equated score of 26 on a different test form.

Processes That Are Related to Equating

There are processes that are similar to equating, which may be more properly referred to as *scaling to achieve comparability*, in the terminology of the *Standards for Educational and Psychological Testing* (AERA, APA, NCME, 1985), or *linking*, in the terminology of Linn (1993) and Mislevy (1992). One of these processes is *vertical scaling* (frequently referred to as vertical *"equating"*), which often is used with elementary school achievement test batteries. In these batteries, students often are administered questions that test content matched to their current grade level. This procedure allows developmental scores (e.g., grade equivalents) of examinees at different grade levels to be compared. Because the content of the tests administered to students at various educational levels is different, however, scores on tests intended for different educational levels cannot be used interchangeably. Other examples of scaling to achieve comparability include relating scores on one test to those on another, and scaling the tests within a battery so that they all have the same distributional characteristics. As with vertical scaling, solutions to these problems do not allow test scores to be used interchangeably, because the content of the tests is different.

Although similar statistical procedures often are used in scaling to achieve comparability and equating, their purposes are different. Whereas tests that are purposefully built to be different are scaled to achieve comparability, equating is used to adjust scores on test forms that are built to be as similar as possible in content and statistical characteristics. When equating is successful, scores on alternate forms can be used interchangeably. Issues in linking tests that are not built to the same specifications are considered further in Chapter 8.

Equating and Score Scales

On a multiple-choice test, the *raw score* an examinee earns is often the number of items the examinee answers correctly. Other raw scores might involve penalties for wrong answers or weighting items differentially. On tests that require ratings by judges, a raw score might be the sum of the numerical ratings made by the judges.

Raw scores often are transformed to *scale scores*. The *raw-to-scale score transformation* can be chosen by test developers to enhance the interpretability of scores by incorporating useful information into the score scale (Petersen *et al.*, 1989). Information based on a nationally representative group of examinees, referred to as a *national norm group*, sometimes is used as a basis for establishing score scales. For example, the number-correct scores for the four tests of the initial form of a revised version of the ACT Assessment were scaled (Brennan, 1989) to have a mean scale score of 18 for a nationally representative sample of college-bound 12th graders. Thus, an examinee who earned a scale score of 22, for example, would know that this score was above the mean scale score for the nationally representative sample of college-bound 12th graders used to develop the score scale. One alternative to using nationally representative norm groups is to base scale score characteristics on a *user norm group*, which is a group of examinees that is administered the test under operational conditions. For example, Cook (1994) and Dorans (1994a) indicated that a rescaled SAT scale will be established in 1995 by setting the mean score equal to 500 for the group of SAT examinees that graduated from high school in 1990.

Scaling and Equating Process. Equating can be viewed as an aspect of a more general *scaling and equating process*. Score scales typically are established using a single test form. For subsequent test forms, the scale is maintained through an equating process that places raw scores from subsequent forms on the established score scale. In this way, a scale score has the same meaning regardless of the test form administered or the group of examinees tested. Typically, raw scores on the new form are equated to raw scores on the old form, and these equated raw scores are then converted to scale scores using the raw-to-scale score transformation for the old form.

Table 1.1. Hypothetical Conversion Tables for Test Forms.

Scale	Form Y Raw	Form X_1 Raw	Form X_2 Raw
.	.	.	.
.	.	.	.
.	.	.	.
13	26	27	28
14	27	28	29
14	28	29	30
15	29	30	31
15	30	31	32
.	.	.	.
.	.	.	.
.	.	.	.

Example of the Scaling and Equating Process. The hypothetical conversions shown in Table 1.1 illustrate the scaling and equating process. The first two columns show the relationship between Form Y raw scores and scale scores. For example, a raw score of 28 on Form Y converts to a scale score of 14. (At this point there is no need to be concerned about what particular method was used to develop the raw-to-scale score transformation.) The relationship between Form Y raw scores and scale scores shown in the first two columns involves scaling—not equating, because Form Y is the only form that is being considered so far.

Assume that an equating process indicates that Form X_1 is 1 point easier than Form Y throughout the score scale. A raw score of 29 on Form X_1 would thus reflect the same level of achievement as a raw score of 28 on Form Y. This relationship between Form Y raw scores and Form X_1 raw scores is displayed in the second and third columns in Table 1.1. What scale score corresponds to a Form X_1 raw score of 29? A scale score of 14 corresponds to this raw score, because a Form X_1 raw score of 29 corresponds to a Form Y raw score of 28, and a Form Y raw score of 28 corresponds to a scale score of 14.

To carry the example one step further, assume that Form X_2 is found to be uniformly 1 raw score point easier than Form X_1. Then, as illustrated in Table 1.1, a raw score of 30 on Form X_2 corresponds to a raw score of 29 on Form X_1, which corresponds to a raw score of 28 on Form Y, which corresponds to a scale score of 14. Later, additional forms could be converted to scale scores by a similar chaining process. The result of a successful scaling and equating process is that scale scores on all forms can be used interchangeably.

Possible Alternatives to Equating. Equating has the potential to improve score reporting and interpretation of tests that have alternate forms when examinees administered different forms are evaluated at the same time, or when score trends are to be evaluated over time. When at least one of these characteristics is present, at least two possible, but typically unacceptable, alternatives to equating exist. One alternative is to report raw scores regardless of the form administered. As was the case with Student 1 and Student 2 considered earlier, this approach could cause problems because examinees who were administered an easier form are advantaged and those who were administered a more difficult form are disadvantaged. As another example, suppose that the mean score on a test increased from 27 one year to 30 another year, and that different forms of the test were administered in the two years. Without additional information, it is impossible to determine whether this 3-point score increase is attributable to differences in the difficulty of the two forms, differences in the achievement level of the groups tested, or some combination of these two factors.

A second alternative to equating is to convert raw scores to other types of scores so that certain characteristics of the score distributions are the same across all test dates. For example, for a test with two test dates per year, say in February and August, the February raw scores might be converted to scores having a mean of 50 among the February examinees, and the August raw scores might be converted to have a mean of 50 among the August examinees. Suppose, given this situation, that an examinee somehow knew that August examinees were higher achieving, on average, than February examinees. In which month should the examinee take the test to earn the highest score? Because the August examinees are higher achieving, a high converted score would be more difficult to get in August than in February. Examinees who take the test in February, therefore, would be advantaged. Under these circumstances, examinees who take the test with a lower achieving group are advantaged, and examinees who take the test with a higher achieving group are disadvantaged. Furthermore, trends in average examinee performance cannot be addressed using this alternative because the average converted scores are the same regardless of the achievement level of the group tested.

Successfully equated scores are not affected by the problems that occur with these two alternatives. Successful equating adjusts for differences in the difficulty of test forms; the resulting equated scores have the same meaning regardless of when or to whom the test was administered.

Equating and the Test Score Decline of the 1960s and 1970s

The importance of equating in evaluating trends over time is illustrated by issues surrounding the substantial decline in test scores in the 1960s and 1970s. A number of studies were undertaken to try to understand the causes for this decline. (See, for example, Advisory Panel on the Scholastic

Aptitude Test Score Decline, 1977; Congressional Budget Office, 1986; and Harnischfeger and Wiley, 1975). One of the potential causes that was investigated was whether the decline was attributable to inaccurate equating. The studies concluded that the equating was adequate. Thus, the equating procedures allowed the investigators to rule out changes in test difficulty as being the reason for the score decline. Next the investigators searched for other explanations. These explanations included changes in how students were being educated, changes in demographics of test takers, and changes in social and environmental conditions. It is particularly important to note that the search for these other explanations was made possible because equating ruled out changes in test difficulty as the reason for the score decline.

Equating in Practice—A Brief Overview of This Book

So far, what equating is and why it is important have been described in general terms. Clearly, equating involves the implementation of statistical procedures. In addition, as has been stressed, equating requires that all test forms be developed according to the same content and statistical specifications. Equating also relies on adequate test administration procedures, so that the collected data can be used to judge accurately the extent to which the test forms differ statistically. In our experience, the most challenging part of equating often is ensuring that the test development, test administration, and statistical procedures are coordinated. The following is a list of steps for implementing equating (the order might vary in practice):

1. *Decide on the purpose for equating.*
2. *Construct alternate forms.* Alternate test forms are constructed in accordance with the same content and statistical specifications.
3. *Choose a design for data collection.* Equating requires that data be collected for providing information on how the test forms differ statistically.
4. *Implement the data collection design.* The test is administered and the data are collected as specified by the design.
5. *Choose one or more operational definitions of equating.* Equating requires that a choice be made about what types of relationships between forms are to be estimated. For example, this choice might involve deciding on whether to implement linear or nonlinear equating methods.
6. *Choose one or more statistical estimation methods.* Various procedures exist for estimating a particular equating relationship. For example, in Chapter 4, linear equating relationships are estimated using the Tucker and Levine methods.

7. *Evaluate the results of equating.* After equating is conducted, the results need to be evaluated. Some evaluation procedures are discussed along with methods described in Chapters 2–6. The test development process, test administration, statistical procedures, and properties of the resulting equating are all components of the evaluation, as is discussed in Chapter 8.

As these steps in the equating process suggest, individuals responsible for conducting equating make choices about designs, operational definitions, statistical techniques, and evaluation procedures. In addition, various practical issues in test administration and quality control are often vital to successful equating.

In practice, equating requires considerable judgment on the part of the individuals responsible for conducting equating. General experience and knowledge about equating, along with experience in equating for tests in a testing program, are vital to making informed judgments. As a statistical process, equating also requires the use of statistical techniques. Therefore, conducting equating involves a mix of practical issues and statistical knowledge. This book treats both practical issues and statistical concepts and procedures.

Many of the changes that have taken place in the literature on equating over the last 10 to 15 years are reflected in this book. Although the vast literature that has developed is impossible to review in a single volume, this book provides many references that should help the reader access the literature on equating. We recommend that the classic work by Angoff (1971) be consulted as a supplement to this book for its treatment of many of the issues in traditional equating methods and for its perspective on equating. Works by Harris (1993), Harris and Crouse (1993), Holland and Rubin (1982), and Petersen *et al.* (1989) also should be consulted as supplements.

This book is intended to describe the concept of test form equating, to distinguish equating from other similar processes, to describe techniques used in equating, and to describe various practical issues involved in conducting equating. These purposes are addressed by describing information, techniques, and resources that are necessary to understand the principles of equating, to design and conduct an informed equating, and to evaluate the results of equating in reasonable ways.

Subsequent sections of this chapter focus on equating properties and equating designs, which are required concepts for Chapters 2–6. Equating error and evaluation of equating methods also are briefly discussed. Specific operational definitions and statistical estimation methods are the focus of Chapters 2–6. Equating error is described in Chapters 7 and 8. Practical issues in equating, along with new directions, are also discussed in Chapter 8.

Properties of Equating

Various desirable properties of equating relationships have been proposed in the literature (Angoff, 1971; Harris and Crouse, 1993; Lord, 1980; Petersen *et al.*, 1989). Some properties focus on individuals' scores, others on distributions of scores. At the individual level, ideally, an examinee taking one form would earn the same reported score regardless of the form taken. At the distribution level, for a group of examinees, the same proportion would earn a reported score at or below, say, 26 on Form X as they would on Form Y. These types of properties have been used as the principal basis for developing equating procedures.

Some properties focus on variables that cannot be directly observed, such as *true scores* in *classical test theory* (Lord and Novick, 1968) and *latent abilities* in *item response theory* (*IRT*) (Lord, 1980). True scores and latent abilities are scores that an examinee would have earned had there been no measurement error. For example, in classical test theory the score that an examinee earns, the examinee's *observed score*, is viewed as being composed of the examinee's true score and measurement error. It is assumed that if the examinee could be measured repeatedly, then measurement error would, on average, equal zero. Statistically, the true score is the expected score over replications. Because the examinee is not measured repeatedly in practice, the examinee's true score is not directly observed. Instead, the true score is modeled using a test theory model.

Other equating properties focus on observed scores. Observed score properties of equating do not rely on test theory models.

Symmetry Property

The *symmetry property* (Lord, 1980), which requires that equating transformations be symmetric, is required for a relationship to be considered an equating relationship. This property requires that the function used to transform a score on Form X to the Form Y scale be the inverse of the function used to transform a score on Form Y to the Form X scale. For example, this property implies that if a raw score of 26 on Form X converts to a raw score of 27 on Form Y, then a raw score of 27 on Form Y must convert to a raw score of 26 on Form X. This symmetry property rules out regression as an equating method, because the regression of Y on X is, in general, different from the regression of X on Y. As a check on this property, an equating of Form X to Form Y and an equating of Form Y to Form X could be conducted. If these equating relationships are plotted, then the symmetry property requires that these plots be indistinguishable. Symmetry is considered again in Chapter 2.

Same Specifications Property

As indicated earlier, test forms must be built to the same content and statistical specifications if they are to be equated. Otherwise, regardless of the statistical procedures used, the scores can not be used interchangeably. This *same specifications property* is essential if scores on alternate forms are to be considered interchangeable.

Equity Properties

Lord (1980, p. 195) proposed *Lord's equity property* of equating, which is based on test theory models. For Lord's equity property to hold, it must be a matter of indifference to each examinee whether Form X or Form Y is administered.

Lord defined this property specifically. Lord's equity property holds if examinees with a given true score have the same distribution of converted scores on Form X as they would on Form Y. To make the description of this property more precise, define

τ as the true score;

Form X as the new form, let X represent the random variable score on Form X, and let x represent a particular score on Form X (i.e., a realization of X);

Form Y as the old form, let Y represent the random variable score on Form Y, and let y represent a particular score on Form Y (i.e., a realization of Y);

G as the cumulative distribution of scores on Form Y for the population of examinees;

eq_Y as an equating function that is used to convert scores on Form X to the scale of Form Y; and

G^* as the cumulative distribution of eq_Y for the same population of examinees.

Lord's equity property holds in the population if

$$G^*[eq_Y(x)|\tau] = G(y|\tau), \text{ for all } \tau. \qquad (1.1)$$

This property implies that examinees with a given true score would have identical observed score means, standard deviations, and distributional shapes of converted scores on Form X and scores on Form Y. In particular, the identical standard deviations imply that the conditional standard error of measurement at any true score are equal on the two forms. If, for example, Form X measured somewhat more precisely at high scores than Form Y, then Lord's equity property would not be met.

Lord (1980) showed that, under fairly general conditions, Lord's equity property specified in equation (1.1) is possible only if Form X and Form Y are essentially identical. However, identical forms typically cannot be constructed in practice. Furthermore, if identical forms could be constructed, then there would be no need for equating. Thus, *using Lord's equity property as the criterion, equating is either impossible or unnecessary.*

Morris (1982) suggested a less restrictive version of Lord's equity property that might be more readily achieved, which is referred to as the *first-order equity property* or *weak equity property* (also see Yen, 1983). Under the first-order equity property, examinees with a given true score have the same mean converted score on Form X as they have on Form Y. Defining E as the expectation operator, an equating achieves the first-order equity property if

$$\mathbf{E}[eq_Y(X)|\tau] = \mathbf{E}(Y|\tau) \text{ for all } \tau. \tag{1.2}$$

The first-order equity property implies that examinees are expected to earn the same equated score on Form X as they would on Form Y. Suppose examinees with a given true score earn, on average, a score of 26 on Form Y. Under the first-order equity property, these examinees also would earn, on average, an equated score of 26 on Form X.

As is described in Chapter 4, linear methods have been developed that are consistent with the first-order equity property. Also, the IRT true score methods that are discussed in Chapter 6 are related to this equity property. The equating methods that are based on equity properties are closely related to other psychometric procedures, such as models used to estimate reliability. These methods make explicit the requirement that the two forms measure the same achievement through the true score.

Observed Score Equating Properties

In observed score equating, the characteristics of score distributions are set equal for a specified *population of examinees* (Angoff, 1971). For the *equipercentile equating property*, the converted scores on Form X have the same distribution as scores on Form Y. More explicitly, this property holds, for the *equipercentile equating function*, e_Y, if

$$G^*[e_Y(x)] = G(y), \tag{1.3}$$

where G^* and G were defined previously. The equipercentile equating property implies that the cumulative distribution of equated scores on Form X is equal to the cumulative distribution of scores on Form Y.

Suppose a passing score was set at a *scale score* of 26. If the equating of the forms achieved the equipercentile equating property, then the proportion of examinees in the population earning a scale score below 26 on Form X would be the same as the proportion of examinees in the population earning a scale score below 26 on Form Y. In addition, in the population, the same proportion of examinees would score below any particular scale score, regardless of the form taken. For example, if a scale score of 26 was chosen as a passing score, then the same proportion of examinees in the population would pass using either Form X or Form Y.

The equipercentile equating property is the focus of the equipercentile equating methods described in Chapters 2, 3, and 5 and the IRT observed score equating method described in Chapter 6. Two other observed score equating properties also may be used sometimes. Under the *mean equating property*, converted scores on the two forms have the same mean. This property is the focus of the mean observed score equating methods described in Chapter 2. Under the *linear equating property*, converted scores on the two forms have the same mean and standard deviation. This property is the focus of the linear observed score methods described in Chapters 2, 4, and 5. When the equipercentile equating property holds, the linear and mean equating properties must also hold. When the linear equating property holds, the mean equating property also must hold.

Observed score equating methods associated with the observed score properties of equating predate other methods, which partially explains why they have been used more often. Observed score methods do not directly consider true scores or other unobservable variables, and in this way they are less complicated. As a consequence, however, nothing in the statistical machinery of observed score equating requires that test forms be built to the same specifications. This requirement is added so that results from equating may be reasonably and usefully interpreted.

Group Invariance Property

Under the *group invariance property*, the equating relationship is the same regardless of the group of examinees used to conduct the equating. For example, if the group invariance property holds, the same equating relationship would be found for females and males. Lord and Wingersky (1984) indicated that methods based on observed score properties of equating are not strictly group invariant. However, research on the group invariance property conducted by Angoff and Cowell (1986) and Harris and Kolen (1986) suggested that the conversions are very similar across various examinee groups, at least in those situations where carefully constructed alternate forms are equated. Lord and Wingersky (1984) indicated that, under certain theoretical conditions, true score equating methods are group invariant. However, group invariance does not necessarily hold for these methods when observed scores are substituted for true scores. Because group invariance cannot be assumed to exist in the strictest sense, the population of examinees on which the equating relationship is developed should be clearly stated and representative of the group of examinees who are administered the test.

Equating Designs

A variety of designs can be used for collecting data for equating. The group of examinees included in an equating study should be reasonably representative of the group of examinees who will be administered the test under typical test administration conditions. The choice of a design involves both practical and statistical issues. Three commonly used designs are illustrated in Figure 1.1. Assume that a conversion from Form Y to scale scores has been developed, and that Form X is a new form to be equated to Form Y.

Random Groups Design

The *random groups design* is the first design shown in Figure 1.1. In this design, examinees are randomly assigned the form to be administered.

A *spiraling* process is one procedure that can be used to randomly assign forms using this design. In one method for spiraling, Form X and Form Y are alternated when the test booklets are packaged. When the booklets are handed out, the first examinee receives Form X, the second examinee Form Y, the third examinee Form X, and so on. This spiraling process typically leads to comparable, *randomly equivalent* groups taking Form X and Form Y. When using this design, the difference between group-level performance on the two forms is taken as a direct indication of the difference in difficulty between the forms.

For example, suppose that the random groups design is used to equate Form X to Form Y using large representative examinee groups. Suppose also that the mean for Form Y is 77 raw score points and the mean for Form X is 72 raw score points. Because the mean for Form Y is 5 points higher than the mean for Form X, Form Y is 5 raw score points easier, on average, than Form X. This example is a simplification of equating in practice. More complete methods for equating using the random groups design are described in detail in Chapter 2.

One practical feature of the random groups design is that each examinee takes only one form of the test, thus minimizing testing time relative to a design in which examinees take more than one form. In addition, more than one new form can be equated at the same time by including the additional new forms in the spiraling process. The random groups design requires that all the forms be available and administered at the same time, which might be difficult in some situations. If there is concern about test form security, administering more than one form could exacerbate these concerns. Because different examinees take the forms to be equated, large sample sizes are typically needed.

When spiraling is used for random assignment, certain practical issues should be considered. First, examinees should not be seated in a way that

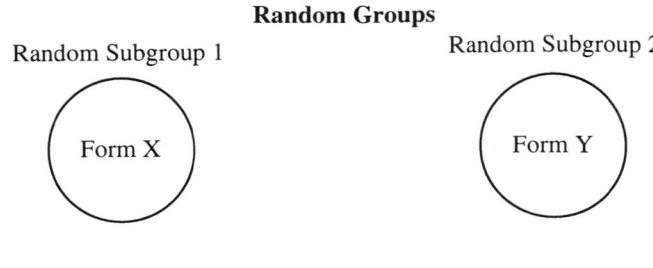

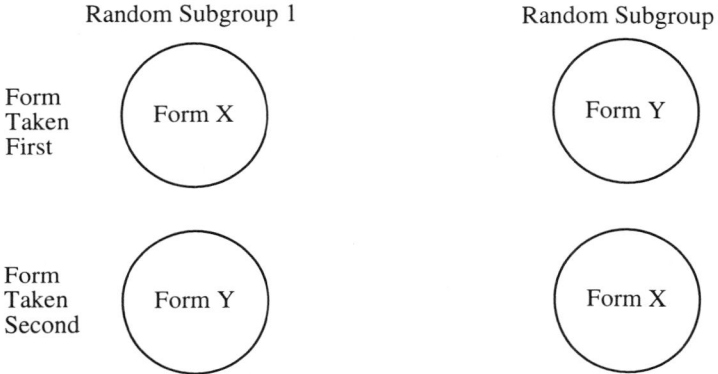

Figure 1.1. Illustration of three data collection designs.

would defeat the process. For example, if examinees were systematically seated boy—girl, boy—girl, then the boys might all be administered Form X and the girls Form Y. Also, suppose that there were many testing rooms. If the first examinee in each room was administered Form X, then more Form X booklets would be administered than Form Y booklets in those rooms with an odd number of examinees.

Single Group Design

In the *single group design* (not shown in Figure 1.1) the same examinees are administered both Form X and Form Y. What if Form X was administered first to all examinees followed by Form Y? If fatigue was a factor in examinee performance, then Form Y could appear relatively more difficult than Form X because examinees would be tired when administered Form Y. On the other hand, if familiarity with the test increased performance, then Form Y could appear to be easier than Form X. Because these *order effects* are typically present, and there is no reason to believe they cancel each other out, this design is rarely used in practice.

Single Group Design with Counterbalancing

Counterbalancing the order of administration of the forms is one way to deal with order effects in the single group design. In one method for counterbalancing, test booklets are constructed that contain Form X and Form Y. One-half of the test booklets are printed with Form X following Form Y, and the other half are printed with Form Y following Form X. In packaging, test booklets having Form X first are alternated with test booklets having Form Y first. When the test booklets are handed out, the first examinee takes Form X first, the second examinee takes Form Y first, the third examinee takes Form X first, and so on. When the booklets are administered, the first and second forms are separately timed. This spiraling process helps to ensure that the examinee group receiving Form Y first is comparable to the examinee group receiving Form X first.

Figure 1.1 provides an illustration of the *single group design with counterbalancing*. The portion of the design labeled "Form Taken First" is identical to the random groups design shown in Figure 1.1. Therefore, Form X could be equated to Form Y using only the data from the form taken first (i.e., Form X data from Subgroup 1 and Form Y data from Subgroup 2). To take full advantage of this design, however, the data from the "Form Taken Second" also could be used. Assume that examinees typically take only one form of the test when the test is later administered operationally to examinees. In this case, the equating relationship of interest would be the relationship between the forms when the forms are administered first. If the effect of taking Form X after taking Form Y is the same as the effect of taking Form Y after taking Form X, then the equating relationship will be the same between the forms taken first as it is between the forms taken second. Otherwise, a *differential order effect* is said to have occurred, and the equating relationships would differ. In this case, the data for the form that is taken second might need to be disregarded, which could lead to instability in the equating (see Chapter 7 for a discussion of equating error) and a waste of examinee time.

Table 1.2. Means for Two Forms of a
Hypothetical Test Administered Using the
Single Group Design with Counterbalancing.

	Subgroup 1	Subgroup 2
Form taken first	Form X 72	Form Y 77
Form taken second	Form Y 75	Form X 71

As an example, Table 1.2 presents a situation in which the effect of taking Form X after taking Form Y differs from the effect of taking Form Y after taking Form X. In this example, alternate forms of a test are to be equated by the single group design with counterbalancing using very large groups of examinees. The raw score means for the form that was taken first are shown in the first line of the table. Subgroup 2 had a mean of 77 on Form Y, which is 5 points higher than the mean of 72 earned by the randomly equivalent Subgroup 1 on Form X. Thus, using only data from the form that was taken first, Form Y appears to be 5 points easier, on average, than Form X. The means for the form that was taken second are shown in the second line of the table. Subgroup 1 had a mean of 75 on Form Y, which is 4 points higher than the mean of 71 earned by randomly equivalent Subgroup 2 on Form X. Thus, using data from the form taken second, Form Y is 4 points easier, on average, than Form X. Because the sample size is very large, this 4- versus 5-point difference suggests that there is a differential order effect. When a differential order effect like this one is present, the data from the form taken second might need to be disregarded. These issues are discussed further in Chapter 2.

In addition to the need to control for differential order effects, other practical problems can restrict the usefulness of the single group design with counterbalancing. Because two forms must be administered to the same students, testing time needs to be doubled, which often is not practically feasible. If fatigue and practice are effectively controlled by counterbalancing and differential order effects are not present, then the primary benefit in using the single group design with counterbalancing is that it typically has smaller sample size requirements than the random groups design, because, by taking both of the forms, each examinee serves as his or her own control.

In practice, the single group design with counterbalancing might be used instead of the random groups design when (1) administering two forms to examinees is operationally possible, (2) differential order effects are not expected to occur, and (3) it is difficult to obtain participation of a sufficient number of examinees in an equating study that uses the random groups design. Relative sample size requirements for these two designs are discussed in Chapter 7.

ASVAB Problems with a Single Group Design

The Armed Services Vocational Aptitude Battery (ASVAB) is a battery of ability tests that is used in the process of selecting individuals for the military. In 1976, new forms of the ASVAB were introduced. Scores on these forms were to be reported on the scale of previous forms through the use of a scaling process. (Because the content of the new forms differed somewhat from the content of the previous forms, the process used to convert scores to the scale of the previous forms is referred to here as scaling rather than as equating.) Maier (1993) indicated that problems occurred in the scaling process, with the result that many individuals entered the military who were actually not eligible to enter under the standards that were intended to be in effect at the time. As a result, Maier estimated that between January 1, 1976 and September 30, 1980, over 350,000 individuals entered the military who should have been judged ineligible. Maier reported that a complicated set of circumstances led to these problems. Most of the problems were a result of how the scaling study was designed and carried out. The effects of one of these problems are discussed here.

The examinees included in the study were applying to the military. In the scaling process, each examinee was administered both the old and new forms. (Supposedly, the order was counterbalanced—see Maier, 1993, for a discussion.) The scores on the old form were used for selection. No decisions about the examinees were made using the scores on the new form. Many examinees were able to distinguish between the old and the new forms. (For example, the content differed and the printing quality of the old form was better than that for the new form.) Also, many examinees knew that only the scores on the old form were to be used for selection purposes. Because the scores on the old form were to be used in the process of making selection decisions, the examinees were likely more motivated when taking the old form than they were when taking the new form. It seems reasonable to assume that scores under conditions of greater motivation would be higher than they would be under lower motivation conditions.

The following hypothetical example demonstrates how this motivation difference might be reflected in the scale scores. Suppose that the following conditions hold:

1. A raw score of 10 on the old form corresponds to a raw score of 10 on the new form under conditions of high motivation.
2. A raw score of 8 on the old form corresponds to a raw score of 8 on the new form under conditions of high motivation.

3. A raw score of 10 on each form corresponds to a scale score of 27 under the conditions of high motivation.
4. A raw score of 8 on each form corresponds to a scale score of 25 under the conditions of high motivation.
5. When either of the forms is administered under conditions of lower motivation the raw scores are depressed by 2 points.

Conditions 1 and 2 imply that the old and new forms are equally difficult at a raw score of 10 under high motivation conditions. The same is true at a raw score of 8.

What would happen in a scaling study if the old form was administered under high motivation and the new form under low motivation, and the motivation differences were not taken into account? In this case, a score of 8 on the new form would appear to correspond to a score of 10 on the old form, because the new form score would be depressed by 2 points. In the scaling process, an 8 on the new form would be considered to be equivalent to a 10 on the old form and to a scale score of 27. That is, an 8 on the new form would correspond to a scale score of 27 instead of the correct scale score of 25. Thus, when the new form is used later under high motivation conditions, scale scores on the new form would be too high.

Reasoning similar to that in this hypothetical example led Maier (1993) to conclude that motivation differences caused the scale scores on the new form to be too high when the new form was used to make selection decisions for examinees. The most direct effect of these problems was that the military selected many individuals using scores on the new form whose skill levels were lower than the intended standards. After the problem was initially detected in 1976, it took until October of 1980 to sort out the causes for the problems and to build new tests and scales that were judged to be sound. It took much effort to resolve the ASVAB scaling problem, including conducting a series of research studies, hiring a panel of outside testing experts, and significantly improving the quality control and oversight procedures for the ASVAB program.

Common-Item Nonequivalent Groups Design

The last design shown in Figure 1.1 is the *common-item nonequivalent groups design*. This design often is used when more than one form per test date cannot be administered because of test security or other practical concerns. In this design, Form X and Form Y have a set of items in common, and different groups of examinees are administered the two forms. For example, a group tested one year might be administered Form X and a group tested another year might be administered Form Y. This design has two variations. When the score on the set of common items contributes to the examinee's score on the test, the set of common items is referred to as *internal.* The internal common items are chosen to represent the content and statistical characteristics of the old form. For this reason, internal common items typically are interspersed among the other items in the test form. When the score on the set of common items does *not* contribute to the examinee's score on the test form, the set of common items is referred to as *external.* Typically, external common items are administered as a separately timed section.

To accurately reflect group differences, the set of common items should be proportionally representative of the total test forms in content and statistical characteristics. That is, the common-item set should be a "mini version" of the total test form. The common items also should behave similarly in the old and new forms. To help ensure similar behavior, each common item should occupy a similar location (item number) in the two forms. In addition, the common items should be exactly the same (e.g., no wording changes or rearranging of alternatives) in the old and new forms. Additional ways to help ensure adequate equating using the common-item nonequivalent groups design are described in Chapter 8.

In this design, the group of examinees taking Form X is *not* considered to be equivalent to the group of examinees taking Form Y. Differences between means (and other score distribution characteristics) on Form X and Form Y can result from a combination of examinee group differences and test form differences. The central task in equating using this design is to separate group differences from form differences.

The hypothetical example in Table 1.3 illustrates how differences might be separated. Form X and Form Y each contain 100 multiple-choice items that are scored number correct, and there is an internal set of 20 items in common between the two forms. The means on the common items suggest that Group 2 is higher achieving than Group 1, because members of Group 2, on average, correctly answered 75% of the common items, whereas members of Group 1 correctly answered only 65% of the common items. That is, on average, Group 2 correctly answered 10% more of the common items than did Group 1.

Which of the two forms is easier? To provide one possible answer, consider the following question: What would have been the mean on Form X for Group 2 had Group 2 taken Form X? Group 2 correctly answered 10%

Table 1.3. Means for Two Forms of a Hypothetical 100-Item Test with an Internal Set of 20 Common Items.

Group	Form X (100 items)	Form Y (100 items)	Common Items (20 items)
1	72	—	13 (65%)
2	—	77	15 (75%)

more of the common items than did Group 1. Therefore, Group 2 might be expected to correctly answer 10% more of the Form X items than would Group 1. Using this line of reasoning (and using the fact that Form X contains 100 items), the mean for Group 2 on Form X would be expected to be $82 = 72 + 10$. Because Group 2 earned a mean of 77 on Form Y and has an expected mean of 82 on Form X, Form X appears to be 5 points easier than Form Y.

This example is an oversimplification of how equating actually would be accomplished, and these results would hold only under very stringent conditions. The equating methods discussed in Chapters 4, 5, and 6 might even lead to the opposite conclusion about which form is more difficult. This example is intended to illustrate that a major task in conducting equating with the nonequivalent groups design is to separate group and form differences.

As indicated earlier, for this design to function well the common items need to represent the content and statistical characteristics of the total test. Table 1.4 provides data for a hypothetical test that is intended to illustrate the need for the set of common items to be content representative. In this example, Group 1 and Group 2 are again nonequivalent groups of examinees. The test consists of items from two content areas, Content I and Content II. As shown near the top of Table 1.4, on average, Group 1 correctly answered 70% of the Content I items and 80% of the Content II items. Group 2 correctly answered 80% of the Content I items and 70% of the Content II items. If the total test contains one-half Content I items and one-half Content II items, then, as illustrated near the middle of Table 1.4, both Group 1 and Group 2 will earn an average score of 75% correct on the whole test. Thus, the two groups have the same average level of achievement for the total test, consisting of one-half Content I and one-half Content II items.

Assume that two forms of the test are to be equated. If, as illustrated near the bottom of Table 1.4, the common-item set contains three-fourths

Table 1.4. Percent Correct for Two Groups on a Hypothetical Test.

	Group 1	Group 2
Content		
I	70%	80%
II	80%	70%
For Total Test	75% =	75% =
1/2(Content I) + 1/2(Content II)	1/2(70%) + 1/2(80%)	1/2(80%) + 1/2(70%)
For Common Items	72.5% =	77.5% =
3/4(Content I) + 1/4(Content II)	3/4(70%) + 1/4(80%)	3/4(80%) + 1/4(70%)

Content I items and one-fourth Content II items, Group 1 will correctly answer 72.5% of the common items, and Group 2 will correctly answer 77.5% of the common items. Thus, for this set of common items, Group 2 appears to be higher achieving than Group 1, even though the two groups are at the same level on the total test. This example illustrates that common items need to be content representative if they are to portray group differences accurately and lead to a satisfactory equating. (See Klein and Jarjoura, 1985, for an illustration of the need for content representativeness for an actual test.)

The common-item nonequivalent groups design is widely used. A major reason for its popularity is that this design requires that only one test form be administered per test date, which is how test forms usually are administered in operational settings. In contrast, the random groups design typically requires different test forms to be administered to random subgroups of examinees, and the single group design requires that more than one form be administered to each examinee. Another advantage of the common-item nonequivalent groups design is that, with external sets of common items, it might be possible for all items that contribute to an examinee's score (the noncommon items) to be disclosed following the test date. The ability to disclose items is important for some testing programs, because some states have mandated disclosure for certain tests, and some test publishers have opted for disclosure. However, common items should not be disclosed if they are to be used to equate subsequent forms. (See Chapter 8 for further discussion.)

The administrative flexibility offered by the use of nonequivalent groups is gained at some cost. As is described in Chapters 4, 5, and 6, strong statistical assumptions are required to separate group and form differences. The larger the differences between examinee groups, the more difficult it becomes for the statistical methods to separate the group and form differences. The only link between the two groups is the common items, so the content and statistical representativeness of the common items are especially crucial when the groups differ. Although a variety of statistical equating methods have been proposed for the common-item nonequivalent groups design, no method has been found that provides completely appropriate adjustments when the examinee groups are very different.

NAEP Reading Anomaly—Problems with Common Items

The National Assessment of Educational Progress (NAEP) is a congressionally mandated survey of the educational achievement of students in American schools. NAEP measures performance trends in many achievement areas, based on representative samples at three grade/age levels. The preliminary results from the 1986 NAEP Assessment in Reading indicated that the reading results "showed a surprisingly large decrease from 1984 at age 17 and, to a lesser degree, at age 9.... Such large changes in reading proficiency were considered extremely unlikely to have occurred in just two years without the awareness of the educational community." (Zwick, 1991, p. 11).

A series of inquiries were conducted to better understand the reasons for the decline. One potential cause that was investigated was the manner in which common items were used in linking the 1984 and 1986 assessments. Zwick (1991) indicated that the following differences existed between the administration:

1. In 1984, the test booklets administered to examinees contained reading and writing sections. In 1986, the booklets administered to examinees contained reading, mathematics, and/or science sections at ages 9 and 13. In 1986, the booklets contained reading, computer science, history, and/or literature at age 17.
2. The composition of the reading sections differed in 1984 and 1986. Items that were common to the two years appeared in different orders, and the time available to complete the common items differed in the two years.

The investigations concluded that these differences in the context in which the common items appeared in the two years, rather than changes in reading achievement, were responsible for much of the difference that was observed (Zwick, 1991). This so-called NAEP reading anomaly illustrates the importance of administering common items in the same context in the old and new forms. Otherwise, context effects can lead to very misleading results.

Error in Estimating Equating Relationships

Estimated equating relationships typically contain estimation error. A major goal in designing and conducting equating is to minimize such equating error.

Random equating error is present whenever samples from populations of examinees are used to estimate parameters (e.g., means, standard deviations, and percentile ranks) that are involved in estimating an equating relationship. Random error is typically indexed by the *standard error of equating*, which is the focus of Chapter 7. Conceptually, the standard error of equating is the standard deviation of score equivalents over replications of the equating procedure. The following situation illustrates the meaning of the standard error of equating when estimating the Form Y score equivalent of a Form X score.

1. Draw a random sample of size 1000 from a population of examinees.
2. Find the Form Y score equivalent of a Form X score of 75 using data from this sample and a given equating method.
3. Repeat steps 1 and 2 a large number of times, which results in a large

number of estimates of the Form Y score equivalent of a Form X score
of 75.
4. The standard deviation of these estimates is an estimate of the standard
   error of equating for a Form X score of 75.

As these steps illustrate, the standard error of equating is defined separately for each score on Form X.

As the sample size becomes larger, the standard error of equating
becomes smaller, and it becomes inconsequential for very large sample
sizes (assuming very large populations, as discussed in Chapter 7). Random
error can be controlled by using large samples of examinees, by choosing
an equating design that reduces such error, or both. Random error is especially troublesome when practical issues dictate the use of small samples of
examinees.

Systematic equating error results from violations of the assumptions and
conditions of equating. For example, in the random groups design, systematic error results if a particular spiraling process is inadequate for achieving group comparability. In the single group design with counterbalancing,
failure to control adequately for differential order effects can be a major
source of systematic error. In the common-item nonequivalent groups design, systematic error results if the assumptions of statistical methods used
to separate form and group differences are not met. These assumptions can
be especially difficult to meet under the following conditions: the groups
differ substantially, the common items are not representative of the total
test form in content and statistical characteristics, or the common items
function differently from one administration to the next. A major problem
with this design is that sufficient data typically are not available to estimate or adjust for systematic error.

Over time, after a large number of test forms are involved in the scaling
and equating process, both random and systematic errors tend to accumulate. Although the amount of random error can be quantified readily using
the standard error of equating, systematic error is much more difficult to
quantify. In conducting and designing equating studies, both types of error
should be minimized to the extent possible. In some practical circumstances the amount of equating error might be so large that equating
would add more error into the scores than if no equating had been done.
Thus, equating is not always defensible. This issue is described further in
Chapter 8.

Evaluating the Results of Equating

In addition to designing an equating study, an operational definition of
equating and a method for estimating an equating relationship need to be
chosen. Then, after the equating is conducted, the results should be eval-

uated. As indicated by Harris and Crouse (1993), such evaluation requires that criteria for equating be identified. Estimating random error using standard errors of equating can be used to develop criteria. Criteria for evaluating equating also can be based on consistency of results with previous results.

The properties of equating that were described earlier also can be used to develop evaluative criteria. The symmetry and same specifications properties always must be achieved. Some aspects of Lord's equity property can be evaluated. For example, procedures are discussed in Chapter 8 that indicate the extent to which examinees can be expected to earn approximately the same score, regardless of the form that they take. Procedures are also considered that can be used to evaluate the extent to which examinees are measured with equal precision across forms. Observed score equating properties are especially important when equating is evaluated from an institutional perspective. An institution that is admitting students needs to know that the particular test form administered would not affect the numbers of students who would be admitted. The group invariance property is important from the perspective of treating subgroups of examinees equitably. The equating relationship should be very similar across subgroups. As a check on the group invariance property, the equating can be conducted on various important subgroups. Procedures for evaluating equating are discussed more fully in Chapter 8.

Testing Situations Considered

In this chapter, equating has been described for testing programs in which alternate forms of tests are administered on various test dates. Equating is very common in this circumstance, especially when tight test security is required, such as when equating professional certification, licensure, and college admissions tests. Another common circumstance is for two or more forms of a test to be developed and equated at one time. The equated forms then are used for a period of years until the content becomes dated. Alternate forms of elementary achievement level batteries, for example, often are administered under these sorts of conditions. The procedures described in this book pertain directly to equating alternate forms of tests under either of these circumstances.

In recent years, test administration on computer has become common. Computer administration is often done by selecting test items to be administered from a pool of items, with each examinee being administered a different set of items. In this case, a clear need exists to use processes to ensure that scores earned by different examinees are comparable to one another. However, as discussed in Chapter 8, such procedures often are different from the equating methods to be discussed in Chapters 2 through 7 of this book.

In this book, equating is presented mainly in the context of dichotomously (right versus wrong) scored tests. Recently, there has been considerable attention given to performance tests, which require judges or a computer to score tasks or items. Many of the concepts of equating for multiple-choice tests also pertain to performance tests. However, the use of judges along with difficulties in representing the domain of content complicate equating for performance tests. Chapter 8 discusses when and how the methods treated in this book can be applied to performance tests.

The procedures used to calculate raw scores on a test affect how equating procedures are implemented. In this book, tests typically are assumed to be scored number-correct, with scores ranging from zero to the number of items on the test. Many of the procedures described can be adapted to other types of scoring, however, such as scores that are corrected for guessing. For example, a generalization of equipercentile equating to scoring which produces scores that are not integers is described in Chapter 2. In Chapter 3, smoothing techniques are referenced which can be used with scores that are not integers. Many of the techniques in Chapters 4 and 5 can be adapted readily to other scoring schemes. In Chapter 6, noninteger IRT scoring is discussed. Issues in performance assessments are described in Chapter 8. To simplify exposition, unless noted otherwise, assume that alternate forms of dichotomously scored tests are being equated. Scores on these tests range from zero to the number of items on the test.

Preview

This chapter has discussed equating properties and equating designs. Chapter 2 treats equating using the random groups design, which, compared to other designs, requires very few statistical assumptions. For this reason, the random groups design is ideal for presenting many of the statistical concepts in observed score equating. Specifically, the mean, linear, and equipercentile equating methods are considered. The topic of Chapter 3 is smoothing techniques that are used to reduce total error in estimated equipercentile relationships.

Linear methods appropriate for the common-item nonequivalent groups design are described in Chapter 4. In addition to considering observed score methods, methods based on test theory models are introduced in Chapter 4. Equipercentile methods for the common-item nonequivalent groups design are presented in Chapter 5.

IRT methods, which are also test theory-based methods, are the topic of Chapter 6. IRT methods are presented that can be used with the equating designs described in this chapter. In addition, IRT methods appropriate for equating using item pools are described.

Equating procedures are all statistical techniques that are subject to random error. Procedures for estimating the standard error of equating are described in Chapter 7 along with discussions of sample sizes required to attain desired levels of equating precision. Chapter 8 focuses on various practical issues in equating. These topics include evaluating the results of equating and choosing among equating methods and results. In addition, current topics, such as linking performance assessments and linking computerized tests, are considered.

Exercises

Exercises are presented at the end of each chapter of this book. Some of the exercises are intended to reinforce important concepts and consider practical issues; others are intended to facilitate learning how to apply statistical techniques.

1.1. A scholarship test is administered twice per year, and different forms are administered on each test date. Currently, the top 1% of the examinees on each test date earn scholarships.
   a. Would equating the two forms affect who was awarded a scholarship? Why or why not?
   b. Suppose the top 1% who took the test during the year (rather than at each test date) were awarded scholarships. Would the use of equating affect who was awarded a scholarship? Why or why not?

1.2. Refer to the example in Table 1.1. Suppose that a new form, Form X_3, was found to be uniformly 1 point easier than Form X_2. What scale score would correspond to a Form X_3 raw score of 29?

1.3. A state passes a law that all items which contribute to an examinee's score on a test will be released to that examinee, on request, following the test date. Assume that the test is to be secure. Which of the following equating designs could be used in this situation: random groups, single group with counterbalancing, common-item nonequivalent groups with an internal set of common items, common-item nonequivalent groups with an external set of common items? Briefly indicate how equating would be accomplished using this (these) design(s).

1.4. Equating of forms of a 45-minute test is to be conducted by collecting data on a group of examinees who are being tested for the purpose of conducting equating. Suppose that it is relatively easy to get large groups of examinees to participate in the study, but it is difficult to get any student to test for more than one 50-minute class period, where 5 minutes are needed to hand out materials, give instructions, and collect materials. Would it be better to use the random groups design or the single group design with counterbalancing in this situation? Why?

1.5. Suppose that only one form of a test can be administered on any given test date. Of the designs discussed, which equating design(s) can be used?

1.6. Refer to the data shown in Table 1.4.
   a. Which group would appear to be higher achieving on a set of common items composed only of Content I items?

    b. Which group would appear to be higher achieving on a set of common items composed only of Content II items?

    c. What is the implication of your answers to a and b?

1.7. Consider the following statements for equated Forms X and Y:

    I. "Examinees A and B are at the same level of achievement, because A scored at the 50th percentile nationally on Form X and B scored at the 50th percentile nationally on Form Y."

    II. "Examinees A and B are at the same level of achievement, because the expected equated score of A on Form X equals the expected score of B on Form Y."

Which of these statements is consistent with an observed score property of equating? Which is consistent with Lord's equity property of equating?

1.8. If a very large group of examinees is used in an equating study, which source of equating error would almost surely be small, random or systematic? Which source of equating error could be large if the very large group examinees used in the equating were not representative of the examinees that are to be tested, random or systematic?

CHAPTER 2

Observed Score Equating Using the Random Groups Design

As was stressed in Chapter 1, the same specifications property is an essential property of equating, which means that the forms to be equated must be built to the same content and statistical specifications. We also stressed that the symmetry property is essential for any equating relationship. The focus of the present chapter is on methods that are designed to achieve the observed score equating property, along with the same specifications and symmetry properties. As was described in Chapter 1, these observed score equating methods are developed with the goal that, after equating, converted scores on two forms have at least some of the same score distribution characteristics in a population of examinees.

In this chapter, these methods are developed in the context of the random groups design. Of the designs discussed thus far, the assumptions required for the random groups design are the least severe and most readily achieved. Thus, very few sources of systematic error are present with the random groups design. Because of the minimal assumptions required with the random groups design, this design is ideal for use in presenting the basic statistical methods in observed score equating, which is the focus of the present chapter.

The definitions and properties of mean, linear, and equipercentile equating methods are described in this chapter. These methods are presented, initially, in terms of population parameters (e.g., population means and standard deviations) for a specific population of examinees. We also discuss the process of estimating equating relationships, which requires that statistics (e.g., sample means and standard deviations) be substituted in place of the parameters. The methods then are illustrated using a real data example. Following the presentation of the methods, issues in using scale scores are described and illustrated. We then briefly discuss equating using single group designs.

An important practical challenge in using the random groups design is to obtain large enough sample sizes so that random error (see Chapter 7 for a discussion of standard errors) is at an acceptable level (rules of thumb for appropriate sample sizes are given in Chapter 8). For the equipercentile equating method, in Chapter 3 we describe statistical smoothing methods that often are used to help reduce random error when conducting equipercentile equating using the random groups design.

For simplicity, the statistical methods in this chapter are developed using a testing situation in which tests consist of test items that are scored correct (1) or incorrect (0), and where the total score is the number of items answered correctly. Near the end of the chapter, a process for equating tests that are scored using other scoring schemes is described.

Mean Equating

In mean equating, Form X is considered to differ in difficulty from Form Y by a constant amount along the score scale. For example, under mean equating, if Form X is 2 points easier than Form Y for high-scoring examinees, it is also 2 points easier than Form Y for low-scoring examinees. Although a constant difference might be overly restrictive in many testing situations, mean equating is useful for illustrating some important equating concepts.

As was done in Chapter 1, define Form X as the new form, let X represent the random variable score on Form X, and let x represent a particular score on Form X (i.e., a realization of X); and define Form Y as the old form, let Y represent the random variable score on Form Y, and let y represent a particular score on Form Y (i.e., a realization of Y). Also, define $\mu(X)$ as the mean on Form X and $\mu(Y)$ as the mean on Form Y for a population of examinees. In mean equating, scores on the two forms that are an equal (signed) distance away from their respective means are set equal:

$$x - \mu(X) = y - \mu(Y). \tag{2.1}$$

Then solve for y and obtain

$$m_Y(x) = y = x - \mu(X) + \mu(Y). \tag{2.2}$$

In this equation, $m_Y(x)$ refers to a score x on Form X transformed to the scale of Form Y using mean equating.

As an illustration of how to apply this formula, consider the situation discussed in Chapter 1, in which the mean on Form X was 72 and the mean on Form Y was 77. Based on this example, equation (2.2) indicates that 5 points would need to be added to a Form X score to transform a score on Form X to the Form Y scale. That is,

$$m_Y(x) = x - 72 + 77 = x + 5.$$

For example, using mean equating, a score of 72 on Form X is considered to indicate the same level of achievement as a score of 77 ($77 = 72 + 5$) on Form Y. And, a score of 75 on Form X is considered to indicate the same level of achievement as a score of 80 on Form Y. Thus, mean equating involves the addition of a constant (which might be negative) to all raw scores on Form X to find equated scores on Form Y.

Linear Equating

Rather than considering the differences between two forms to be a constant, linear equating allows for the differences in difficulty between the two test forms to vary along the score scale. For example, linear equating allows Form X to be more difficult than Form Y for low achieving examinees but less difficult for high achieving examinees.

In linear equating, scores that are an equal (signed) distance from their means in standard deviation units are set equal. Thus, linear equating can be viewed as allowing for the scale units, as well as the means, of the two forms to differ. Define $\sigma(X)$ and $\sigma(Y)$ as the standard deviations of Form X and Form Y scores, respectively. The linear conversion is defined by setting standardized deviation scores (z-scores) on the two forms to be equal such that

$$\frac{x - \mu(X)}{\sigma(X)} = \frac{y - \mu(Y)}{\sigma(Y)}. \qquad (2.3)$$

If the standard deviations for the two forms were equal, equation (2.3) could be simplified to equal the mean equating equation (2.2). Thus, if the standard deviations of the two forms are equal, then mean and linear equating produce the same result. Solving for y in equation (2.3),

$$l_Y(x) = y = \sigma(Y) \left[\frac{x - \mu(X)}{\sigma(X)} \right] + \mu(Y), \qquad (2.4)$$

where $l_Y(x)$ is the linear conversion equation for converting observed scores on Form X to the scale of Form Y. By rearranging terms, an alternate expression for $l_Y(x)$ is,

$$l_Y(x) = y = \frac{\sigma(Y)}{\sigma(X)} x + \left[\mu(Y) - \frac{\sigma(Y)}{\sigma(X)} \mu(X) \right]. \qquad (2.5)$$

This expression is a linear equation of the form $slope(x) + intercept$ with

$$slope = \frac{\sigma(Y)}{\sigma(X)}, \text{ and } intercept = \mu(Y) - \frac{\sigma(Y)}{\sigma(X)}\mu(X). \quad (2.6)$$

What if the standard deviations in the mean equating example were $\sigma(X) = 10$ and $\sigma(Y) = 9$? The slope is $9/10 = .9$, and the intercept is $77 - (9/10)72 = 12.2$. The resulting conversion equation is $l_Y(x) = .9x + 12.2$. What is $l_Y(x)$ if $x = 75$?

$$l_Y(75) = .9(75) + 12.2 = 79.7.$$

How about if $x = 77$ or $x = 85$?

$$l_Y(77) = .9(77) + 12.2 = 81.5, \text{ and}$$

$$l_Y(85) = .9(85) + 12.2 = 88.7.$$

These equated values illustrate that the difference in test form difficulty varies with score level. For example, the difference in difficulty between Form X and Form Y for a Form X score of 75 is 4.7 ($79.7 - 75$), whereas the difference for a Form X score of 85 is 3.7 ($88.7 - 85$).

Properties of Mean and Linear Equating

In general, what are the properties of the equated scores? From Chapter 1, E is the expectation operator. The mean of a variable is found by taking the expected value of that variable. Using equation (2.2), the mean converted score $m_Y(x)$, for mean equating is

$$\mathbf{E}[m_Y(X)] = \mathbf{E}[X - \mu(X) + \mu(Y)] = \mu(X) - \mu(X) + \mu(Y) = \mu(Y). \quad (2.7)$$

That is, for mean equating the mean of the Form X scores equated to the Form Y scale is equal to the mean of the Form Y scores. In the example described earlier, the mean of the equated Form X scores is 77 [recall that $m_Y(x) = x + 5$ and $\mu(X) = 72$], the same value as the mean of the Form Y scores. Note that standard deviations were not shown in equation (2.7). What would be the standard deviation of Form X scores converted using the mean equating equation (2.2)? Because the Form X scores are converted to Form Y by adding a constant, the standard deviation of the converted scores would be the same as the standard deviation of the scores prior to conversion. That is, under mean equating, $\sigma[m_Y(X)] = \sigma(X)$.

Using equation (2.5), the mean equated score for linear equating can be found as follows:

$$E[l_Y(X)] = E\left[\frac{\sigma(Y)}{\sigma(X)}X + \mu(Y) - \frac{\sigma(Y)}{\sigma(X)}\mu(X)\right]$$

$$= \frac{\sigma(Y)}{\sigma(X)}E(X) + \mu(Y) - \frac{\sigma(Y)}{\sigma(X)}\mu(X)$$

$$= \mu(Y), \qquad\qquad\qquad (2.8)$$

because $E(X) = \mu(X)$.

The standard deviation of the equated scores is found by first substituting equation (2.5) for $l_Y(X)$ as follows:

$$\sigma[l_Y(X)] = \sigma\left[\frac{\sigma(Y)}{\sigma(X)}X + \mu(Y) - \frac{\sigma(Y)}{\sigma(X)}\mu(X)\right]$$

To continue, the standard deviation of a score plus a constant is equal to the standard deviation of the score. That is, $\sigma(X + constant) = \sigma(X)$. By recognizing in the linear equating equation that the terms to the right of the addition sign are a constant, the following holds:

$$\sigma[l_Y(X)] = \sigma\left[\frac{\sigma(Y)}{\sigma(X)}X\right].$$

Also note that the standard deviation of a score multiplied by a constant equals the standard deviation of the score multiplied by the constant. That is, $\sigma(constant\ X) = constant\ \sigma(X)$. Noting that the ratio of standard deviations in the large parentheses is also a constant that multiplies X,

$$\sigma[l_Y(X)] = \frac{\sigma(Y)}{\sigma(X)}\sigma(X) = \sigma(Y). \qquad\qquad (2.9)$$

Therefore, the mean and standard deviation of the Form X scores equated to the Form Y scale are equal to the mean and standard deviation, respectively, of the Form Y scores. In the example described earlier for linear equating, the mean of the equated Form X scores is 77 and the standard deviation is 9; these are the same values as the mean and standard deviation of the Form Y scores.

Consider the equation for mean equating, equation (2.2), and the equation for linear equating (2.5). If either of the equations were solved for x, rather than for y, the equation for equating Form Y scores to the scale of Form X would result. These conversions would be symbolized by $m_X(y)$ and $l_X(y)$, respectively. Equating relationships are defined as being *symmetric* because the equation used to convert Form X scores to the Form Y scale is the inverse of the equation used to convert Form Y scores to the Form X scale.

The equation for linear equating (2.5), is deceptively like a linear regression equation. The difference is that, for linear regression, the $\sigma(Y)/\sigma(X)$ terms are multiplied by the correlation between X and Y. However, a linear regression equation does not qualify as an equating function because the

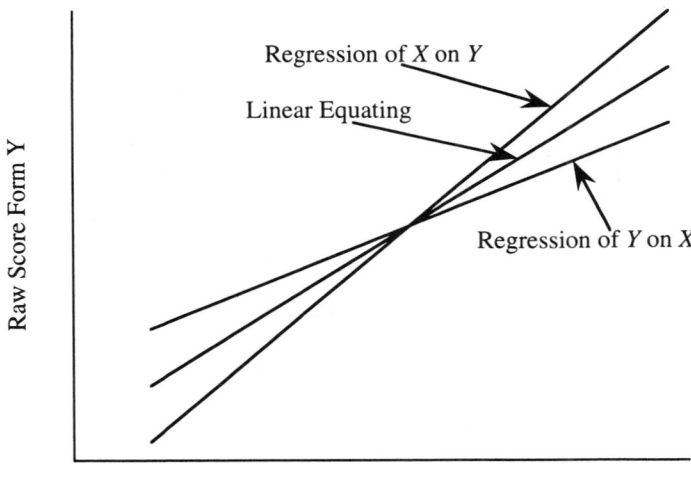

Figure 2.1. Comparison of linear regression and linear equating.

regression of X on Y is different from the regression of Y on X, unless the correlation coefficient is 1. For this reason, regression equations cannot, in general, be used as equating functions. The comparison between linear regression and linear equating is illustrated in Figure 2.1. The regression of Y on X is different from the regression of X on Y. Also note that there is only one linear equating relationship graphed in the figure. This relationship can be used to transform Form X scores to the Form Y scale, or to transform Form Y scores to the Form X scale.

Comparison of Mean and Linear Equating

Figure 2.2 illustrates the equating of Form X and Form Y using the hypothetical test forms already discussed. The equations for equating scores on Form X to the Form Y scale are plotted in this figure. Also plotted in this figure are the results from the "identity equating." In the *identity equating*, a score on Form X is considered to be equivalent to the identical score on Form Y; for example, a 40 on Form X is considered to be equivalent to a 40 on Form Y. Identity equating would be the same as mean and linear equating if the two forms were identical in difficulty all along the score scale.

To find a Form Y equivalent of a Form X score using the graph, find the Form X value of interest on the horizontal axis, go up to the function, and then go over to the vertical axis to read off the Form Y equivalent.

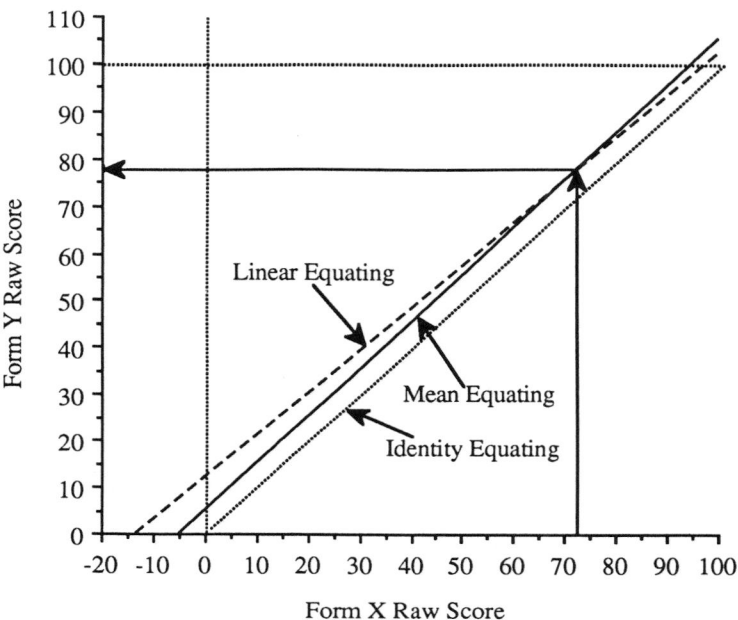

Figure 2.2. Graph of mean and linear equating for a hypothetical 100-item test.

How to find the Form Y equivalent of a Form X score of 72 is illustrated in the figure using the arrows. This equivalent is 77, using either mean or linear equating. The score 72 is the mean score on Form X. As indicated earlier, both mean and linear equating will produce the same result at the mean.

Now refer to the identity equating line in the figure, and note that the line for mean equating is parallel to the line for the identity equating. The lines for these two methods will always be parallel. As can be seen, the line for mean equating is uniformly 5 points vertically above the line for the identity equating, because Form Y is, on average, 5 points less difficult than Form X. Refer to the line for linear equating. This line is not parallel to the identity equating line. The linear equating line is further above the identity equating line at the low scores than at the high scores. This observation is consistent with the earlier discussion in which the difference in difficulty between Form X and Form Y was shown to be greater at the lower scores than at the higher scores.

Assume that the test in this example is scored number-correct. Number-correct scores for this 100-item test can range from 0 to 100. Figure 2.2 illustrates that equated scores from mean and linear equating can sometimes be out of the range of possible observed scores. The dotted lines at 0 on Form X and at 100 illustrate the boundaries of possible observed scores. For example, using linear equating, a score of 100 on Form X equates to a score of approximately 102 on Form Y. Also, using linear equating, a score of 0 on Form Y equates to a score of approximately -14 on Form X. There are a variety of ways to handle this problem. One way is to allow the top and bottom to "float." For example, the highest equated score might be allowed to exceed the highest raw score. An alternative is to truncate the conversion at the highest and lowest scores. In the example, truncation involves setting all converted scores greater than 100 equal to 100 and setting all converted scores less than 0 equal to 0. That is, all Form Y scores that equate to Form X scores below 0 would be set to 0 and all Form X scores that equate to Form Y scores above 100 would be set to 100. In practice, the decision about how to handle equated scores outside the range typically interacts with the score scale that is used for reporting scores. Sometimes this issue is effectively of no consequence, because no one achieves the extreme raw scores on Form X that equate to unobtainable scores on Form Y.

In summary, in mean equating the conversion is derived by setting the deviation scores on the two forms equal, whereas in linear equating the standardized deviation scores (z-scores) on the two forms are set equal. In mean equating, scores on Form X are adjusted by a constant amount that is equal to the difference between the Form Y and Form X means. In linear equating, scores on Form X are adjusted using a linear equation that allows for the forms to be differentially difficult along the score scale. In mean equating, the mean of the Form X scores equated to the Form Y scale is equal to the mean of the Form Y scores; whereas in linear equating, the standard deviation as well as the mean are equal. In general, mean equating is less complicated than linear equating, but linear equating provides for more flexibility in the conversion than does mean equating.

Equipercentile Equating

In equipercentile equating, a curve is used to describe form-to-form differences in difficulty, which makes equipercentile equating even more general than linear equating. Using equipercentile equating, for example, Form X could be more difficult than Form Y at high and low scores, but less difficult at the middle scores.

The equating function is an equipercentile equating function if the distribution of scores on Form X converted to the Form Y scale is equal to the distribution of scores on Form Y in the population. The equipercentile equating function is developed by identifying scores on Form X that have the same percentile ranks as scores on Form Y.

The definition of equipercentile equating developed by Braun and Holland (1982) is adapted for use here. Consider the following definitions of terms, some of which were presented previously:

X is a random variable representing a score on Form X, and x is a particular value (i.e., a realization) of X.

Y is a random variable representing a score on Form Y, and y is a particular value (i.e., a realization) of Y.

F is the cumulative distribution function of X in the population.

G is the cumulative distribution function of Y in the same population.

e_Y is a symmetric equating function used to convert scores on Form X to the Form Y scale.

G^* is the cumulative distribution function of e_Y in the same population. That is, G^* is the cumulative distribution function of scores on Form X converted to the Form Y scale.

The function e_Y is defined to be the equipercentile equating function in the population if

$$G^* = G. \tag{2.10}$$

That is, the function e_Y is the equipercentile equating function in the population if the cumulative distribution function of scores on Form X converted to the Form Y scale is equal to the cumulative distribution function of scores on Form Y.

Braun and Holland (1982) indicated that the following function is an equipercentile equating function when X and Y are continuous random variables:

$$e_Y(x) = G^{-1}[F(x)], \tag{2.11}$$

where G^{-1} is the inverse of the cumulative distribution function G.

As previously indicated, to be an equating function, e_Y must be symmetric. Define

e_X as a symmetric equating function used to convert scores on Form Y to the Form X scale, and

F^* as the cumulative distribution function of e_X in the population. That is, F^* is the cumulative distribution function of scores on Form Y converted to the Form X scale.

By the symmetry property,

$$e_X^{-1}(x) = e_Y(x) \text{ and } e_Y^{-1}(y) = e_X(y). \tag{2.12}$$

Also,

$$e_X(y) = F^{-1}[G(y)], \tag{2.13}$$

is the equipercentile equating function for converting Form Y scores to the Form X scale. In this equation, F^{-1} is the inverse of the cumulative distribution function F.

Following the definitions in equations (2.10)–(2.13), an equipercentile equivalent for the population of examinees can be constructed in the following manner: For a given Form X score, find the percentage of examinees earning scores at or below that Form X score. Next, find the Form Y score that has the same percentage of examinees at or below it. These Form X and Form Y scores are considered to be equivalent. For example, suppose that 20% of the examinees in the population earned a Form X score at or below 26 and 20% of the examinees in the population earned a Form Y score at or below 27. Then a Form X score of 26 would be considered to represent the same level of achievement as a Form Y score of 27. Using equipercentile equating, a Form X score of 26 would be equated to a Form Y score of 27.

The preceding discussion was based on an assumption that test scores are continuous random variables. Typically, however, test scores are discrete. For example, number-correct scores take on only integer values. With discrete test scores, the definition of equipercentile equating is more complicated than the situation just described. Consider the following situation. Suppose that a test is scored number-correct and that the following is true of the population distributions:

1. 20% of the examinees score at or below 26 on Form X.
2. 18% of the examinees score at or below 27 on Form Y.
3. 23% of the examinees score at or below 28 on Form Y.

What is the Form Y equipercentile equivalent of a Form X score of 26? No Form Y score exists that has precisely 20% of the scores at or below it. Strictly speaking, no Form Y equivalent of a Form X score of 26 exists. Thus, the goal of equipercentile equating stated in equation (2.10) cannot be met strictly when test scores are discrete.

How can equipercentile equating be conducted when scores are discrete? A tradition exists in educational and psychological measurement to view discrete test scores as being continuous by using percentiles and percentile ranks as defined in many educational and psychological measurement textbooks (e.g., Blommers and Forsyth, 1977; Ebel and Frisbie, 1991). In this approach, an integer score of 28, for example, is considered to represent scores in the range 27.5–28.5. Examinees with scores of 28 are conceived of being uniformly distributed in this range. The percentile rank of a score of 28 is defined as being the percentage of scores *below* 28. However, because only 1/2 of the examinees who score 28 are considered to be below 28 (the remainder being between 28 and 28.5), the percentile rank of 28 is the percentage of examinees who earned integer scores of 27 and below, plus 1/2 the percentage of examinees who earned an integer score of 28. Placing the preceding example in the context of percentile ranks, 18% of the examinees earned a Form Y score below 27.5 and 5% (23–18%) of the examinees earned a score between 27.5 and 28.5. So the percentile rank of a Form Y score of 28 would be 18% + 1/2(5%) = 20.5%. In the terminology typically used, the percentile rank of an integer score is the percentile rank at the midpoint of the interval that contains that score.

Holland and Thayer (1989) presented a statistical justification for using percentiles and percentile ranks. In their approach, they use what they refer to as a *continuization* process. Given a discrete integer-valued random variable X and a random variable U that is uniformly distributed over the range $-1/2$ to $+1/2$, they define a new random variable

$$X^* = X + U.$$

This new random variable is continuous. The cumulative distribution function of this new random variable corresponds to the percentile rank function. The inverse of the cumulative distribution of this new function exists and is the percentile function. Holland and Thayer (1989) also generalized their approach to incorporate continuization processes that are based on distributions other than the uniform.

In the present chapter, the traditional approach to percentiles and percentile ranks is followed. The Holland and Thayer (1989) paper should be consulted if a more in-depth discussion of continuization approaches is desired.

The equipercentile methods presented next assume that the observed scores on the tests to be equated are integer scores that range from zero through the number of items on the test, as would be true of tests scored number-correct. Generalizations to other scoring schemes are discussed as well.

Graphical Procedures

Equipercentile equating using graphical methods provides a conceptual framework for subsequent consideration of analytic methods. A hypothetical four-item test is used to illustrate the graphical process for equipercentile equating. Data for Form X are presented in Table 2.1. In this table, x refers to test score and $f(x)$ to the proportion of examinees earning the score x. For example, the proportion of examinees earning a score of 0 is .20. $F(x)$ is the cumulative proportion at or below x. For example, the proportion of examinees scoring 3 or below is .9. $P(x)$ refers to the percen-

Table 2.1. Form X Score Distribution
for a Hypothetical Four-Item Test.

x	$f(x)$	$F(x)$	$P(x)$
0	.2	.2	10
1	.3	.5	35
2	.2	.7	60
3	.2	.9	80
4	.1	1.0	95

tile rank, and for an integer score it equals the percentage of examinees below x plus 1/2 the percentage of examinees at x—i.e., for integer score x, $P(x) = 100[F(x-1) + f(x)/2]$.

To be consistent with traditional definitions of percentile ranks, the percentile rank function is plotted as points at the upper limit of each score interval. For example, the percentile rank of a score of 3.5 is 90, which is 100 times the cumulative proportion at or below 3. Therefore, to plot the percentile ranks, plot the percentile ranks at each integer score plus .5. The percentile ranks at an integer score plus .5 can be found from Table 2.1 by taking the cumulative distribution function values, $F(x)$, at an integer and multiplying them by 100 to make them percentages. Figure 2.3 illustrates how to plot the percentile rank distribution for Form X. A percentile rank of 0 is also plotted at a Form X score of −.5. The points are then connected with straight lines. An example is presented for finding the percentile rank of a Form X integer score of 2 using the arrows in Figure 2.3. As can be seen, the percentile rank of a score of 2 is 60, which is the same result found in Table 2.1.

In Figure 2.3, percentile ranks of scores between −.5 and 0.0 are greater than zero. These nonzero percentile ranks result from using the traditional definition of percentile ranks, in which scores of 0 are assumed to be uniformly distributed from −.5 to .5. Also, scores of 4 are considered to be uniformly distributed between 3.5 to 4.5, so that scores above 4 have percentile ranks less than 100. Under this conceptualization, the range of pos-

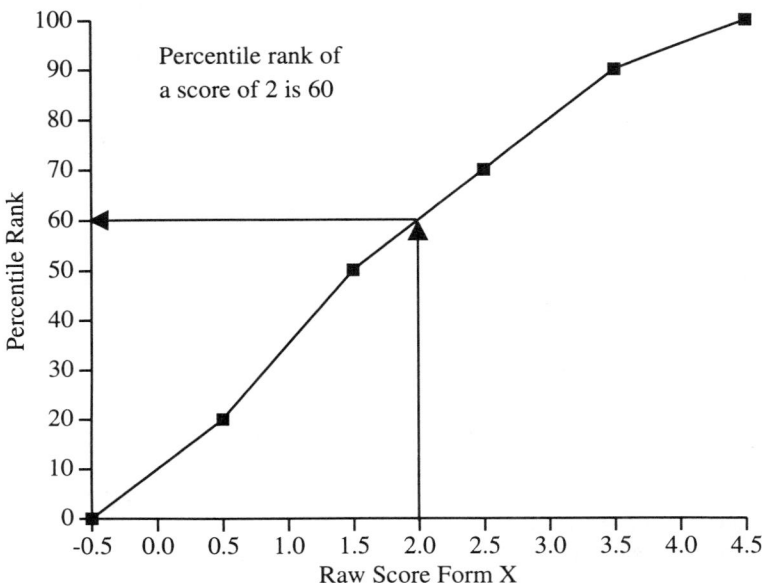

Figure 2.3. Form X percentile ranks on a hypothetical four-item test.

Table 2.2. Form X and Form Y Distributions for a Hypothetical Four-Item Test.

y	g(y)	G(y)	Q(y)	x	f(x)	F(x)	P(x)
0	.1	.1	5	0	.2	.2	10
1	.2	.3	20	1	.3	.5	35
2	.2	.5	40	2	.2	.7	60
3	.3	.8	65	3	.2	.9	80
4	.2	1.0	90	4	.1	1.0	95

sible scores is treated as being between $-.5$ and the highest integer score $+.5$.

Data from Form Y also need to be used in the equating process. The data for Form Y are presented along with the Form X data in Table 2.2. In this table, y refers to Form Y scores, $g(y)$ to the proportion of examinees at each score, $G(y)$ to the proportion at or below each score, and $Q(y)$ to the percentile rank at each score. Percentile ranks for Form Y are plotted in the same manner as they were for Form X. To find the equipercentile equivalent of a particular score on Form X, find the Form Y score with the same percentile rank. Figure 2.4 illustrates this process for finding the equipercentile equivalent of a Form X score of 2. As indicated by the arrows, a Form X score of 2 has a percentile rank of 60. Following the arrows, it can be seen that the Form Y score of about 2.8 (actually 2.83) is equivalent to the Form X score of 2.

The equivalents can also be plotted. To construct such a graph, plot, as points, Form Y equivalents of Form X scores at each integer plus .5. Then plot Form X equivalents of Form Y scores at each integer plus .5. To handle scores below the lowest integer scores $+.5$, a point is plotted at the (x, y) pair $(-.5, -.5)$. The plotted points are then connected by straight lines. This process is illustrated for the example in Figure 2.5. As indicated by the arrows in the figure, a Form X score of 2 is equivalent to a Form Y score of 2.8 (actually 2.83), which is consistent with the result found earlier. This plot of equivalents displays the Form Y equivalents of Form X scores.

In summary, the graphical process of finding equipercentile equivalents is as follows: Plot percentile ranks for each form on the same graph. To find a Form Y equivalent of a Form X score, start by finding the percentile rank of the Form X score. Then find the Form Y score that has that same percentile rank. Equivalents can be plotted in a graph that shows the equipercentile relationship between the two forms.

One issue that arises in equipercentile equating is how to handle situations in which no examinees earn a particular score. When this occurs, the score that corresponds to a particular percentile rank might not be unique. Suppose for example that x has a percentile rank of 20. To find the equipercentile equivalent, the Form Y score that has a percentile rank of 20

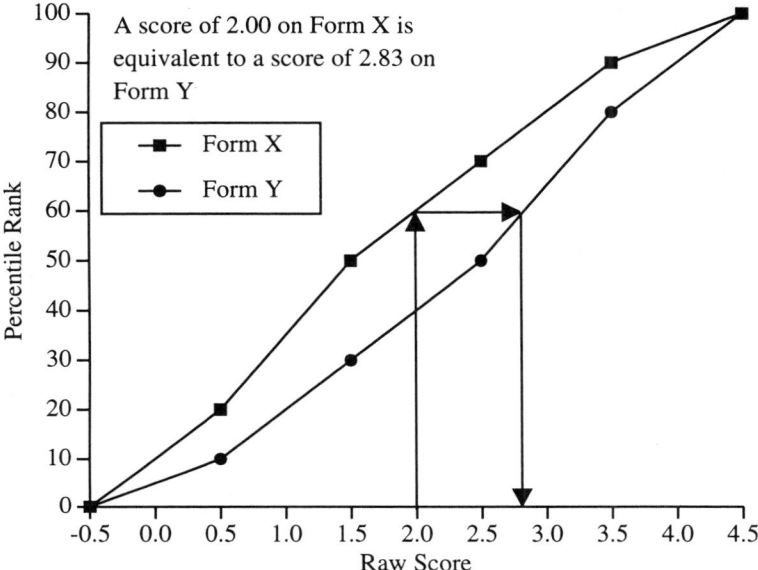

Figure 2.4. Graphical equipercentile equating for a hypothetical four-item test.

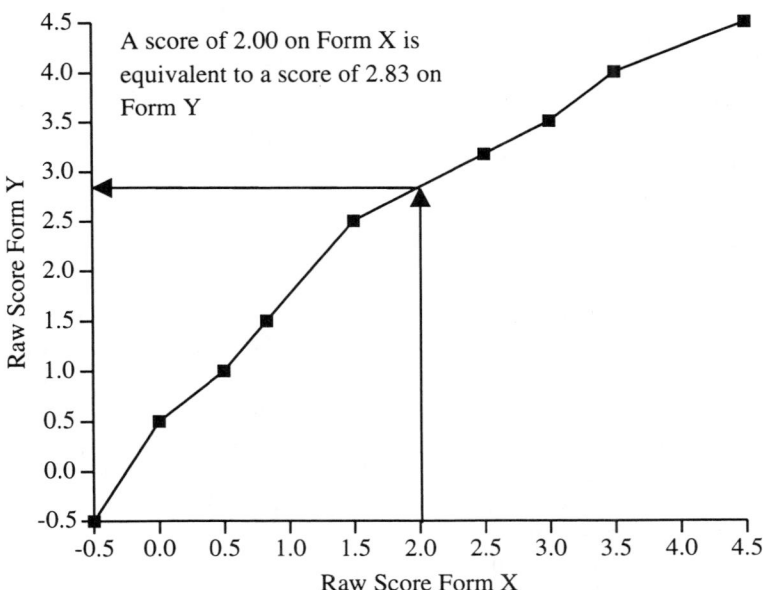

Figure 2.5. Equipercentile equivalents for a hypothetical four-item test.

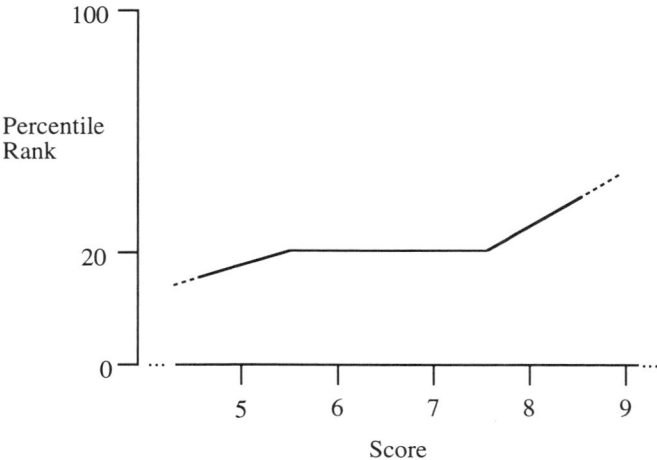

Figure 2.6. Illustration of percentile ranks when no examinees earn a particular score.

needs to be found. Suppose, however, that there is no unique score on Form Y that has a percentile rank of 20, as illustrated in Figure 2.6. The percentile ranks shown in Figure 2.6 could occur if no examinees earned scores of 6 and 7. In this case, the graph indicates that scores in the range 5.5 to 7.5 all have percentile ranks of 20. The choice of the Form Y score that has a percentile rank of 20 is arbitrary. In this situation, usually the middle score would be chosen. So, in the example the score with a percentile rank of 20 would be designated as 6.5. Choosing the middle score is arbitrary, technically, but doing so seems sensible.

Analytic Procedures

The graphical method discussed in the previous section is not likely to be viable for equating a large number of real forms in real time. In addition, equating using graphical procedures can be inaccurate. What is needed are formulas that provide more formal definitions of percentile ranks and equipercentile equivalents. The following discussion provides such formulas. The result of applying these formulas is to produce percentile ranks and equipercentile equivalents that are equal to those that would result using the graphical procedures.

To define percentile ranks, let K_X represent the number of items on Form X of a test. Define X as a random variable representing test scores on Form X that can take on the integer values $0, 1, \ldots, K_X$. Define $f(x)$ as the discrete density function for $X = x$. That is,

$f(x) \geq 0$ for integer scores $x = 0, 1, \ldots, K_X$;
$f(x) = 0$ otherwise; and
$\sum f(x) = 1$.

Define $F(x)$ as the discrete cumulative distribution function. That is, $F(x)$ is the proportion of examinees in the population earning a score *at or below* x. Therefore,

$0 \leq F(x) \leq 1$ for $x = 0, 1, \ldots, K_X$;
$F(x) = 0$ for $x < 0$; and
$F(x) = 1$ for $x > K_X$.

Consider a possible noninteger value of x. Define x^* as that integer that is closest to x such that $x^* - .5 \leq x < x^* + .5$. For example, if $x = 5.7$, $x^* = 6$; if $x = 6.4$, $x^* = 6$; and if $x = 5.5$, $x^* = 6$. The percentile rank function for Form X is

$$P(x) = 100\{F(x^* - 1) + [x - (x^* - .5)][F(x^*) - F(x^* - 1)]\},$$

$$-.5 \leq x < K_X + .5,$$

$$= 0, \qquad x < -.5,$$

$$= 100, \qquad x \geq K_X + .5. \tag{2.14}$$

To illustrate how this equation functions, consider the following example based on the data in Table 2.1. To calculate the percentile rank for a score of $x = 1.3$, using equation (2.14)

$$P(1.3) = 100\{F(0) + [1.3 - (1 - .5)][F(1) - F(0)]\}$$

$$= 100\{.2 + [.8][.5 - .2]\} = 100\{.2 + .24\} = 44.$$

In this case, $x^* = 1.0$, because 1 is the integer score that is closest to 1.3. The term $[F(1) - F(0)] = .5 - .2 = .3$ represents the proportion of examinees earning a score of 1. These scores are considered to range from .5 to 1.5. The term $[1.3 - (1 - .5)] = .8$ indicates that the score of 1.3 is, proportionally, .8 of the distance between .5 and 1.5. So, $[.8][.3] = .24$ represents the probability of scoring between .5 and 1.3. The probability of scoring below .5 is represented by $F(0) = .2$. Therefore, the percentile rank of a score of 1.3 equals 44.

The inverse of the percentile rank function, which often is referred to as the percentile function, is symbolized as P^{-1}. Two alternate percentile functions are given as follows. These functions produce the same result, unless some of the probabilities are zero. Given a percentile rank (e.g., the 10th percentile rank), this inverse function is used to find the score corresponding to that percentile rank. To find this function, solve equation (2.14) for x. Specifically, for a given percentile rank P^*, the percentile is

$$x_U(P^*) = P^{-1}[P^*] = \frac{P^*/100 - F(x_U^* - 1)}{F(x_U^*) - F(x_U^* - 1)} + (x_U^* - .5), \quad 0 \leq P^* < 100,$$

$$= K_X + .5, \qquad\qquad\qquad P^* = 100 \qquad (2.15)$$

In equation (2.15), for $0 \leq P^* < 100$, x_U^* is the *smallest* integer score with a cumulative percent $[100F(x)]$ that is *greater than* P^*. An alternate expression for the percentile is

$$x_L(P^*) = P^{-1}[P^*] = \frac{P^*/100 - F(x_L^*)}{F(x_L^* + 1) - F(x_L^*)} + (x_L^* + .5), \quad 0 < P^* \leq 100,$$

$$= -.5. \qquad\qquad\qquad P^* = 0. \qquad (2.16)$$

In equation (2.16), for $0 < P^* \leq 100$, x_L^* is the *largest* integer score with a cumulative percent $[100F(x)]$ that is *less than* P^*. If the $f(x)$ are nonzero at all score points $0, 1, \ldots, K_X$, then $x = x_U = x_L$, and either expression can be used. If some of the $f(x)$ are zero, then $x_U \neq x_L$ for at least some percentile ranks. In this case, the convention $x = (x_U + x_L)/2$ is used. This convention produces the same results as the one described in association with Figure 2.6 using the graphical procedures. In most situations, it seems reasonable to assume that the $f(x)$ are all nonzero over the integer score range $0, 1, \ldots, K_X$. For this reason, and to simplify issues, when considering population distributions in the following discussion, only equation (2.15) is used with $x_U = x$. When considering estimates of population distributions, estimated probabilities of zero are often encountered (i.e., when no examinees in a sample earn a particular score).

As an example of how to use equation (2.15), find the score corresponding to a percentile rank of 62 using the inverse of the percentile rank function using the data in Table 2.1. In this case $x_U^* = 2$ because, in Table 2.1, it is the *smallest* integer score with $F(x)$ that is *greater than* .62. Then

$$P^{-1}(62) = \frac{62/100 - F(1)}{F(2) - F(1)} + (2 - .5)$$

$$= \frac{.62 - .5}{.7 - .5} + (2 - .5) = .12/.20 + 1.5 = .60 + 1.5 = 2.1.$$

In equipercentile equating, the interest is in finding a score on Form Y that has the same percentile rank as a score on Form X. Referring to y as a score on Form Y, let K_Y refer to the number of items on Form Y, let $g(y)$ refer to the discrete density of y, let $G(y)$ refer to the discrete cumulative distribution of y, let $Q(y)$ refer to the percentile rank of y, and let Q^{-1} refer to the inverse of the percentile rank function for Form Y. Then the Form Y equipercentile equivalent of score x on Form X is

$$e_Y(x) = y = Q^{-1}[P(x)], \quad -.5 \leq x \leq K_X + .5. \qquad (2.17)$$

This equation indicates that, to find the equipercentile equivalent of score

x on the scale of Form Y, first find the percentile rank of x in the Form X distribution. Then find the Form Y score that has that same percentile rank in the Form Y distribution. Equation (2.17) is symmetric. That is, to find the Form X equivalent of a Form Y score, equation (2.17) is solved for y, giving $e_X(y) = P^{-1}[Q(y)]$.

Analytically, to find $e_Y(x)$ given by equation (2.17), use the analog of equation (2.15) for the Form Y distribution. That is, use

$$e_Y(x) = Q^{-1}[P(x)]$$

$$= \frac{P(x)/100 - G(y_U^* - 1)}{G(y_U^*) - G(y_U^* - 1)} + (y_U^* - .5), \qquad 0 \le P(x) < 100,$$

$$= K_Y + .5, \qquad\qquad\qquad\qquad P(x) = 100. \qquad (2.18)$$

[Note that, to use this equation when some Form Y scores have zero probabilities, it also is necessary to use y_L^* as described in the discussion following equation (2.16).] Refer to Table 2.2. As an example of finding equipercentile equivalents, find the Form Y equipercentile equivalent of a Form X score of 2. The percentile rank of a Form X score of 2 is $P(2) = 60$, as is shown in Table 2.2. To find the equipercentile equivalent, the Form Y score that has a percentile rank of 60 must be found. Because 3 is the score with the *smallest* $G(y)$ that is *greater than* .60, $y_U^* = 3$. Thus, using equation (2.18),

$$e_Y(x) = Q^{-1}[60] = \frac{60/100 - .5}{.8 - .5} + (3 - .5) = .1/.3 + 2.5 = 2.8333.$$

The raw score equipercentile equivalents that result typically are noninteger. Noninteger scores arise through the continuization process used to define percentiles and percentile ranks. Issues related to rounding to integers are considered later in the discussion of scale scores.

Properties of Equated Scores in Equipercentile Equating

Conducting equipercentile equating using equation (2.18) always results in equated scores in the range $-.5 \le e_Y(x) \le K_Y + .5$. Thus, equipercentile equating has the desirable property that the equated scores will always be within the range of possible scores under the traditional conceptualization of percentiles and percentile ranks. The problem of having equated scores that are out of the range of possible scores which occur with mean and linear equating does not occur with equipercentile equating.

Ideally, in equipercentile equating the equated scores on Form X would have the same distribution as the scores on Form Y. As was previously indicated, if test scores were continuous, then these distributions would be the same. However, test scores are discrete. A continuization process in-

Table 2.3. Form Y Equivalents of Form X Scores for a Hypothetical Four-Item Test.

x	$f(x)$	$e_Y(x)$
0	.2	.50
1	.3	1.75
2	.2	2.8333
3	.2	3.50
4	.1	4.25

Table 2.4. Moments for Equating Form X and Form Y of a Hypothetical Four-Item Test.

Score	μ	σ	sk	ku
y	2.3000	1.2689	−.2820	1.9728
x	1.7000	1.2689	.2820	1.9728
$e_Y(x)$	2.3167	1.2098	−.0972	1.8733

volving percentiles and percentile ranks was used to render the problem mathematically tractable. However, when the results of equating are applied to discrete scores, the equated Form X score distribution will differ from the Form Y distribution.

Consider the following illustration. Using the hypothetical four-item test from Table 2.2, Table 2.3 provides the Form Y equivalents of scores resulting from the use of equation (2.18). The moments that result are shown in Table 2.4, where skewness and kurtosis are defined for Form X, respectively, as

$$sk(X) = \frac{\mathbf{E}[X - \mu(X)]^3}{[\sigma(X)]^3}, \text{ and} \tag{2.19}$$

$$ku(X) = \frac{\mathbf{E}[X - \mu(X)]^4}{[\sigma(X)]^4}, \tag{2.20}$$

Central moments for other variables are defined similarly. To arrive at the moments of the equated scores, $e_Y(x)$, in Table 2.4, the Form X scores were equated to Form Y scores. For example, as indicated in Table 2.3, the proportion of examinees earning an $e_Y(x)$ of .50 is .20. Moments of these equated scores then were found. Ideally, the moments for $e_Y(x)$ in Table 2.4 would be equal to those for y. As can be seen, however, there are departures. These departures are a result of the discreteness of the scores. The departures in Table 2.4 are relatively large because the test is so short. Departures likely would be considerably less with longer, more realistic tests. For tests of realistic lengths, not being able to achieve the equal distribution goal precisely often is more of a theoretical concern than a practical one.

Estimating Observed Score Equating Relationships

So far, the methods have been described using population parameters. In practice, sample statistics are all that are available, and these sample statistics are substituted for the parameters in the preceding equations.

One estimation problem that occurs in equipercentile equating is how to calculate the function P^{-1} when the frequency at some score points is zero. The conventions associated with equations (2.15) and (2.16) for averaging the results is one procedure for producing a unique result. Another procedure is to add a very small relative frequency to each score, and then adjust the relative frequencies so they sum to one. If adj is taken as this small quantity, then the adjusted relative frequencies on Form Y are

$$\hat{g}_{adj}(y) = \frac{\hat{g}(y) + adj}{1 + (K_Y + 1) \cdot adj},$$

where $\hat{g}(y)$ is the relative frequency that was observed. For example, if $K_Y = 10, adj = 10^{-6}$, and $\hat{g}(2) = .02$, then

$$\hat{g}_{adj}(2) = \frac{.02 + 10^{-6}}{1 + (10 + 1) \cdot 10^{-6}} = .02000078.$$

A similar procedure could be used for Form X. The equating then can be done using the adjusted relative frequencies. Experience has shown that a value around $adj = 10^{-6}$ can be used without creating a serious bias in the equating. A third solution to the zero frequency problem is to use smoothing methods, which are the subject of Chapter 3.

Data for an example of an equating of Form X and Form Y of the Original ACT Mathematics test are presented in Table 2.5. This test contains 40 multiple-choice items scored incorrect (0) or correct (1). Form X was administered to 4329 examinees and Form Y to 4152 examinees in a spiral administration, which resulted in random groups of examinees being administered Form X and Form Y. The sample sizes for the two forms differ, in part, because Form X always preceded Form Y in the distribution of booklets in each testing room. Thus, one more Form X than Form Y booklet was administered in some testing rooms. In the table, a "^" is used to indicate an estimate of a population parameter, and N_X and N_Y refer to sample sizes for the forms. Consider, for example, a score of 10 on Form Y.

Table 2.5. Data for Equating Form X and Form Y of the Original ACT Mathematics Test.

| | Form Y | | | | | Form X | | | | |
Score	$N_Y \cdot \hat{g}(y)$	$N_Y \cdot \hat{G}(y)$	$\hat{g}(y)$	$\hat{G}(y)$	$\hat{Q}(y)$	$N_X \cdot \hat{f}(x)$	$N_X \cdot \hat{f}(x)$	$\hat{f}(x)$	$\hat{F}(x)$	$\hat{P}(x)$
0	0	0	.0000	.0000	.00	0	0	.0000	.0000	.00
1	1	1	.0002	.0002	.01	1	1	.0002	.0002	.01
2	3	4	.0007	.0010	.06	1	2	.0002	.0005	.03
3	13	17	.0031	.0041	.25	3	5	.0007	.0012	.08
4	42	59	.0101	.0142	.92	9	14	.0021	.0032	.22
5	59	118	.0142	.0284	2.13	18	32	.0042	.0074	.53
6	95	213	.0229	.0513	3.99	59	91	.0136	.0210	1.42
7	131	344	.0316	.0829	6.71	67	158	.0155	.0365	2.88
8	158	502	.0381	.1209	10.19	91	249	.0210	.0575	4.70
9	161	663	.0388	.1597	14.03	144	393	.0333	.0908	7.42
10	194	857	.0467	.2064	18.30	149	542	.0344	.1252	10.80
11	164	1021	.0395	.2459	22.62	192	734	.0444	.1696	14.74
12	166	1187	.0400	.2859	26.59	192	926	.0444	.2139	19.17
13	197	1384	.0474	.3333	30.96	192	1118	.0444	.2583	23.61
14	177	1561	.0426	.3760	35.46	201	1319	.0464	.3047	28.15
15	158	1719	.0381	.4140	39.50	204	1523	.0471	.3518	32.83
16	169	1888	.0407	.4547	43.44	217	1740	.0501	.4019	37.69
17	132	2020	.0318	.4865	47.06	181	1921	.0418	.4438	42.28
18	158	2178	.0381	.5246	50.55	184	2105	.0425	.4863	46.50
19	151	2329	.0364	.5609	54.28	170	2275	.0393	.5255	50.59
20	134	2463	.0323	.5932	57.71	201	2476	.0464	.5720	54.87
21	137	2600	.0330	.6262	60.97	147	2623	.0340	.6059	58.89
22	122	2722	.0294	.6556	64.09	163	2786	.0377	.6436	62.47
23	110	2832	.0265	.6821	66.88	147	2933	.0340	.6775	66.05

24	116	2948	.0279	.7100	69.61	140	3073	.0323	.7099	69.37
25	132	3080	.0318	.7418	72.59	147	3220	.0340	.7438	72.68
26	104	3184	.0250	.7669	75.43	126	3346	.0291	.7729	75.84
27	104	3288	.0250	.7919	77.94	113	3459	.0261	.7990	78.60
28	114	3402	.0275	.8194	80.56	100	3559	.0231	.8221	81.06
29	97	3499	.0234	.8427	83.10	106	3665	.0245	.8466	83.44
30	107	3606	.0258	.8685	85.56	107	3772	.0247	.8713	85.90
31	88	3694	.0212	.8897	87.91	91	3863	.0210	.8924	88.18
32	80	3774	.0193	.9090	89.93	83	3946	.0192	.9115	90.19
33	79	3853	.0190	.9280	91.85	73	4019	.0169	.9284	92.00
34	70	3923	.0169	.9448	93.64	72	4091	.0166	.9450	93.67
35	61	3984	.0147	.9595	95.22	75	4166	.0173	.9623	95.37
36	48	4032	.0116	.9711	96.53	50	4216	.0116	.9739	96.81
37	47	4079	.0113	.9824	97.68	37	4253	.0085	.9824	97.82
38	29	4108	.0070	.9894	98.59	38	4291	.0088	.9912	98.68
39	32	4140	.0077	.9971	99.33	23	4314	.0053	.9965	99.39
40	12	4152	.0029	1.0000	99.86	15	4329	.0035	1.000	99.83

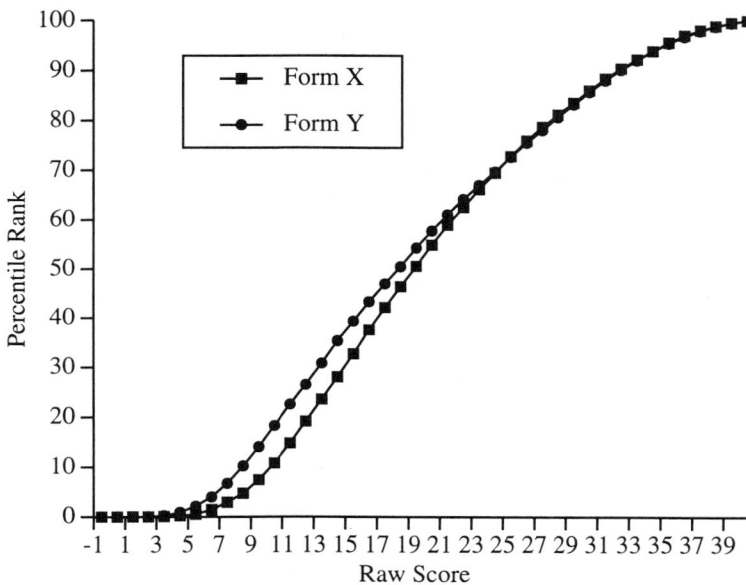

Figure 2.7. Percentile ranks for equating Form X and Form Y of the original ACT Mathematics test.

From Table 2.5, 194 examinees earned a score of 10, and 857 examinees earned a score of 10 or below; the proportion of examinees earning a score of 10 is .0467, the proportion of examinees at or below a score of 10 is .2064, and the estimated percentile rank of a score of 10 is 18.30.

Percentile ranks for Forms X and Y are plotted in Figure 2.7. The percentile ranks are plotted for each score point plus .5. Form X appears to be somewhat easier than Form Y, because the Form X distribution is shifted to the right. The relative frequency distributions are shown in Figure 2.8. Both score distributions are positively skewed, and Form X again appears to be somewhat easier than Form Y. Estimates of central moments for Form X and Form Y are given in the upper portion of Table 2.6. Both forms have means, $\hat{\mu}$, less than 20 (which is 50% of the 40 items), so it appears that the tests are somewhat difficult for these examinees. Form X is, on average, nearly 1 point easier than Form Y. Based on the standard deviations, $\hat{\sigma}$, the distribution for Form X is less variable than the distribution for Form Y. As indicated by the skewness values, $\widehat{sk}$ the distributions are positively skewed, where skewness for the population is defined in equation (2.19). Based on the kurtosis estimates, $\widehat{ku}$, the distributions have lower kurtosis than a normal distribution, which would have a kurtosis value of 3, where kurtosis for the population is defined in equation (2.20).

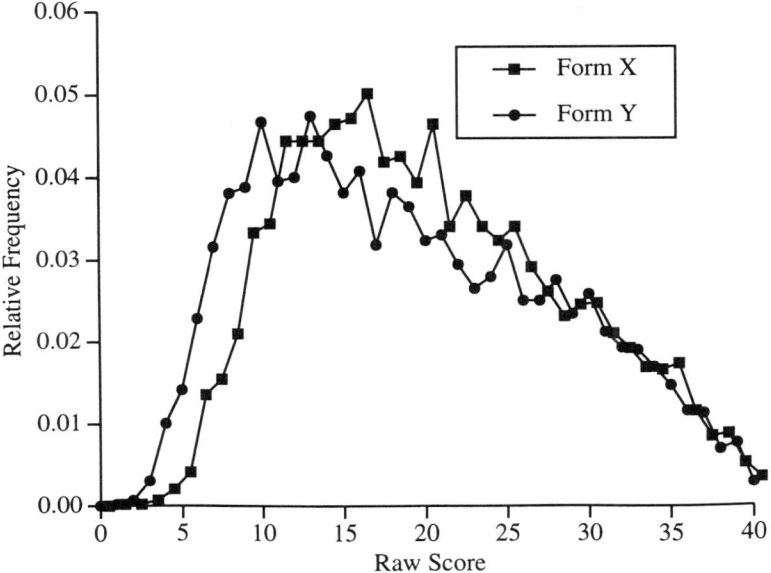

Figure 2.8. Relative frequency distributions for Form X and Form Y of the original ACT Mathematics test.

Table 2.6. Moments for Equating Form X and Form Y.

Test Form	$\hat{\mu}$	$\hat{\sigma}$	$\widehat{sk}$	$\widehat{ku}$
Form Y	18.9798	8.9393	.3527	2.1464
Form X	19.8524	8.2116	.3753	2.3024
Form X Equated to Form Y Scale for Various Methods				
Mean	18.9798	8.2116	.3753	2.3024
Linear	18.9798	8.9393	.3753	2.3024
Equipercentile	18.9799	8.9352	.3545	2.1465

The conversions for mean, linear, and equipercentile equating are shown in Table 2.7 and are graphed in Figure 2.9. The linear and equipercentile results were calculated using the *RAGE* computer program described in Appendix B. The moments for converted scores are shown in the bottom portion of Table 2.6. As expected, the mean converted scores for mean equating are the same as the mean for Form Y. For linear equating, the mean and standard deviation of the converted scores agree with the mean and standard deviation of Form Y. The first four moments of converted scores for equipercentile equating are very similar to those for Form Y. In Table 2.7, it can be seen that mean and linear equating produce results that

Table 2.7. Raw-to-Raw Score Conversion Tables.

Form X Score	Form Y Equivalent Using Equating Method		
	Mean	Linear	Equipercentile
0	−.8726	−2.6319	.0000
1	.1274	−1.5432	.9796
2	1.1274	−.4546	1.6462
3	2.1274	.6340	2.2856
4	3.1274	1.7226	2.8932
5	4.1274	2.8112	3.6205
6	5.1274	3.8998	4.4997
7	6.1274	4.9884	5.5148
8	7.1274	6.0771	6.3124
9	8.1274	7.1657	7.2242
10	9.1274	8.2543	8.1607
11	10.1274	9.3429	9.1827
12	11.1274	10.4315	10.1859
13	12.1274	11.5201	11.2513
14	13.1274	12.6088	12.3896
15	14.1274	13.6974	13.3929
16	15.1274	14.7860	14.5240
17	16.1274	15.8746	15.7169
18	17.1274	16.9632	16.8234
19	18.1274	18.0518	18.0092
20	19.1274	19.1405	19.1647
21	20.1274	20.2291	20.3676
22	21.1274	21.3177	21.4556
23	22.1274	22.4063	22.6871
24	23.1274	23.4949	23.9157
25	24.1274	24.5835	25.0292
26	25.1274	25.6722	26.1612
27	26.1274	26.7608	27.2633
28	27.1274	27.8494	28.1801
29	28.1274	28.9380	29.1424
30	29.1274	30.0266	30.1305
31	30.1274	31.1152	31.1297
32	31.1274	32.2039	32.1357
33	32.1274	33.2925	33.0781
34	33.1274	34.3811	34.0172
35	34.1274	35.4697	35.1016
36	35.1274	36.5583	36.2426
37	36.1274	37.6469	37.1248
38	37.1274	38.7355	38.1321
39	38.1274	39.8242	39.0807
40	39.1274	40.9128	39.9006

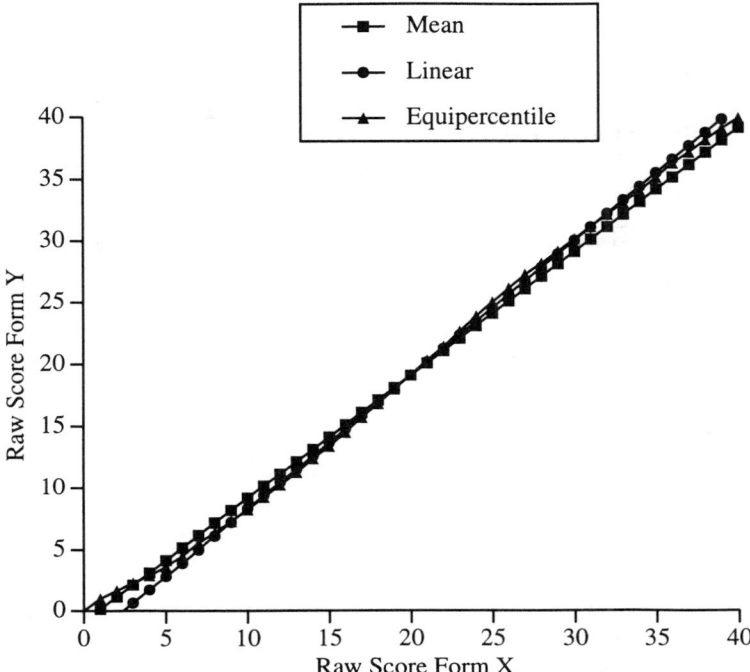

Figure 2.9. Results for equating Form X and Form Y of the original ACT Mathematics test.

are outside the range of possible raw scores. Because of the large number of values in Table 2.7 and the considerable similarity of equating functions in Figure 2.9, differences between the functions are difficult to ascertain. The use of considerably larger graph paper would help in such a comparison. Alternatively, difference-type plots can be used, as in Figure 2.10. In this graph, the difference between the results for each method and the results for the identity equating are plotted. To find the Form Y equivalent of a Form X score, just add the vertical axis value to the horizontal axis value. For example, for equipercentile equating a Form X score of 10 has a vertical axis value of approximately -1.8. Thus, the Form Y equivalent of a Form X score of 10 is approximately $8.2 = 10 - 1.8$. This value is the same as the one indicated in Table 2.7 (8.1607), apart from error inherent in trying to read values from a graph.

In Figure 2.10, the horizontal line for the identity equating is at a vertical axis value of 0, which will always be the case with difference plots constructed in the manner of Figure 2.10. The results for mean equating are displayed by a line that is parallel to, but nearly 1 point below, the line for the identity equating. The line for linear equating crosses the identity

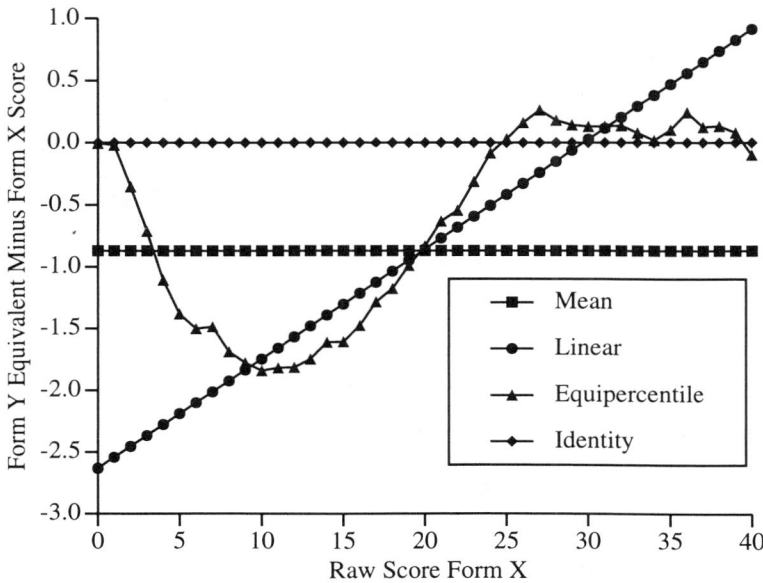

Figure 2.10. Results expressed as differences for equating Form X and Form Y of the original ACT Mathematics test.

equating and mean equating lines. The equipercentile equating relationship appears to be definitely nonlinear. Referring to the equipercentile relationship, Form X appears to be nearly 2 points easier around a Form X score of 10, and the two forms appear to be similar in difficulty at scores in the range of 25 to 40.

The plot in Figure 2.10 for equipercentile equating is somewhat irregular (bumpy). These irregularities are a result of random error in estimating the equivalents. Smoothing methods are introduced in Chapter 3, which lead to more regular plots and less random error.

Scale Scores

When equating is conducted in practice, raw scores typically are converted to scale scores. Scale scores are constructed to facilitate score interpretation, often by incorporating normative or content information. For example, scale scores might be constructed to have a particular mean in a nationally representative group of examinees. The effects of equating on scale scores are crucial to the interpretation of equating results, because scale scores are the scores typically reported to examinees. A discussion of methods for developing score scales is provided by Petersen et al. (1989). The use of scale scores in the equating context are described next.

Linear Conversions

The least complicated raw-to-scale score transformations that typically are used in practice are linear in form. For example, suppose that a national norming study was conducted using Form Y of the 100-item test that was used earlier in this chapter to illustrate mean and linear equating. Assume that the mean raw score, $\mu(Y)$, was 70 and the standard deviation, $\sigma(Y)$, was 10 for the national norm group. Also assume that the mean scale score, $\mu(sc)$, was intended to be 20 and the standard deviation of the scale scores, $\sigma(sc)$, 5. Then by equation (3) in Petersen *et al.* (1989), whose derivation follows logic similar to the derivation of linear equating (2.5), the raw-to-scale score transformation (sc) for converting raw scores on the old form, Form Y, to scale scores is

$$sc(y) = \frac{\sigma(sc)}{\sigma(Y)} y + \left[\mu(sc) - \frac{\sigma(sc)}{\sigma(Y)} \mu(Y) \right]$$

$$= \frac{5}{10} y + \left[20 - \frac{5}{10} 70 \right]$$

$$= .5y - 15.$$

Now assume that scores on Form X are to be converted to scale scores based on the equating used in the earlier linear equating example. As was found earlier, the linear conversion equation for equating raw scores on Form X to raw scores on Form Y was $l_Y(x) = .9x + 12.2$. To find the raw-to-scale score transformation for Form X, substitute $l_Y(x)$ for y in the raw-to-scale score transformation for Form Y. This gives

$$sc[l_Y(x)] = .5[l_Y(x)] - 15$$

$$= .5[.9x + 12.2] - 15$$

$$= .45x - 8.9.$$

For example, a raw score of 74 on Form X converts to a scale score of $.45(74) - 8.9 = 24.4$. In this manner, raw-to-scale score conversions for all Form X raw scores can be found. When another new form is constructed and equated to Form X, a similar process can be used to find the scale score equivalents of scores on this new form.

Truncation of Linear Conversions

When linear transformations are used as scaling transformations, the score scale transformation often needs to be truncated at the upper and/or lower extremes. For example, the Form Y raw-to-scale score transformation, $sc(y) = .5y - 15$, produces scale scores below 1 for raw scores below 32. Suppose that scale scores are intended to be 1 or greater. The transformation for this form then would be as follows:

$$sc(y) = .5y - 15, \quad y \geq 32,$$

$$= 1, \quad\quad\quad y < 32.$$

Also, a raw score of 22 on Form X is equivalent to a raw score of $32 = .9(22) + 12.2$ on Form Y. So, the raw-to-scale score conversion for Form X is

$$sc[l_Y(x)] = .45x - 8.9, \quad x \geq 22,$$

$$= 1, \quad\quad\quad x < 22.$$

Truncation can also occur at the top end. For example, truncation would be needed at the top end for Form X but not for Form Y if the highest scale score was set to 35 on this 100-item test (the reader should verify this fact).

Scale scores are typically rounded to integers for reporting purposes. Define sc_{int} as the scale score rounded to an integer. Then, for example, $sc_{int}[l_Y(x = 74)] = 24$, because a scale score of 24.4 rounds to a scale score of 24.

Nonlinear Conversions

Nonlinear raw-to-scale score transformations are often used in practice. Examples of nonlinear transformations include the following: normalized scales, grade equivalents, and scales constructed to stabilize measurement error variability (Petersen et al., 1989). The use of nonlinear transformations complicates the process of converting raw scores to scale scores. The nonlinear function could be specified as a continuous function. However, when using discrete test scores (e.g., number-correct scores) the function is often defined at selected raw score values, and linear interpolation is used to compute scale score equivalents at other raw score values. The scheme for nonlinear raw-to-scale score transformations that is presented here is designed to be consistent with the definitions of equipercentile equating described earlier.

The first step in describing the process is to specify $sc(y)$, the raw-to-scale score function for Form Y. In the present approach, the conversions of Form Y raw scores to scale scores are specified at Form Y raw scores of $-.5, K_Y + .5$, and all integer score points through and including 0 to K_Y. The first two columns of Table 2.8 present an example. As can be seen, each integer raw score on Form Y has a scale score equivalent. For example, the scale score equivalent of a Form Y raw score of 24 is 22.3220. These equivalents resulted from an earlier equating of Form Y.

When Form X is equated to Form Y, the Form Y equivalents are typically noninteger. These noninteger equivalents need to be converted to scale scores, so a procedure is needed to find scale score equivalents of noninteger scores on Form Y. Linear interpolation is used in the present approach. For example, to find the scale score equivalent of a Form Y

Table 2.8. Raw-to-Scale Score Conversion Tables.

| Raw Score | Form Y Scale Scores | | Form X Scale Scores | | | | | |
| | | | Mean Equating | | Linear Equating | | Equipercentile | |
	sc	sc_{int}	sc	sc_{int}	sc	sc_{int}	sc	sc_{int}
−.5	.5000	1	.5000	1	.5000	1	.5000	1
0	.5000	1	.5000	1	.5000	1	.5000	1
1	.5000	1	.5000	1	.5000	1	.5000	1
2	.5000	1	.5000	1	.5000	1	.5000	1
3	.5000	1	.5000	1	.5000	1	.5000	1
4	.5000	1	.5000	1	.5000	1	.5000	1
5	.6900	1	.5242	1	.5000	1	.5000	1
6	1.6562	2	.8131	1	.5000	1	.5949	1
7	3.1082	3	1.8412	2	.6878	1	1.1874	1
8	4.6971	5	3.3106	3	1.7681	2	2.1098	2
9	6.1207	6	4.8784	5	3.3715	3	3.4645	3
10	7.4732	7	6.2930	6	5.0591	5	4.9258	5
11	8.9007	9	7.6550	8	6.5845	7	6.3678	6
12	10.3392	10	9.0839	9	8.0892	8	7.7386	8
13	11.6388	12	10.5047	11	9.6489	10	9.2622	9
14	12.8254	13	11.7899	12	11.1303	11	10.8456	11
15	14.0157	14	12.9770	13	12.4663	12	12.1050	12
16	15.2127	15	14.1682	14	13.7610	14	13.4491	13
17	16.3528	16	15.3579	15	15.0626	15	14.8738	15
18	17.3824	17	16.4839	16	16.3109	16	16.1515	16
19	18.3403	18	17.5044	18	17.4321	17	17.3912	17
20	19.2844	19	18.4606	18	18.4729	18	18.4958	18
21	20.1839	20	19.3990	19	19.4905	19	19.6151	20
22	20.9947	21	20.2872	20	20.4415	20	20.5533	21
23	21.7000	22	21.0845	21	21.2813	21	21.4793	21
24	22.3220	22	21.7792	22	22.0078	22	22.2695	22
25	22.9178	23	22.3979	22	22.6697	23	22.9353	23
26	23.5183	24	22.9943	23	23.3214	23	23.6171	24
27	24.1314	24	23.5964	24	23.9847	24	24.2949	24
28	24.7525	25	24.2105	24	24.6590	25	24.8496	25
29	25.2915	25	24.8212	25	25.2581	25	25.3538	25
30	25.7287	26	25.3472	25	25.7400	26	25.7841	26
31	26.1534	26	25.7828	26	26.2104	26	26.2176	26
32	26.6480	27	26.2164	26	26.7684	27	26.7281	27
33	27.2385	27	26.7232	27	27.4343	27	27.2908	27
34	27.9081	28	27.3238	27	28.2070	28	27.9216	28
35	28.6925	29	28.0080	28	29.1886	29	28.7998	29
36	29.7486	30	28.8270	29	30.5595	31	30.1009	30
37	31.2010	31	29.9336	30	32.1652	32	31.3869	31
38	32.6914	33	31.3908	31	33.7975	34	32.8900	33
39	34.1952	34	32.8830	33	35.2388	35	34.2974	34
40	35.4615	35	34.3565	34	36.5000	36	35.3356	35
40.5	36.5000	36	34.9897	35	36.5000	36	36.5000	36

score of 24.5 in Table 2.8, find the scale score that is halfway between the scale score equivalents of Form Y raw scores of 24 (22.3220) and 25 (22.9178). The reader should verify that this value is 22.6199.

Note that scale score equivalents are provided in the table for raw scores of −.5 and 40.5. These values provide minimum and maximum scale scores when equipercentile equating is used. (As was indicated earlier, the minimum equated raw score in equipercentile equating is −.5 and the maximum is $K_Y + .5$.)

To make the specification of conversion for Form Y to scale scores more explicit, let y_i refer to the i-th point that is tabled. For $-.5 \le y \le K_Y + .5$, define y_i^* as the tabled raw score that is the *largest* among the tabled scores that are *less than or equal* to y. In this case, the linearly inter-polated raw-to-scale score transformation is defined as

$$sc(y) = sc(y_i^*) + \frac{y - y_i^*}{y_{i+1}^* - y_i^*} [sc(y_{i+1}^*) - sc(y_i^*)], \quad -.5 \le y \le K_Y + .5,$$

$$= sc(-.5), \qquad\qquad\qquad\qquad y < -.5,$$

$$= sc(K_Y + .5), \qquad\qquad\qquad y > K_Y + .5, \qquad (2.21)$$

where y_{i+1}^* is the *smallest* tabled raw score that is *greater than or equal to* y_i^*. Note that $sc(-.5)$ is the minimum scale score and that $sc(K_Y + .5)$ is the maximum scale score.

To illustrate how this equation works, refer again to Table 2.8. How would the scale score equivalent of a raw score of $y = 18.3$ be found using equation (2.21)? Note that $y_i^* = 18$, because this score is the *largest* tabled score that is *less than or equal to* y. Using equation (2.21),

$$sc(y) = sc(18) + \frac{18.3 - 18}{19 - 18} [sc(19) - sc(18)]$$

$$= 17.3824 + \frac{18.3 - 18}{19 - 18} [18.3403 - 17.3824]$$

$$= 17.6698.$$

To illustrate that equation (2.21) is a linear interpolation expression, note that the scale score equivalent of 18 is 17.3842. The scale score 18.3 is, proportionally, .3 of the way between 18 and 19. This .3 value is multiplied by the difference between the scale score equivalents at 19 (18.3403) and at 18 (17.3824). Then .3 times this difference is .3[18.3403 − 17.3824] = .2874. Adding .2874 to 17.3824 gives 17.6988.

Typically, the tabled scores used to apply equation (2.21) will be integer raw scores along with −.5 and $K_Y + .5$. Equation (2.21), however, allows for more general schemes. For example, scale score equivalents could be tabled at each half raw score, such as −.5, .0, .5, 1.0, etc.

In practice, integer scores, which are found by rounding $sc(y)$, are re-ported to examinees. The third column of the table provides these integer scale score equivalents for integer raw scores (sc_{int}). A raw score $-.5$ was set equal to a scale score value of .5 and a raw score of 40.5 was set equal to a scale score value of 36.5. These values were chosen so that the minimum possible rounded scale score would be 1 and the maximum 36. In rounding, a convention is used where a scale score that precisely equals an integer score plus .5 rounds up to the next integer score. The exception to this convention is that the scale score is rounded down for the highest scale score, so that 36.5 rounds to 36.

To find the scale score equivalents of the Form X raw scores, the raw scores on Form X are first equated to raw scores on Form Y using equation (2.18). Then, substituting $e_Y(x)$ for y in equation (2.21),

$$sc[e_Y(x)] = sc(y_i^*) + \frac{e_Y(x) - y_i^*}{y_{i+1}^* - y_i^*}[sc(y_{i+1}^*) - sc(y_i^*)], -.5 \le e_Y(x) \le K_X + .5.$$

$$(2.22)$$

In this equation, y_i^* is defined as the *largest* tabled raw score that is *less than or equal to* $e_Y(x)$. This definition of y_i^* as well as the definition of y_{i+1}^* are consistent with their definitions in equation (2.21). The transformation is defined only for the range of Form X scores, $-.5 \le x \le K_X + .5$. There is no need to define this function outside this range, because $e_Y(x)$ is defined only in this range in equation (2.17). The minimum and maximum scale scores for Form X are identical to those for Form Y, which occur at $sc[e_Y(x = -.5)]$ and at $sc[e_Y(x = K_X + .5)]$, respectively.

As an example, equation (2.22) is used with the ACT Mathematics equating example. Suppose that the scale score equivalent of a Form X raw score of 24 is to be found using equipercentile equating. In Table 2.7, a Form X raw score of 24 is shown to be equivalent to a Form Y raw score of 23.9157. To apply equation (2.22), note that the largest Form Y raw score in Table 2.8 that is less than 23.9157 is 23. So, $y_i^* = 23$, and $y_{i+1}^* = 24$. From Table 2.8, $sc(23) = 21.7000$ and $sc(24) = 22.3220$. Applying equation (2.22),

$$sc[e_Y(x = 24)] = sc(23.9157)$$

$$= sc(23) + \frac{23.9157 - 23}{24 - 23}[sc(24) - sc(23)]$$

$$= 21.7000 + \frac{23.9157 - 23}{24 - 23}[22.3220 - 21.7000]$$

$$= 22.2696.$$

For a Form X raw score of 24, this value agrees with the value using equipercentile equating in Table 2.8, apart from rounding. Rounding to an integer, $sc_{int}[e_Y(x = 24)] = 22$.

Mean and linear raw score equating results can be converted to nonlinear scale scores by substituting $m_Y(x)$ or $l_Y(x)$ for y in equation (2.21).

The raw score equivalents from either the mean or linear methods might fall outside the range of possible Form Y scores. This problem is handled in equation (2.21) by truncating the scale scores. For example, if $l_Y(x) < -.5$, then $sc(y) = sc(-.5)$ by equation (2.21). The unrounded and rounded raw-to-scale score conversions for the mean and linear equating results are presented in Table 2.8.

Inspecting the central moments of scale scores can be useful in judging the accuracy of equating. Ideally, after equating, the scale score moments for converted Form X scores would be identical to those for Form Y. However, the moments typically are not identical, in part because the scores are discrete. If equating is successful, then the scale score moments for converted Form X scores should be very similar (say agree to at least one decimal place) to the scale score moments for Form Y. Should the Form X moments be compared to the rounded or unrounded Form Y moments? The answer is not entirely clear. However, the approach taken here is to compare the Form X moments to the Form Y unrounded moments. The rationale for this approach is that the unrounded transformation for Form Y most closely defines the score scale for the test, whereas rounding is used primarily to facilitate score interpretability. Following this logic, the use of Form Y unrounded moments for comparison purposes should lead to greater score scale stability when, over time, many forms become involved in the equating process.

Moments are shown in Table 2.9 for Form Y and for Form X using mean, linear, and equipercentile equating. Moments are shown for the unrounded (sc) and rounded (sc_{int}) score transformations. Note that the process of rounding affects the moments for Form Y. Also, the Form X scale

Table 2.9. Scale Score Moments.

Test Form	$\hat{\mu}_{SC}$	$\hat{\sigma}_{SC}$	$\widehat{sk}_{SC}$	$\widehat{ku}_{SC}$
Form Y				
unrounded	16.5120	8.3812	−.1344	2.0557
rounded	16.4875	8.3750	−.1025	2.0229
Form X Equated to Form Y Scale for Various Methods				
Mean				
unrounded	16.7319	7.6474	−.1868	2.1952
rounded	16.6925	7.5965	−.1678	2.2032
Linear				
unrounded	16.5875	8.3688	−.1168	2.1979
rounded	16.5082	8.3065	−.0776	2.1949
Equipercentile				
unrounded	16.5125	8.3725	−.1300	2.0515
rounded	16.4324	8.3973	−.1212	2.0294

score mean for mean equating (both rounded and unrounded) is much larger than the unrounded scale score mean for Form Y. Presumably, the use of a nonlinear raw-to-scale score transformation for Form Y is responsible. When the raw-to-scale score conversion for Form Y is nonlinear, the mean scale score for Form X is typically not equal to the mean scale score for Form Y for mean and linear equating. Similarly, when the raw-to-scale score conversion for Form Y is nonlinear, the standard deviation of the Form X scale scores typically is not equal to the standard deviation of Form Y scale scores for linear equating.

For equipercentile equating, the unrounded moments for Form X are similar to the unrounded moments for Form Y. The rounding process results in the mean of Form X being somewhat low. Is there anything that can be done to raise the mean of the rounded scores? Refer to Table 2.8. In this table, a raw score of 23 converts to an unrounded scale score of 21.4793 and a rounded scale score of 21. If the unrounded converted score had been only .0207 points higher, then the rounded converted score would have been 22. This observation suggests that the rounded conversion might be adjusted to make the moments more similar. Consider adjusting the conversion so that a raw score of 23 converts to a scale score of 22 (instead of 21) and a raw score of 16 converts to a scale score of 14 (instead of 13). The moments for the adjusted conversion are as follows: $\hat{\mu}_{sc} = 16.5165$, $\hat{\sigma}_{sc} = 8.3998$, $\widehat{sk}_{sc} = -.1445$, and $\widehat{ku}_{sc} = 2.0347$. Overall, the moments of the adjusted conversion seem closer to the moments of the original unrounded conversion. For this reason, the adjusted conversion might be used in practice.

Should the rounded conversions actually be adjusted in practice? To the extent that moments for the Form X rounded scale scores are made more similar to the unrounded scale score moments for Form Y, adjusting the conversions would seem advantageous. However, adjusting the conversions might lead to greater differences between the cumulative distributions of scale scores for Form X and Form Y at some scale score points. That is, adjusted conversions lead to less similar percentile ranks of scale scores across the two forms. In addition, adjusted conversions affect the scores of individual examinees.

Because adjusting can lead to less similar scale score distributions, and because it adds a subjective element into the equating process, we typically take a conservative approach to adjusting conversions. A rule of thumb that we often follow is to consider adjusting the conversions only if the moments are closer after adjusting than before adjusting, and the unrounded conversion is within .1 point of rounding to the next higher or lower value (e.g., 21.4793 in the example is within .1 point of rounding to 22). Smoothing methods are considered in Chapter 3, which might eliminate the need to consider subjective adjustments.

In the examples, scale score equivalents of integer raw scores were specified and linear interpolation was used between the integer scores. If more precision is desired, scale score equivalents of fractional raw scores could be specified. The procedures associated with equations (2.21) and (2.22) are expressed in sufficient generality to handle this additional precision. Procedures using nonlinear interpolation also could be developed, although linear interpolation is likely sufficient for practical purposes.

When score scales are established, the highest and lowest possible scale scores are often fixed at particular values. For example, the ACT score scale is said to range from 1 to 36. The approach taken here to scaling when using nonlinear conversions is to fix the ends of the score scale at specific points. Over time, if forms become easier or more difficult, the end points could be adjusted. However, such adjustments would require careful judgment. An alternative procedure involves leaving enough room at the top and bottom of the score scale to handle these problems. For example, suppose that the rounded score scale for an original form is to have a high score of 36 for the first form developed. However, there is a desire to allow scale scores on subsequent forms to go as high as 40 if the forms become more difficult. For the initial Form Y, a scale score of 36 could be assigned to a raw score equal to K_Y and a scale score of 40.5 could be assigned to a raw score equal to $K_Y + .5$. If subsequent forms are more difficult than Form Y, the procedures described here could lead to scale scores as high as 40.5. Of course, alternate interpolation rules could lead to different properties. Rules for nonlinear scaling and equating also might be developed that would allow the highest and lowest scores to float without limit. The approach taken here is to provide a set of equations to be used for nonlinear equating and scaling that can adequately handle, in a consistent manner, many of the situations we have encountered in practice.

One practical problem sometimes occurs when the highest possible raw score does not equate to the highest possible scale score. For the ACT, for example, the highest possible raw score is assigned a scale score value of 36, regardless of the results of the equating. For the SAT (Donlon, 1984, p. 19), the highest possible raw score is assigned a scale score value of 800, and other converted scores are sometimes adjusted, as well.

Equating Using Single Group Designs

If practice, fatigue, and other order effects do not have an effect on scores, then the statistical process for mean, linear, and equipercentile equating using the single group design (without counterbalancing) is essentially the same as with the random groups design. However, order typically has an affect, and for this reason the single group design (without counterbalancing) is not recommended.

When the single group design with counterbalancing is used, the following four equatings can be conducted:

1. Equate Form X and Form Y using the random groups design for examinees who were administered Form X first and Form Y first.
2. Equate Form X and Form Y using the random groups design for examinees who were administered Form X second and Form Y second.
3. Equate Form X and Form Y using the single group design for examinees who were administered Form X first and Form Y second.
4. Equate Form X and Form Y using the single group design for examinees who were administered Form X second and Form Y first.

Compare equatings 1 and 2. Standard errors of equating described in Chapter 7 can be used as a baseline for comparing the equatings. If the equatings give different results, apart from sampling error, then Forms X and Y are differentially affected by appearing second. In this case, only the first equating should be used. Note that the first equating is a random groups equating, so it is unaffected by order. The problem with using the first equating only is that the sample size might be quite small. However, when differential order effects occur, then equating 1 might be the only equating that would not be biased.

If equatings 1 and 2 give the same results, apart from sampling error, then Forms X and Y are similarly affected by appearing second. In this case, all of the data can be used. One possibility would be to pool all of the Form X data and all of the Form Y data, and equate the pooled distributions. Angoff (1971) and Petersen *et al.* (1989) provide procedures for linear equating. Holland and Thayer (1990) presented a systematic scheme that is based on statistical tests using log-linear models for equipercentile equating under the single group counterbalanced design.

Equating Using Alternate Scoring Schemes

The presentation of equipercentile equating and scale scores assumed that the tests to be equated are scored number-correct with the observed scores ranging from 0 to the number of items. Although this type of scoring scheme is the one that is used most often with educational tests, alternative scoring procedures are becoming much more popular, and the procedures described previously can be generalized to other scoring schemes. For example, whenever raw scores are integer scores that range from 0 to a positive integer value, the procedures can be used directly by defining K as the maximum score on a form, rather than as the number of items on a form as has been done.

Some scoring schemes might produce discrete scores that are not necessarily integers. For example, when tests are scored using a correction for guessing, a fractional score point often is subtracted from the total score whenever an item is answered incorrectly. In this case, raw scores are not integers. However, the discrete score points that can possibly occur are specifiable and equally spaced. One way to conduct equating in this situation is to transform the raw scores. The lowest possible raw score is transformed to a score of 0, the next lowest raw score is transformed to a score of 1, and so on through K, which is defined as the transformed value of the highest possible raw score. The procedures described in this chapter then can be applied and the scores transformed back to their original units.

Equipercentile equating also can be conducted when the scores are considered to be continuous, which might be the case when equating forms of a computerized adaptive test. In many ways, equating in this situation is more straightforward than with discrete scores, because the definitional problems associated with continuization do not need to be considered. Still, difficulties might arise in trying to define score equivalents in portions of the score scale where few examinees earn scores. In addition, even if the range of scores is potentially infinite, the range of scores for which equipercentile equivalents are to be found needs to be considered.

Preview of What Follows

In this chapter, we described many of the issues associated with observed score equating using the random groups design, including defining methods, describing their properties, and estimating the relationships. We also discussed the relationships between equating and score scales. One of the major relevant issues not addressed in this chapter is the use of smoothing methods to reduce random error in estimating equipercentile equivalents. Smoothing is the topic of Chapter 3. Also, as we show in Chapters 4 and 5, the implementation of observed score equating methods becomes much more complicated when the groups administered the two forms are not randomly equivalent. Observed score methods associated with IRT are described in Chapter 6. Estimating random error in observed score equating is discussed in detail in Chapter 7, and practical issues are discussed in Chapter 8.

Exercises

2.1. From Table 2.2 find $P(2.7)$, $P(.2)$, $P^{-1}(25)$, $P^{-1}(97)$.

2.2. From Table 2.2, find the linear and mean conversion equation for converting scores on Form X to the Form Y scale.

2.3. Find the mean and standard deviation of the Form X scores converted to the Form Y scale for the equipercentile equivalents shown in Table 2.3.

2.4. Fill in Table 2.10 and Table 2.11.

Table 2.10. Score Distributions for Exercise 2.4.

x	$f(x)$	$F(x)$	$P(x)$	y	$g(y)$	$G(y)$	$Q(y)$
0	.00			0	.00		
1	.01			1	.02		
2	.02			2	.05		
3	.03			3	.10		
4	.04			4	.20		
5	.10			5	.25		
6	.20			6	.20		
7	.25			7	.10		
8	.20			8	.05		
9	.10			9	.02		
10	.05			10	.01		

Table 2.11. Equated Scores for Exercise 2.4.

x	$m_Y(x)$	$l_Y(x)$	$e_Y(x)$
0			
1			
2			
3			
4			
5			
6			
7			
8			
9			
10			

2.5. If the standard deviations on Form X and Y are equal, which methods, if any, among mean, linear, and equipercentile will produce the same results? Why?

2.6. Suppose that a raw score of 20 on Form W was found to be equivalent to a raw score of 23.15 on Form X. What would be the scale score equivalent of a Form W raw score of 20 using the Form X equipercentile conversion shown in Table 2.8?

2.7. Suppose that the linear raw-to-scale score conversion equation for Form Y was $sc(y) = 1.1y + 10$. Also suppose that the linear equating of Form X to Form Y was $l_Y(x) = .8x + 1.2$. What is the linear conversion of Form X scores to scale scores?

2.8. In general, how would the shape of the distribution of Form X raw scores equated to the Form Y raw scale compare to the shape of the original Form X raw score distribution using mean, linear, and equipercentile equating?

CHAPTER 3

Random Groups—Smoothing in Equipercentile Equating

As described in Chapter 2, sample statistics are used to estimate equating relationships. For mean and linear equating, the use of sample means and standard deviations in place of the parameters typically leads to adequate equating precision, even when the sample size is fairly small. However, when sample percentiles and percentile ranks are used to estimate equipercentile relationships, equating often is not sufficiently precise for practical purposes because of sampling error.

One indication that considerable error is present in estimating equipercentile equivalents is that score distributions and equipercentile relationships appear irregular when graphed. For example, the equating shown in Figure 2.10 was based on over 4000 examinees per form. Even with these large sample sizes, the equipercentile relationship is somewhat irregular. Presumably, if very large sample sizes or the entire population were available, score distributions and equipercentile relationships would be reasonably smooth.

Smoothing methods have been developed that produce estimates of the empirical distributions and the equipercentile relationship which will have the smoothness property that is characteristic of the population. In turn, it is hoped that the resulting estimates will be more precise than the unsmoothed relationships. However, the danger in using smoothing methods is that the resulting estimates of the population distributions, even though they are smooth, might be poorer estimates of the population distributions or equating relationship than the unsmoothed estimates. The quality of analytic smoothing methods is an empirical issue and has been the focus of research (Cope and Kolen, 1990; Fairbank, 1987; Hanson et al., 1994; Kolen, 1984). Also, when there are very few score points, the equating relationships can appear irregular, even after smoothing, because of the discreteness issues discussed in Chapter 2. Two general types of smoothing can be conducted: In *presmoothing*, the score distributions are smoothed; in *postsmoothing*, the equipercentile equivalents are smoothed. Smoothing can be conducted by hand or by using analytical methods. Various analytic smoothing techniques are described in this chapter. In addition, we consider various practical issues in choosing among various equating relationships.

A Conceptual Statistical Framework for Smoothing

A conceptual statistical framework is developed in this section which is intended to provide a framework for distinguishing random error in equipercentile equating from systematic error that is introduced by smoothing. The following discussion considers different sources of equating errors. To be clear that the focus is on a Form X raw score, define x_i as a particular score on Form X. Define $e_Y(x_i)$ as the population equipercentile equivalent at that score, and define $\hat{e}_Y(x_i)$ as the sample estimate. Also assume that $E[\hat{e}_Y(x_i)] = e_Y(x_i)$, where E is the expectation over random samples. Equating error at a particular score is defined as the difference between the sample equipercentile equivalent and the population equipercentile equivalent. That is, equating error at score x_i for a given equating is

$$[\hat{e}_Y(x_i) - e_Y(x_i)]. \tag{3.1}$$

Conceive of replicating the equating a large number of times; for each replication the equating is based on two random samples of examinees from a population of examinees who take Form X and Form Y, respectively. Equating error variance at score point x_i is

$$var[\hat{e}_Y(x_i)] = E[\hat{e}_Y(x_i) - e_Y(x_i)]^2, \tag{3.2}$$

where the variance is taken over replications. The standard error of equating is defined as the square root of the error variance,

$$se[\hat{e}_Y(x_i)] = \sqrt{var[\hat{e}_Y(x_i)]} = \sqrt{E[\hat{e}_Y(x_i) - e_Y(x_i)]^2}. \tag{3.3}$$

The error indexed in equations (3.1)–(3.3) is random error that is due to the sampling of examinees to estimate the population quantity.

A graphic depiction is given in Figure 3.1. In this figure, the Form X equivalents of Form Y scores, indicated by $e_Y(x)$, are graphed. Also, a particular score, x_i, is shown on the horizontal axis. Above x_i, a distribution is plotted that represents estimated Form Y equivalents of x_i over replications of the equating. As can be seen, the mean equivalent falls on the

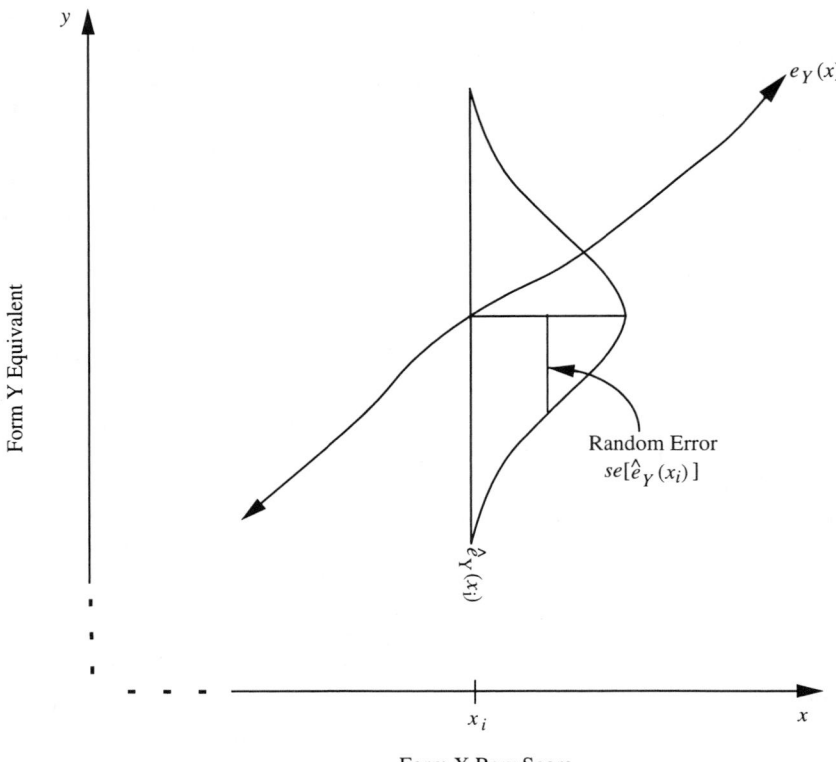

Form X Raw Score

Figure 3.1. Schematic plot illustrating random equating error in unsmoothed equipercentile equating.

$e_Y(x)$ curve. Random variability, due to the sampling of examinees, is indexed by $se[\hat{e}_Y(x_i)]$.

Smoothing methods often can be used to reduce the error variability. Define $\hat{t}_Y(x_i)$ as an alternative estimator of $e_Y(x_i)$ that results from using a smoothing method. Define

$$t_Y(x_i) = \mathbf{E}[\hat{t}_Y(x_i)], \tag{3.4}$$

which is the expected value over replications of the smoothed equating. Defining total error at score x_i as $\hat{t}_Y(x_i) - e_Y(x_i)$, the mean-squared error (*mse*) in equating at score x_i using the smoothing method is

$$mse[\hat{t}_Y(x_i)] = \mathbf{E}[\hat{t}_Y(x_i) - e_Y(x_i)]^2. \tag{3.5}$$

Random error variability in the smoothed equating relationships is indexed by

$$var[\hat{t}_Y(x_i)] = \mathbf{E}[\hat{t}_Y(x_i) - t_Y(x_i)]^2, \tag{3.6}$$

and

$$se[\hat{t}_Y(x_i)] = \sqrt{var[\hat{t}_Y(x_i)]}.$$

Systematic error, or bias, in equating using smoothing is defined as

$$bias[t_Y(x_i)] = t_Y(x_i) - e_Y(x_i). \tag{3.7}$$

Total error can be partitioned into random error and systematic error components as follows:

$$\hat{t}_Y(x_i) - e_Y(x_i) = [\hat{t}_Y(x_i) - t_Y(x_i)] \quad + [t_Y(x_i) - e_Y(x_i)].$$

$$\{\text{Total Error}\} \quad \{\text{Random Error}\} \quad \{\text{Systematic Error}\}$$

In terms of squared quantities,

$$mse[\hat{t}_Y(x_i)] = var[\hat{t}_Y(x_i)] \quad + \{bias[t_Y(x_i)]\}^2$$
$$= \mathbf{E}[\hat{t}_Y(x_i) - t_Y(x_i)]^2 + [t_Y(x_i) - e_Y(x_i)]^2. \tag{3.8}$$

Thus, when using a smoothing method, total error in equating is the sum of random error and systematic error components. Smoothing methods are designed to produce smooth functions which contain less random error than that for unsmoothed equipercentile equating. However, smoothing methods can introduce systematic error. The intent in using a smoothing method is for the increase in systematic error to be more than offset by the decrease in random error. Then the total error using the smoothing method is less than that for the unsmoothed equivalents. That is, smoothing at score point x_i is useful to the degree that $mse[\hat{t}_Y(x_i)]$ is less than $var[\hat{e}_Y(x_i)]$.

Refer to Figure 3.2 for a graphic description. In this figure, the Form X equivalents of Form Y scores, indicated by $e_Y(x)$, are graphed as they were in Figure 3.1. Also, $t_Y(x)$ is graphed and differs from $e_Y(x)$. This difference at x_i is referred to as "Systematic Error" in the graph. The distribution plotted above x_i represents Form Y equivalents of x_i over replications of the smoothed equating. The random variability due to sampling of examinees is indexed by $se[\hat{t}_Y(x_i)]$. Compare the random error component in Figure 3.2 to that in Figure 3.1, which presents random equating error without smoothing. This comparison suggests that the smoothing method results in less random equating error at score x_i than does the unsmoothed equipercentile equating. Thus, the smoothing method reduces random error but introduces systematic error.

The preceding discussion focused on equating error at a single score point. Overall indexes of error can be obtained by summing each of the error components over score points. In this case, the goal of smoothing can be viewed as reducing mean-squared (total) error in estimating the population equipercentile equivalents over score points.

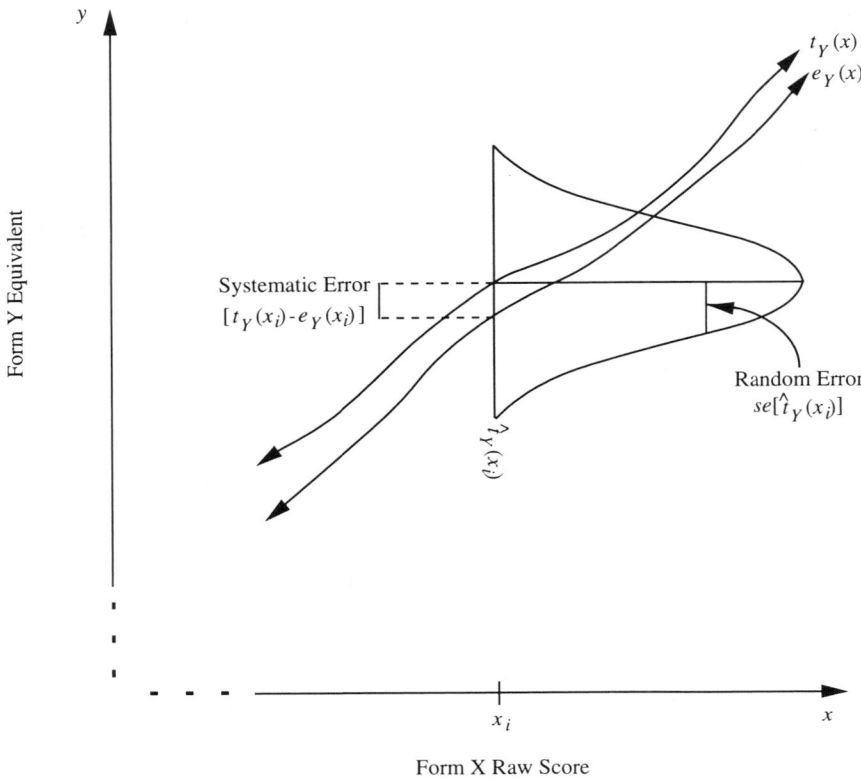

Figure 3.2. Schematic plot illustrating systematic and random equating error in smoothed equipercentile equating.

Properties of Smoothing Methods

Mean and linear equating methods can be viewed as smoothing methods that estimate the equipercentile relationship. In some situations, these methods can lead to less total error in estimating the equipercentile equivalents than equipercentile equating. For example, what if the score distributions for Form X and Form Y are identical in shape (i.e., they differ only in mean and standard deviation)? In this case, the population linear equating and equipercentile equating relationships are identical. For samples of typical size, linear equating will produce less total error in estimating equipercentile equivalents than equipercentile equating when the distributions are the same shape, because less random error is associated with linear equating than with equipercentile equating (see Chapter 7). Even if the distribution shapes are only similar, linear methods might still produce less total error in estimating equipercentile equivalents than equipercentile equating for small samples.

A smoothing method should possess certain desirable characteristics for it to be useful in practice. First, the method should produce *accurate* estimates of the population distributions or equipercentile equivalents. That is, the method should not systematically distort the relationship in a manner that has negative practical consequences. Second, the method should be *flexible* enough to handle the variety of distributions and equipercentile relationships that are found in practice. Third, there should be a *statistical framework* for studying fit. Fourth, the method should improve estimation, as shown by an *empirical research* base. Fortunately, there are analytic smoothing methods that share these characteristics.

Three types of presmoothing methods have been considered for use in equating. One type of method estimates the relative frequency at a score point by averaging the relative frequency at a score point with relative frequencies at surrounding score points. These *rolling average* or *kernel methods* were reviewed by Kolen (1991) and include the Cureton and Tukey (1951) method. Kolen (1991) indicated that these methods often lead to estimated distributions that appear bumpy or are systematically distorted. Although these methods have been found to improve estimation, the improvement is less than for some other methods. For these reasons, rolling average methods are not described further here. Rather, a *log-linear* method and a *strong true score method* for presmoothing are described next. As will be described later in this chapter, these methods have been studied empirically through simulation and have been shown to improve estimation of test score distributions (Kolen, 1991). A postsmoothing method that uses *cubic splines* also is described. The postsmoothing method also has been shown, empirically, to improve estimation (Fairbank, 1987; Kolen, 1984). In addition, Hanson et al. (1994) demonstrated, empirically, that the presmoothing and postsmoothing methods described here improve estimation of equipercentile equivalents to a similar extent. The methods that are described possess the four characteristics of smoothing methods that were described earlier: they have been shown to produce accurate results, they are flexible, they are associated with a statistical framework for studying fit, and they can improve estimation as shown by an empirical research base.

Presmoothing Methods

In presmoothing methods, the score distribution is smoothed. In smoothing the distributions, accuracy in estimating the distributions is crucial. One important property that relates closely to accuracy is *moment pres-*

ervation. In moment preservation, the smoothed distribution has at least some of the same central moments as the observed distribution. For example, a method preserves the first two central moments if the mean and standard deviation of the smoothed distribution are the same as the mean and standard deviation of the unsmoothed distribution.

One presmoothing method uses a polynomial log-linear model with polynomial contrasts to smooth score distributions. The second method is a strong true score model. In strong true score models, a general distributional form is specified for true scores. A distributional form is also specified for error given true score. For both methods, after the distributions are smoothed, Form X is equated to Form Y using the smoothed distributions and equipercentile equating. This equating relationship along and the raw-to-scale score transformation for Form Y are used to convert Form X scores to scale scores.

Polynomial Log-linear Method

Log-linear models that take into account the ordered property of test scores can be used to estimate test score distributions. The method considered here fits polynomial functions to the log of the sample density. This model was described by Darroch and Ratcliff (1972), Haberman (1974a, 1974b, 1978), and Rosenbaum and Thayer (1987). Holland and Thayer (1987) presented a thorough description of this model including algorithms for estimation, properties of the estimates, and applications to fitting test score distributions. The polynomial log-linear method fits a model of the following form to the distribution:

$$\log[N_X f(x)] = \omega_0 + \omega_1 x + \omega_2 x^2 + \cdots + \omega_C x^C. \tag{3.9}$$

In this equation, the log of the density is expressed as a lower order polynomial of degree C. For example, if $C = 2$, then $\log[N_X]f(x) = \omega_0 + \omega_1 x + \omega_2 x^2$, and the model is a polynomial of degree 2 (quadratic). The ω parameters in the model can be estimated by the method of maximum likelihood. Note that the use of logarithms allows for log-linear models to be additive as in equation (3.9).

The resulting fitted distribution has the moment preservation property that the first C moments of the fitted distribution are identical to those of the sample distribution. For example, if $C = 2$, then the mean and standard deviation of the fitted distribution are identical to the mean and standard deviation of the sample distribution. Holland and Thayer (1987) described algorithms for maximum likelihood estimation with this method. Some statistical packages for log-linear models (e.g., the LOGLINEAR procedure of SPSS-X) can also be used. The *RG Equate* and *Usmooth* Macintosh computer programs described in Appendix B also can be used.

The choice of C is an important consideration when using this method. The fitted distribution can be compared, subjectively, to the empirical distribution. Because this method uses a log-linear model, the statistical methods that have been developed for assessing the fit of these models also can be used. In one procedure, likelihood ratio chi-square goodness-of-fit statistics are calculated for each C that is fit. The overall likelihood ratio chi-square for each model also can be tested for significance. In addition, because the models are hierarchical, likelihood ratio difference chi-squares can be tested for significance. For example, the difference between the overall likelihood ratio chi-square statistics for $C = 2$ and $C = 3$ can be compared to a chi-square table with one degree of freedom. A significant difference would suggest that the model with more terms (e.g., $C = 3$) fits the data better than the model with fewer terms (e.g., $C = 2$). In model selection, preference might be given to the simplest model that adequately fits the distribution, under the presumption that models which are more complicated than necessary might lead to excess equating error. A structured procedure that uses the likelihood ratio difference chi-squares to choose C is described by Haberman (1974a) and Hanson (1990).

Because multiple significance tests are involved, these procedures should be used in combination with the inspection of graphs, the inspection of central moments, and previous experience in choosing a degree of smoothing. When inspecting graphs, the investigator tries to judge whether the fitted distribution is smooth and does not depart too much from the empirical distribution. Refer to Bishop *et al.* (1975) for a general description of model fitting procedures for log-linear models.

Strong True Score Method

Unlike the log-linear method, strong true score methods require the use of a parametric model for true scores. Lord (1965) developed a procedure, referred to here as the *beta4* method, to estimate the distribution of true scores. This procedure also results in a smooth distribution of observed scores, which is the primary reason that Lord's (1965) method is of interest here. In the development of the beta4 procedure, a parametric form is assumed for the population distribution of proportion-correct true scores, $\psi(\tau)$. Also, a conditional parametric form is assumed for the distribution of observed score given true score, $f(x|\tau)$. Then the observed score distribution can be written as follows:

$$f(x) = \int_0^1 f(x|\tau)\psi(\tau)\, d\tau. \tag{3.10}$$

In the beta4 method proposed by Lord (1965) the true score distribution, $\psi(\tau)$, was assumed to be four-parameter beta. The four-parameter beta has two parameters that allow for a wide variety of shapes for the distribution.

For example, the four-parameter beta can be skewed positively or negatively, and it can even be U-shaped. The four-parameter beta also has parameters for the high– and low–proportion-correct true scores that are within the range of zero to one. The conditional distribution of observed score given true score, $f(x|\tau)$, was assumed by Lord (1965) to be either binomial or compound binomial. Lord (1965) provided a two-term approximation to the compound binomial method that is usually used in implementing the method. The score distribution, $f(x)$, that results from the use of equation (3.10) in combination with the model assumptions just described is referred to as the *four-parameter beta compound binomial distribution* or the *beta4 distribution*. This distribution can take on a wide variety of forms.

Lord (1965) presented a procedure for estimating this distribution and the associated true score distribution by the method of moments. This estimation procedure uses the number of items, the first four central moments (mean, standard deviation, skewness, and kurtosis) of the sample distribution, and a parameter Lord referred to as k. Lord's k can be estimated directly from the coefficient alpha reliability coefficient. Hanson (1991b) also described the estimation procedure in detail. He described situations in which the method of moments leads to invalid parameter values, such as an upper limit for proportion-correct true scores above 1, and provided procedures for dealing with them.

One important property of this method is that the first four central moments of the fitted distribution agree with those of the sample distribution, provided there are no invalid parameter estimates. Otherwise, fewer than four central moments agree. For example, suppose that the method of moments using the first four central moments produces invalid parameter values. Then the method described by Hanson (1991b) fits the distribution using the method of moments so that the first three central moments agree, and the fourth moment of the fitted distribution is as close as possible to the fourth moment of the observed distribution.

As with the log-linear model, the fit of the model can be evaluated by comparing plots and central moments of the sample and fitted distributions. Statistical methods also can be used. A standard chi-square goodness-of-fit statistic can be calculated, as suggested by Lord (1965). Assuming that all score points are included in the calculation of the chi-square statistic, the degrees of freedom are the number of score points ($K + 1$, to account for a score of 0), minus 1, minus the number of parameters fit. For the beta4 method, the degrees of freedom are $K - 4 = (K + 1) - 1 - 4$.

There are some other strong true score methods that are related to the beta4 method. One simplification of the beta4 method is the *beta-binomial* or *negative hypergeometric distribution* described by Keats and Lord (1962). One difference between this model and the Lord (1965) model is that the two-parameter beta distribution is used for true scores. The two-parameter beta distribution is identical to a four-parameter beta distribution with the highest and lowest proportion-correct true scores set at 1 and 0, respectively. The beta-binomial model uses a binomial distribution for the distribution of observed score given true score. The beta-binomial distribution fits a narrower range of distributions than the beta4 distribution. For example, the beta-binomial distribution cannot be negatively skewed if the mean is less than one-half the items correct. Kolen (1991) concluded that the beta-binomial is not flexible enough to be used in typical equating applications. Carlin and Rubin (1991) studied a special case of the beta4 method that fits three moments, and found that it fit considerably better than the beta-binomial model.

Lord (1969) generalized the beta4 distribution. In this generalization, the parametric form of the true score distribution was not specified. Lord (1969, 1980) referred to the resulting procedure as *Method 20*. Method 20 is more flexible than the beta4 method. For example, Method 20 can produce a variety of multimodal distributions. However, Lord (1969) indicated that Method 20 requires sample sizes of at least 10,000 examinees per form, which makes it impractical in most equating situations.

Another related approach fits continuous distributions to discrete test score distributions. Brandenburg and Forsyth (1974) fit a (continuous) four-parameter beta distribution directly to sample observed test score distributions and concluded that it fit very well. Use of this procedure eliminates further need for a continuization step in equipercentile equating. Although not presented in detail here, fitting of continuous distributions in equating deserves further investigation.

Illustrative Example

The ACT Mathematics example that was considered in the previous chapter is used to illustrate the presmoothing methods. The *RG Equate* Macintosh computer program described in Appendix B was used to conduct the equating. The first step in applying these methods is to fit the raw score distributions. The smoothed distributions (indicated by solid symbols) for Form Y are shown in Figure 3.3 along with the unsmoothed distributions. The distributions for Form X are shown in Figure 3.4. The beta4 and selected log-linear smoothed distributions are shown. In fitting the beta4 method for Form X, fitting all 4 moments resulted in invalid parameter estimates, so only the first 3 moments were fit. The beta4 model was fit setting Lord's $k = 0$.

Visual inspection suggests that the beta4 method fits the distributions of both forms very well. The log-linear method with $C = 2$ appears to fit both distributions poorly. For Form X and Form Y, $C = 6$ appears to fit the distributions well. The $C = 10$ smoothings appear to slightly overfit the distributions for both forms in the score range of 23–30, in that the fitted distributions are a bit irregular. These irregularities, along with the observation that the statistical tests indicate that values of C greater than 6 do

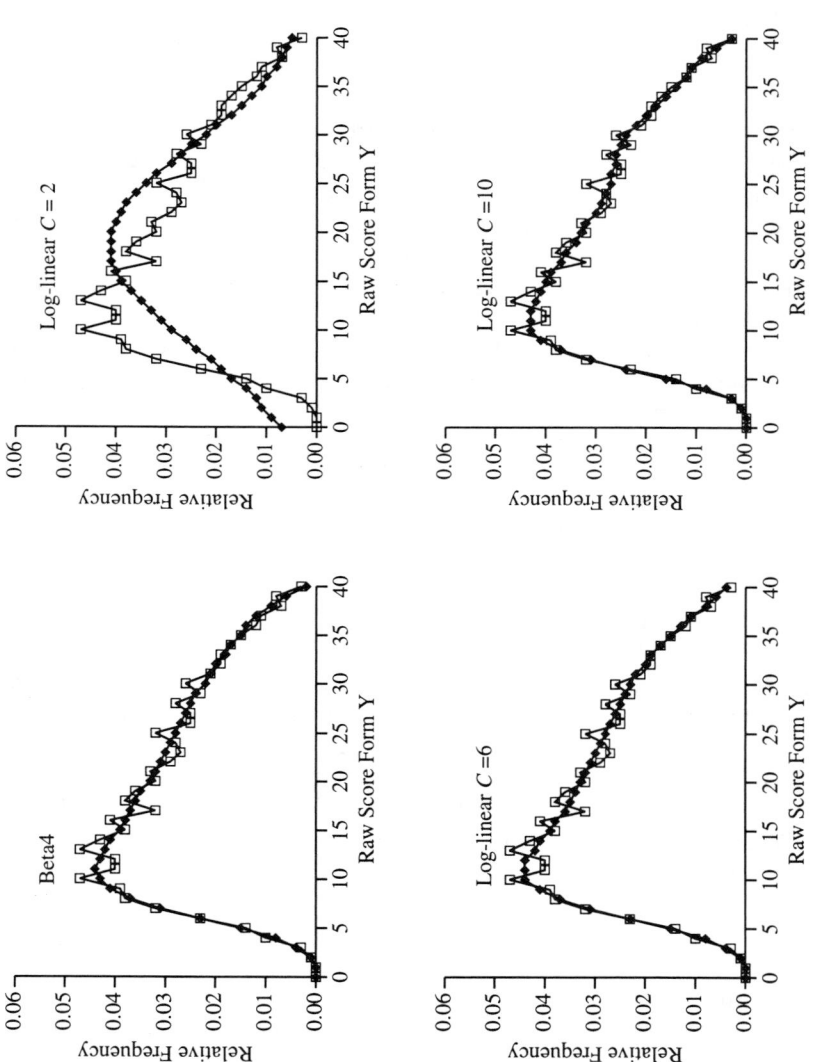

Figure 3.3. Presmoothing Form Y distribution.

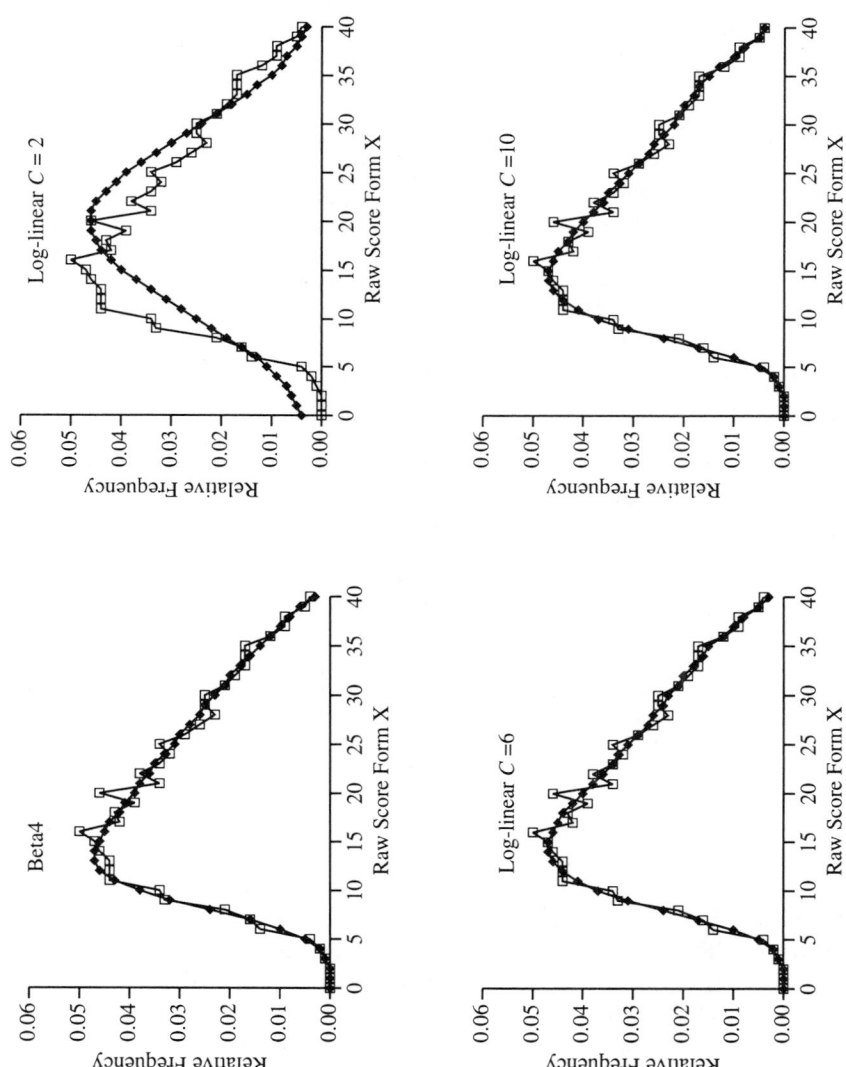

Figure 3.4. Presmoothing Form X distribution.

Table 3.1. Moments and Fit Statistics for Presmoothing.

Form	Method	$\hat{\mu}$	$\hat{\sigma}$	$\widehat{sk}$	$\widehat{ku}$	$\chi^2(df)$	$\chi^2_C - \chi^2_{C+1}$
Y	Sample	18.9798	8.9393	.3527	2.1464		
	Beta4	18.9798	8.9393	.3527	2.1464	31.64(36)	
	Log-linear						
	$C = 10$	18.9798	8.9393	.3527	2.1464	25.92(30)	
	$C = 9$	18.9798	8.9393	.3527	2.1464	26.38(31)	.46
	$C = 8$	18.9798	8.9393	.3527	2.1464	27.00(32)	.62
	$C = 7$	18.9798	8.9393	.3527	2.1464	28.30(33)	1.30
	$C = 6$	18.9798	8.9393	.3527	2.1464	29.45(34)	1.15
	$C = 5$	18.9798	8.9393	.3527	2.1464	39.31(35)	9.86
	$C = 4$	18.9798	8.9393	.3527	2.1464	61.53(36)	22.22
	$C = 3$	18.9798	8.9393	.3527	2.5167	318.66(37)	257.13
	$C = 2$	18.9798	8.9393	.0709	2.3851	489.47(38)	170.81
	$C = 1$	18.9798	11.8057	.1037	1.8134	1579.99(39)	1090.52
X	Sample	19.8524	8.2116	.3753	2.3024		
	Beta4[a]	19.8524	8.2116	.3753	2.2806	33.97(37)	
	Log-linear						
	$C = 10$	19.8524	8.2116	.3753	2.3024	29.68(30)	
	$C = 9$	19.8524	8.2116	.3753	2.3024	29.91(31)	.23
	$C = 8$	19.8524	8.2116	.3753	2.3024	29.94(32)	.03
	$C = 7$	19.8524	8.2116	.3753	2.3024	30.40(33)	.46
	$C = 6$	19.8524	8.2116	.3753	2.3024	30.61(34)	.20
	$C = 5$	19.8524	8.2116	.3753	2.3024	35.78(35)	5.18
	$C = 4$	19.8524	8.2116	.3753	2.3024	40.80(36)	5.01
	$C = 3$	19.8524	8.2116	.3753	2.6565	212.82(37)	172.02
	$C = 2$	19.8524	8.2116	.0082	2.5420	445.19(38)	232.36
	$C = 1$	19.8524	11.8316	.0150	1.7989	2215.02(39)	1769.83

[a] Only 3 moments could be fit using the beta4 method with Form X.

not improve the fit, suggest that $C = 10$ might be fitting aspects of the distributions that are due to sampling error.

Summary statistics for the fitted distributions are shown in Table 3.1 for Form Y and Form X. Because of the moment preservation property of the beta4 method, the first 3 or 4 moments of the fitted distribution for this method agree with those for the sample distribution. Only 3 moments could be fit using the beta4 method with Form X, so the kurtosis for the beta4 method differs from the kurtosis for the sample data. However, this difference in kurtosis values is small (2.3024 for the sample distribution and 2.2806 for the fitted distribution). For both distributions, the chi-square statistic, $\chi^2(df)$ for the beta4 method is less than its degrees of freedom, indicating a reasonable fit.

The log-linear method was fit using values of C ranging from 1 to 10 for both forms. Because of the moment preservation property of the log-linear method, the first 4 moments of the fitted distribution for $C \geq 4$ agree with those for the sample distribution, 3 moments agree for $C = 3$, and fewer moments agree for lower values of C.

Two likelihood ratio χ^2 statistics are presented in the table. The model selection strategy here is to use the significance tests to rule out models. Models that are not ruled out by any of the significance tests then are considered further. Preference is given to the simplest model that is not ruled out, under the presumption that the simplest model which adequately fits the distribution leads to less estimation error than more complicated models. The inspection of plots of the smoothed distributions also is considered in the process of choosing models for further consideration. In model fitting situations such as those considered here, significance tests are viewed more as providing a guide for choosing a model than as a means for providing a definitive decision rule.

The difference statistic, $\chi^2_C - \chi^2_{C+1}$, is a one degree of freedom χ^2 that is the difference between the overall χ^2 at C and the overall χ^2 at $C + 1$. A significant difference suggests that the model with parameter $C + 1$ improves the fit over the model with parameter C. This type of difference χ^2 often is used in fitting hierarchical log-linear models like the one described here. In using the $\chi^2_C - \chi^2_{C+1}$ statistic, a value of C is chosen that is one greater than the largest value of C that has a significant χ^2. For both distributions, using a nominal .05 level, the value at $C = 5$ is the highest value with a significant χ^2 (i.e., $\chi^2 > 3.84$), suggesting that $C \geq 6$ should be considered further. Note that, according to Haberman (1974), the experiment-wise significance level over 9 significance tests with a nominal .05 level is less than or equal to $.3698 = 1 - (1 - .05)^9$.

The term $\chi^2(df)$ test is an overall goodness-of-fit test that compares the fitted model to the empirical distribution. A significant χ^2 suggests that the model does not fit the observed data. For Form Y, the overall χ^2 is significant at the nominal .05 level for $C \leq 4$, suggesting that $C \geq 5$ should be considered further. For Form X, the overall χ^2 is significant at the .05 level for $C \leq 3$, suggesting that $C \geq 4$ should be considered further. (Note that, at the .05 level, the χ^2 critical values range from 43.8 to 54.6, approximately.)

The fit for $C = 6$ is the minimum value of C (the simplest model) that appears to meet both of the significance test criteria for both Form X and Form Y. The smoothed distributions shown in Figures 3.3 and 3.4 for $C = 6$ also appear to fit well. The models using $C = 6$ for Form X and $C = 6$ for Form Y are examined further.

After fitting the distributions, equipercentile methods are used to equate Form X and Form Y. The equipercentile relationships are presented in Table 3.2 and are graphed in Figure 3.5 for the beta4 method and the log-linear method with $C = 6$ in the same format that was used in Figure 2.10. Figure 3.5 also shows the identity equating and unsmoothed relationships. In addition, ± 1 standard error bands are shown. These bands were calcu-

Table 3.2. Raw-to-Raw Score Conversions for Presmoothing.

Form X Score	Standard Error	Form Y Equivalent Using Equating Method		
		Unsmoothed	Beta4	Log-Linear $C = 6$
0	1.9384	.0000	−.4581	−.4384
1	.8306	.9796	.1063	.1239
2	.5210	1.6462	.8560	.9293
3	.8210	2.2856	1.7331	1.8264
4	.2950	2.8932	2.6380	2.7410
5	.1478	3.6205	3.5517	3.6573
6	.2541	4.4997	4.4434	4.5710
7	.1582	5.5148	5.3311	5.4725
8	.1969	6.3124	6.2572	6.3577
9	.1761	7.2242	7.2121	7.2731
10	.1731	8.1607	8.1931	8.2143
11	.1952	9.1827	9.2010	9.1819
12	.1800	10.1859	10.2367	10.1790
13	.2311	11.2513	11.3003	11.2092
14	.2431	12.3896	12.3892	12.2750
15	.2138	13.3929	13.4985	13.3764
16	.2764	14.5240	14.6263	14.5111
17	.2617	15.7169	15.7633	15.6784
18	.3383	16.8234	16.9047	16.8638
19	.2826	18.0092	18.0470	18.0566
20	.2947	19.1647	19.1880	19.2469
21	.3299	20.3676	20.3258	20.4262
22	.3183	21.4556	21.4589	21.5911
23	.3865	22.6871	22.5890	22.7368
24	.3555	23.9157	23.7131	23.8595
25	.3013	25.0292	24.8287	24.9594
26	.3683	26.1612	25.9347	26.0374
27	.3532	27.2633	27.0296	27.0954
28	.3069	28.1801	28.1124	28.1357
29	.3422	29.1424	29.1817	29.1606
30	.2896	30.1305	30.2362	30.1729
31	.3268	31.1297	31.2743	31.1749
32	.3309	32.1357	32.2945	32.1691
33	.3048	33.0781	33.2951	33.1576
34	.3080	34.0172	34.2741	34.1424
35	.3044	35.1016	35.2296	35.1250
36	.3240	36.2426	36.1603	36.1064
37	.2714	37.1248	37.0669	37.0873
38	.3430	38.1321	37.9553	38.0676
39	.2018	39.0807	38.8442	39.0462
40	.2787	39.9006	39.7984	40.0202

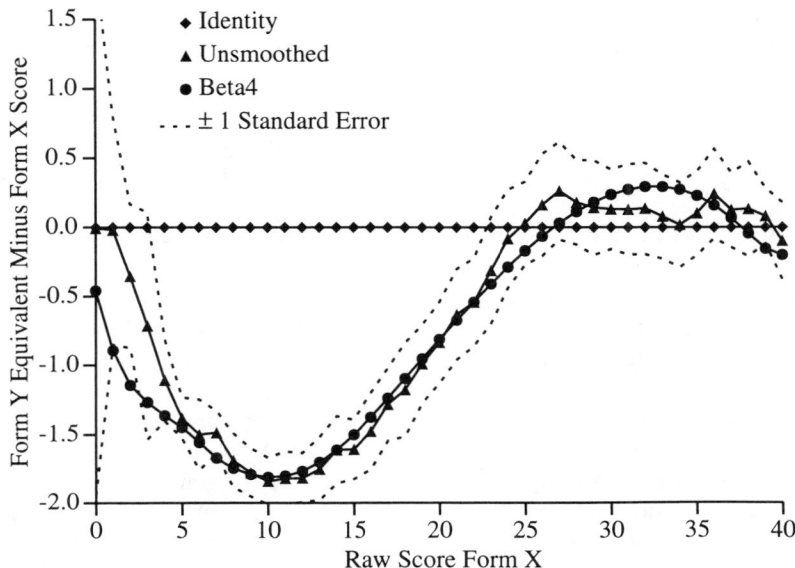

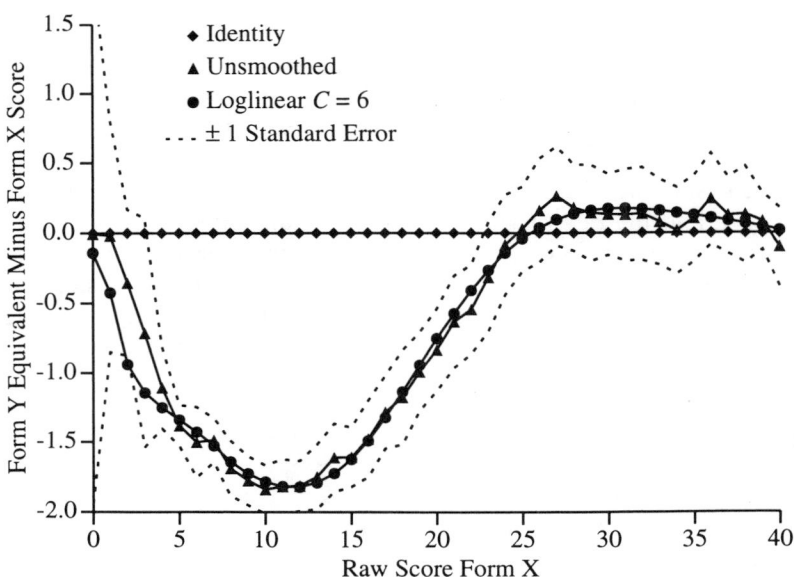

Figure 3.5. Raw-to-raw score equivalents for presmoothing.

lated using standard errors of unsmoothed equipercentile equating that are described in Chapter 7. The upper part of the bands were formed by adding one standard error of equipercentile equating to the unsmoothed relationship. The lower part of the bands were formed by subtracting one standard error. For equating to be adequate, a sensible standard is that the smoothed relationship should lie predominantly within the standard error band.

The equipercentile relationship shown for the beta4 method falls within the standard error band except at Form X raw scores of 1, 2, 7, and 39. These scores are extreme, with few examinees earning any of the scores. Because there are few examinees at these scores, and standard errors of equipercentile equating are poorly estimated at the extremes, these scores can be disregarded and the fit for the beta4 method appears to be adequate. The equipercentile relationship shown for the log-linear method with $C = 6$ is within the standard error band at all scores except at a Form X raw score of 2. The log-linear equivalents are, in general, closer to the unsmoothed relationship than those for the beta4 method. Because the log-linear method results in a smooth curve that is closer to the unsmoothed relationship, it might be viewed as somewhat superior to that for the beta4 method in this case. Because the relationship for both methods appears smooth without departing too far from the unsmoothed relationship, equating using either method seems adequate.

Summary statistics for the raw-to-raw equipercentile equating using these two presmoothing methods are presented in Table 3.3. The moments for the two smoothed methods are very similar to those for Form Y, again suggesting that both of the smoothings were adequate.

The next step in equating is to convert the raw scores on Form X to scale scores, as was done in Table 2.8. Scale score moments are shown in Table 3.4, and the raw-to-scale score conversions are shown in Table 3.5. The unsmoothed moments and equivalents are identical to the values shown previously in Chapter 2. The moments for the unrounded scale scores all appear to be very similar to those for the unrounded scale scores for Form Y. Also, the moments for the rounded scale scores for the beta4

Table 3.3. Raw Score Moments for Presmoothing.

Test Form	$\hat{\mu}$	$\hat{\sigma}$	$\widehat{sk}$	$\widehat{ku}$
Form Y	18.9798	8.9393	.3527	2.1464
Form X	19.8524	8.2116	.3753	2.3024
Form X Equated to Form Y Scale				
Unsmoothed	18.9799	8.9352	.3545	2.1465
Beta4	18.9805	8.9307	.3556	2.1665
Log-linear $C = 6$	18.9809	8.9354	.3541	2.1464

Table 3.4. Scale Score Moments for Presmoothing.

Test Form	$\hat{\mu}_{SC}$	$\hat{\sigma}_{SC}$	$\widehat{sk}_{SC}$	$\widehat{ku}_{SC}$
Form Y				
unrounded	16.5120	8.3812	−.1344	2.0557
rounded	16.4875	8.3750	−.1025	2.0229
Form X Equated to Form Y Scale				
Unsmoothed				
unrounded	16.5125	8.3725	−.1300	2.0515
rounded	16.4324	8.3973	−.1212	2.0294
Beta4				
unrounded	16.5230	8.3554	−.1411	2.0628
rounded	16.4999	8.3664	−.1509	2.0549
Log-linear $C = 6$				
unrounded	16.5126	8.3699	−.1294	2.0419
rounded	16.5461	8.3772	−.1289	2.0003

method appear to be similar to those for the unrounded scale scores for Form Y. However, the mean for the rounded log-linear method (16.5461) appears to be somewhat larger than the mean for the form Y unrounded equivalents (16.5120). This observation suggests that it might be desirable to consider adjusting the rounded raw-to-scale score conversion for the log-linear method, as was done in Chapter 2. Refer to Table 3.5. For the log-linear method, a raw score of 23 converts to a scale score of 21.5143, which rounds to a 22. If the raw score of 23 is converted to a scale score of 21 instead of a scale score of 22, then the moments are as follows: $\hat{\mu}_{sc} = 16.5121$, $\hat{\sigma}_{sc} = 8.3570$, $\widehat{sk}_{sc} = -.1219$, and $\widehat{ku}_{sc} = 2.0142$. After adjustment, the mean is closer to the unrounded mean for Form Y. However, the standard deviation and skewness are farther away. Because the mean is more often given primary attention and the other moments are still reasonably close to the Form Y unrounded moments, the adjustment appears to improve the equating. However, the results without adjustment also appear to be acceptable. As was indicated in Chapter 2, adjustment of conversions should be done conservatively, because it affects score distributions and individual scores.

Postsmoothing Methods

In postsmoothing methods, the equipercentile equivalents, $\hat{e}_Y(x)$, are smoothed directly. Postsmoothing methods fit a curve to the equipercentile relationship. In implementing postsmoothing methods, the smoothed relationship should appear smooth without departing too much from the ob-

Table 3.5. Raw-to-Scale Score Conversions for Presmoothing.

Raw Score	Form Y Scale Scores		Form X Scale Scores					
			Unsmoothed		Beta4		Log-linear $C=6$	
	sc	sc_{int}	sc	sc_{int}	sc	sc_{int}	sc	sc_{int}
−.5	.5000	1	.5000	1	.5000	1	.5000	1
0	.5000	1	.5000	1	.5000	1	.5000	1
1	.5000	1	.5000	1	.5000	1	.5000	1
2	.5000	1	.5000	1	.5000	1	.5000	1
3	.5000	1	.5000	1	.5000	1	.5000	1
4	.5000	1	.5000	1	.5000	1	.5000	1
5	.6900	1	.5000	1	.5000	1	.5000	1
6	1.6562	2	.5949	1	.5842	1	.6084	1
7	3.1082	3	1.1874	1	1.0100	1	1.1465	1
8	4.6971	5	2.1098	2	2.0296	2	2.1756	2
9	6.1207	6	3.4645	3	3.4451	3	3.5421	4
10	7.4732	7	4.9258	5	4.9720	5	5.0022	5
11	8.9007	9	6.3678	6	6.3925	6	6.3667	6
12	10.3392	10	7.7386	8	7.8111	8	7.7287	8
13	11.6388	12	9.2622	9	9.3327	9	9.2016	9
14	12.8254	13	10.8456	11	10.8450	11	10.6965	11
15	14.0157	14	12.1050	12	12.2303	12	12.0855	12
16	15.2127	15	13.4491	13	13.5709	14	13.4337	13
17	16.3528	16	14.8738	15	14.9294	15	14.8277	15
18	17.3824	17	16.1515	16	16.2441	16	16.1975	16
19	18.3403	18	17.3912	17	17.4274	17	17.4367	17
20	19.2844	19	18.4958	18	18.5178	19	18.5734	19
21	20.1839	20	19.6151	20	19.5775	20	19.6678	20
22	20.9947	21	20.5533	21	20.5560	21	20.6631	21
23	21.7000	22	21.4793	21	21.4101	21	21.5143	22
24	22.3220	22	22.2695	22	22.1436	22	22.2346	22
25	22.9178	23	22.9353	23	22.8158	23	22.8936	23
26	23.5183	24	23.6171	24	23.4791	23	23.5412	24
27	24.1314	24	24.2949	24	24.1498	24	24.1906	24
28	24.7525	25	24.8496	25	24.8131	25	24.8256	25
29	25.2915	25	25.3538	25	25.3710	25	25.3617	25
30	25.7287	26	25.7841	26	25.8290	26	25.8021	26
31	26.1534	26	26.2176	26	26.2891	26	26.2399	26
32	26.6480	27	26.7281	27	26.8219	27	26.7479	27
33	27.2385	27	27.2908	27	27.4361	27	27.3441	27
34	27.9081	28	27.9216	28	28.1230	28	28.0198	28
35	28.6925	29	28.7998	29	28.9350	29	28.8245	29
36	29.7486	30	30.1009	30	29.9815	30	29.9032	30
37	31.2010	31	31.3869	31	31.3006	31	31.3312	31
38	32.6914	33	32.8900	33	32.6247	33	32.7931	33
39	34.1952	34	34.2974	34	33.9609	34	34.2539	34
40	35.4615	35	35.3356	35	35.2062	35	35.4871	35
40.5	36.5000	36	36.5000	36	36.5000	36	36.5000	36

served relationship. The method to be described was presented by Kolen (1984) and makes use of cubic smoothing splines described by Reinsch (1967). The spline fitting algorithm was also described by de Boor (1978, pp. 235–243) and is subroutine CSSMH in the IMSL (1991) Math/Library. Polynomials also could be used, but cubic splines are used instead because they appear to provide greater flexibility.

For integer scores, x_i, the spline function is,

$$\hat{d}_Y(x) = v_{0i} + v_{1i}(x - x_i) + v_{2i}(x - x_i)^2 + v_{3i}(x - x_i)^3, \quad x_i \le x < x_i + 1. \tag{3.11}$$

The weights $(v_{0i}, v_{1i}, v_{2i}, v_{3i})$ change from one score point to the next, so that there is a different cubic equation defined between each integer score. At each score point, x_i, the cubic spline is continuous (continuous second derivatives). The spline is fit over the range of scores x_{low} to x_{high}, $0 \le x_{low} \le x \le x_{high} \le K_X$, where x_{low} is the lower integer score in the range and x_{high} is the upper integer score in the range.

The function, over score points, is minimized subject to having minimum curvature and satisfying the following constraint:

$$\frac{\sum_{i=low}^{high} \left[\frac{\hat{d}_Y(x_i) - \hat{e}_Y(x_i)}{\widehat{se}[\hat{e}_Y(x_i)]} \right]^2}{x_{high} - x_{low} + 1} \le S. \tag{3.12}$$

In this equation, the summation is over those points for which the spline is fit. The term $\widehat{se}[\hat{e}_Y(x_i)]$ is the estimated standard error of equipercentile equating, which is defined specifically in Chapter 7. The standard error of equating is used to standardize the differences between the unsmoothed and smoothed relationships. The use of the standard error results in the smoothed and unsmoothed relationships being closer where the standard error is small, and allows them to be farther apart when the standard error is large. The parameter S (where $S \ge 0$) is set by the investigator and controls the degree of smoothing. Several values of S typically are tried and the results compared.

If $S = 0$, then the fitted spline equals the unsmoothed equivalents at all integer score points. If S is very large, then the spline function is a straight line. Intermediate values of S produce a nonlinear function that deviates from the unsmoothed equipercentile relationship by varying degrees. If $S = 1$ then the average squared standardized difference between the smoothed and unsmoothed equivalents is 1.0. Values of S between 0 and 1 have been found to produce adequate results in practice.

The spline is fit over a restricted range of score points so that scores with few examinees and large or poorly estimated standard errors do not unnecessarily influence the spline function. Kolen (1984) recommended that x_{low} and x_{high} be chosen to exclude score points with percentile ranks below .5 and above 99.5.

A linear interpolation procedure that is consistent with the definition of equipercentile equating in Chapter 2 can be used to obtain equivalents outside the range of the spline function. The following equations can be used for linear interpolation outside the range:

$$\hat{d}_Y(x) = \left\{ \frac{[\hat{d}_Y(x_{low}) + .5]}{x_{low} + .5} \right\} x$$

$$+ \left\{ -.5 + \frac{.5[\hat{d}_Y(x_{low}) + .5]}{x_{low} + .5} \right\}, \quad -.5 \le x < x_{low},$$

$$\hat{d}_Y(x) = \left\{ \frac{[\hat{d}_Y(x_{high}) - (K_Y + .5)]}{x_{high} - (K_X + .5)} \right\} x$$

$$+ \left\{ \hat{d}_Y(x_{high}) - \frac{x_{high}[\hat{d}_Y(x_{high}) - (K_Y + .5)]}{x_{high} - (K_X + .5)} \right\}, \quad x_{high} < x \le (K_X + .5).$$

$$(3.13)$$

At the lower end, linear interpolation is between the point $(-.5, -.5)$ and $[x_{low}, \hat{d}_Y(x_{low})]$. At the upper end, linear interpolation is between the point $[x_{high}, \hat{d}_Y(x_{high})]$ and $(K_X + .5, K_Y + .5)$.

Table 3.6 illustrates a cubic spline function that was fit to the ACT Mathematics data using $S = .20$. For this example, the spline function is defined over the Form X raw score range from 5 to 39. The second column shows the spline conversion at Form X integer scores. Equation (3.11) is used to find smoothed values at noninteger scores that are needed for equating. For example, to find the estimated Form Y equivalent of a Form X score of 6.3, note that $x_i = 6$ and $(x - x_i) = (6.3 - 6.0) = .3$. Then,

$$\hat{d}_Y(6.3) = 4.4379 + .9460(.3) + .0013(.3)^2 + .0005(.3)^3 = 4.7218.$$

To illustrate that the spline is continuous, note that the tabled value for a score of $x_i = 7$ is 5.3857. This spline function at 7 also can be obtained using $x = 7$ and $x_i = 6$ as follows. In this case, $(x - x_i) = (7 - 6.0) = 1$. Applying the cubic equation,

$$\hat{d}_Y(7) = 4.4379 + .9460(1) + .0013(1) + .0005(1) = 5.3856,$$

which equals the tabled value for $x_i = 7$, apart from a rounding error in the last digit. Also, the sum of the coefficients in any row equals the value of $\hat{d}_Y(x_i)$ shown in the next row. This equality property is necessary if the spline is to be continuous.

In addition, the spline has continuous second derivatives evaluated at all score points. The second derivative of the spline function evaluated at x in equation (3.11) can be shown to equal $2v_{2i} + 6v_{3i}(x - x_i)$. The second derivative evaluated at a score of 7 using the coefficients at $x_i = 6$ is

Table 3.6. Spline Coefficients for Converting Form X Scores to the Form Y Scale for $S = .20$.

x	$\hat{d}_Y(x) = \hat{v}_0$	$\hat{v}_1$	$\hat{v}_2$	$\hat{v}_3$	$\widehat{se}[\hat{e}_Y(x)]$	$\left[\dfrac{\hat{d}_Y(x) - \hat{e}_Y(x)}{\widehat{se}[\hat{e}_Y(x)]}\right]^2$
5	3.4927	.9447	.0000	.0004	.1478	.7418
6	4.4379	.9460	.0013	.0005	.2541	.0597
7	5.3857	.9502	.0028	.0009	.1582	.6680
8	6.3397	.9585	.0055	.0008	.1969	.0198
9	7.3046	.9721	.0081	.0006	.1761	.2095
10	8.2854	.9902	.0100	.0003	.1731	.5165
11	9.2859	1.0112	.0110	.0001	.1952	.2779
12	10.3082	1.0336	.0114	−.0001	.1800	.4609
13	11.3531	1.0560	.0110	−.0003	.2311	.1952
14	12.4197	1.0770	.0101	−.0003	.2431	.0149
15	13.5066	1.0963	.0091	−.0005	.2138	.2823
16	14.6114	1.1129	.0076	−.0006	.2764	.1000
17	15.7313	1.1263	.0058	−.0006	.2617	.0030
18	16.8627	1.1359	.0039	−.0006	.3383	.0138
19	18.0019	1.1419	.0020	−.0006	.2826	.0006
20	19.1451	1.1439	.0001	−.0006	.2947	.0046
21	20.2885	1.1423	−.0018	−.0006	.3299	.0581
22	21.4285	1.1370	−.0035	−.0005	.3183	.0075
23	22.5615	1.1285	−.0051	−.0005	.3865	.1054
24	23.6844	1.1169	−.0065	−.0003	.3555	.4244
25	24.7945	1.1028	−.0076	−.0002	.3013	.6057
26	25.8895	1.0872	−.0080	.0000	.3683	.5434
27	26.9687	1.0712	−.0080	.0002	.3532	.6943
28	28.0321	1.0557	−.0075	.0003	.3069	.2322
29	29.0806	1.0416	−.0066	.0003	.3422	.0322
30	30.1159	1.0294	−.0056	.0003	.2896	.0024
31	31.1401	1.0192	−.0046	.0003	.3268	.0010
32	32.1551	1.0111	−.0036	.0003	.3309	.0033
33	33.1630	1.0050	−.0026	.0003	.3048	.0778
34	34.1657	1.0006	−.0018	.0001	.3080	.2331
35	35.1646	.9974	−.0014	.0001	.3044	.0423
36	36.1607	.9949	−.0011	.0001	.3240	.0645
37	37.1547	.9931	−.0007	.0001	.2714	.0120
38	38.1473	.9921	−.0003	.0001	.3430	.0020
39	39.1392				.2018	.0832

$$2(.0013) + 6(.0005)(7 - 6) = .0056.$$

The second derivative evaluated at a score of 7 using the coefficients at $x_i = 7$ is

$$2(.0028) + 6(.0009)(7 - 7) = .0056.$$

The equality of these two expressions illustrates the continuous second derivative property of the cubic spline. This property can be shown to hold at the other score points as well.

The rightmost column in Table 3.6 shows the squared standardized difference at each score point. The mean of the values in this column is .20, because $S = .20$.

One problem with the spline expression in equations (3.11) and (3.12) is that it is a regression function, so it is not symmetric. That is, the spline that is used for converting Form X to Form Y is different from the spline that is used for converting Form Y to Form X. To arrive at a function that is more nearly symmetric, define $\hat{d}_X(y)$ as the spline function that converts Form Y scores to Form X scores using the same procedures and the same value of S. Assuming that the inverse function exists, define the inverse of this function as $\hat{d}_X^{-1}(x)$. (Note that the inverse is not guaranteed to exist, although the lack of an inverse has not been known to cause problems in practice.) This inverse can be used to transform Form X scores to the Form Y scale. A more nearly symmetric equating function then can be defined as the average of two splines: the spline developed for converting Form X to the Form Y scale and the inverse of the spline developed for converting Form Y to the Form X scale. For a particular S, define this quantity as

$$\hat{d}_Y^*(x) = \frac{\hat{d}_Y(x) + \hat{d}_X^{-1}(x)}{2}, \quad -.5 \le x \le K_X + .5. \tag{3.14}$$

The expression in equation (3.14) is the final estimate of the equipercentile equating function. [See Wang and Kolen (1994) for a further discussion of symmetry and for an alternative postsmoothing method to the one described here.]

To implement the method, the equating is conducted using a variety of values of S. Graphs of the resulting equivalents can be examined for smoothness and compared to the unsmoothed equivalents. Standard errors of equating can be very useful for evaluating various degrees of smoothing. Ideally, the procedure results in a smooth function that does not depart too much from the unsmoothed equivalents. In addition, the central moments for the Form X scores equated to the Form Y scale using smoothing should be compared to those for the Form Y scores. Central moments for the scale scores that result from the equating also should be inspected.

Illustrative Example

Because there are no statistical tests associated with the postsmoothing method described here, inspection of graphs and moments is even more crucial for choosing a degree of smoothing than in the presmoothing methods. For the ACT Mathematics example, equating was conducted using eight different values for S ranging from .01 to 1.0. The *RAGE* Macintosh computer program described in Appendix B was used to conduct the analyses. The equipercentile relationships using these methods are presented in Table 3.7 and graphed in Figures 3.6 and 3.7.

As can be seen in the figures, the equivalents deviate more from the unsmoothed equivalents as the values of S increase. For $S = .01$, the smoothed and unsmoothed equivalents are very close, and the smoothed equivalents appear to be bumpy. However, the smoothed equivalents are within the standard error bands. For $S = .05$, the equivalents appear to be smooth and are within the standard error bands at all points. As S increases, the smoothed relationship continues to deviate more from the unsmoothed relationship. For $S \geq .75$, the smoothed relationship is outside the standard error bands at many score points. The relationship for $S = .05$ appears to be the one for which there is the least amount of smoothing required to achieve a smooth function of the values tried. The relationship for $S = .10$ also seems acceptable.

Moments for the smoothed relationships are shown in Table 3.8. As S increases, the moments for the smoothed equipercentile equating depart more from the Form Y moments. This result suggests that lower values of S are to be preferred for this example.

Now consider Form X scale score equivalents. Scale score moments are shown in Table 3.9 for the scale score equivalents shown in Tables 3.10 and 3.11. An asterisk indicates the moment that is closest, among the smoothed results, to the Form Y unrounded equivalents. The rounded mean and standard deviation are closest for the $S = .05$ conversion, and the other moments also are fairly close.

As indicated in Chapter 2, scale scores that are reported to examinees are rounded. The rounded conversion is shown in Table 3.11. Triple asterisks in this table indicate score points where adjacent smoothing values convert to different scale scores. For example, a Form X raw score of 9 converts to a scale score of 3 for $S = .01$ and to a scale score of 4 for $S = .05$. As can be seen, this is the only difference in the rounded conversions between these two degrees of smoothing. Sometimes, there are gaps in the conversion table that can be removed by adjusting the conversion. Other times, adjustments can be used to improve the scale score moments. In this example, adjustment of conversions does not seem warranted.

Table 3.7. Raw-to-Raw Score Conversions for Postsmoothing.

Form X Score	No Smooth				Form Y Equivalent				
		$S = .01$	$S = .05$	$S = .10$	$S = .20$	$S = .30$	$S = .50$	$S = .75$	$S = 1.00$
0	.000	-.129	-.129	-.133	-.138	-.141	-.146	-.150	-.154
1	.980	.614	.612	.600	.586	.577	.563	.550	.539
2	1.646	1.356	1.353	1.333	1.311	1.295	1.272	1.250	1.232
3	2.286	2.098	2.094	2.067	2.035	2.013	1.981	1.950	1.925
4	2.893	2.841	2.835	2.800	2.759	2.731	2.690	2.650	2.618
5	3.620	3.583	3.576	3.534	3.484	3.449	3.398	3.350	3.311
6	4.500	4.480	4.440	4.400	4.354	4.322	4.273	4.225	4.185
7	5.515	5.443	5.372	5.349	5.323	5.305	5.277	5.249	5.226
8	6.312	6.324	6.306	6.302	6.296	6.292	6.284	6.276	6.269
9	7.224	7.218	7.252	7.265	7.278	7.286	7.297	7.306	7.313
10	8.161	8.168	8.216	8.243	8.271	8.290	8.317	8.342	8.362
11	9.183	9.166	9.205	9.241	9.281	9.308	9.347	9.385	9.415
12	10.186	10.195	10.221	10.262	10.309	10.342	10.390	10.436	10.474
13	11.251	11.260	11.266	11.307	11.357	11.392	11.445	11.496	11.538
14	12.390	12.345	12.338	12.375	12.424	12.460	12.513	12.565	12.607
15	13.393	13.419	13.434	13.467	13.511	13.544	13.594	13.642	13.683
16	14.524	14.541	14.553	14.579	14.616	14.643	14.686	14.728	14.763
17	15.717	15.695	15.692	15.710	15.736	15.756	15.788	15.820	15.848
18	16.823	16.846	16.846	16.855	16.868	16.879	16.898	16.918	16.936
19	18.009	18.005	18.011	18.010	18.008	18.009	18.013	18.020	18.026
20	19.165	19.171	19.183	19.170	19.153	19.143	19.132	19.123	19.118
21	20.368	20.337	20.356	20.330	20.298	20.278	20.251	20.228	20.211
22	21.456	21.499	21.525	21.485	21.439	21.409	21.368	21.331	21.303

23	22.687	22.695	22.685	22.630	22.572	22.534	22.480	22.432	22.393
24	23.916	23.890	23.826	23.761	23.694	23.650	23.586	23.528	23.481
25	25.029	25.045	24.945	24.873	24.802	24.754	24.685	24.619	24.566
26	26.161	26.160	26.037	25.966	25.894	25.846	25.774	25.704	25.648
27	27.263	27.214	27.101	27.038	26.971	26.924	26.853	26.783	26.725
28	28.180	28.197	28.140	28.091	28.033	27.990	27.922	27.855	27.798
29	29.142	29.161	29.160	29.127	29.080	29.042	28.982	28.920	28.867
30	30.130	30.138	30.166	30.150	30.115	30.084	30.033	29.979	29.932
31	31.130	31.126	31.162	31.162	31.139	31.117	31.076	31.032	30.994
32	32.136	32.107	32.154	32.166	32.156	32.141	32.113	32.081	32.052
33	33.078	33.075	33.144	33.165	33.166	33.160	33.144	33.125	33.108
34	34.017	34.065	34.136	34.161	34.171	34.173	34.171	34.167	34.161
35	35.102	35.112	35.130	35.155	35.174	35.183	35.195	35.205	35.213
36	36.243	36.165	36.126	36.148	36.174	36.191	36.217	36.242	36.263
37	37.125	37.156	37.120	37.140	37.172	37.197	37.237	37.278	37.313
38	38.132	38.125	38.114	38.131	38.169	38.202	38.256	38.313	38.362
39	39.081	39.092	39.103	39.117	39.155	39.188	39.243	39.297	39.341
40	39.901	40.031	40.034	40.039	40.052	40.063	40.081	40.099	40.114

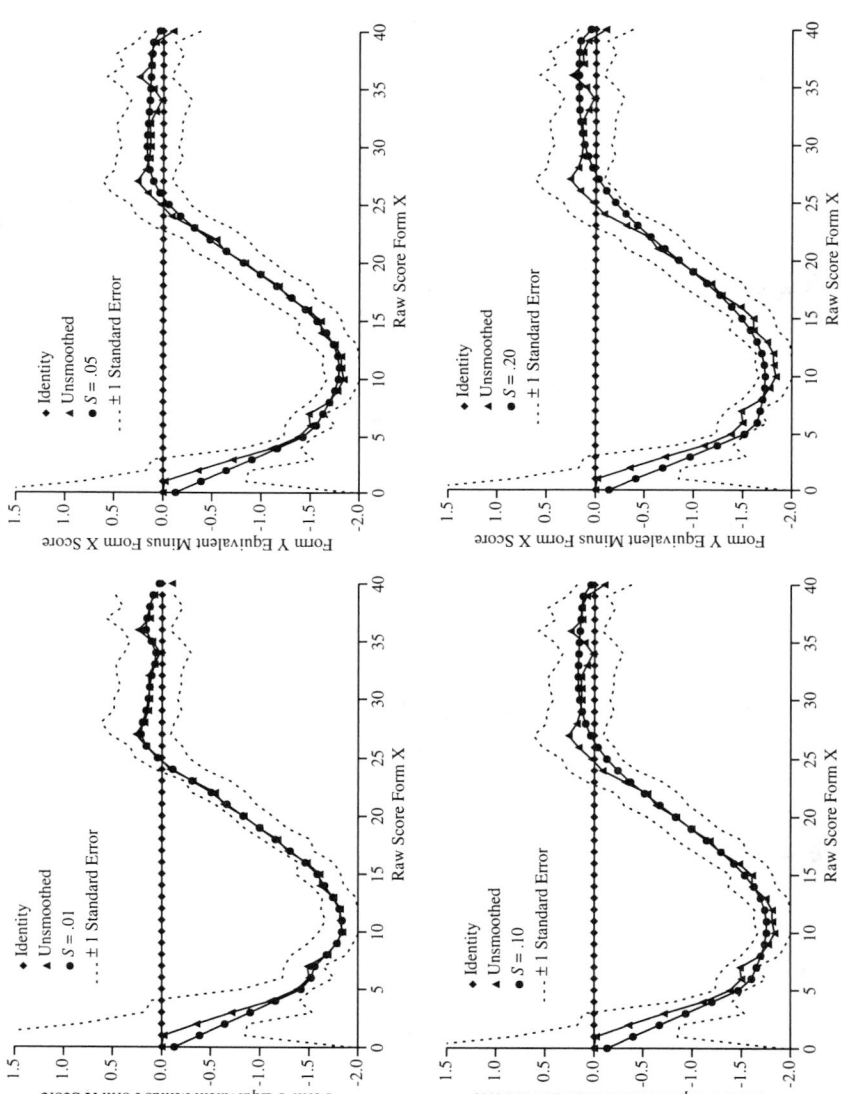

Figure 3.6. Raw-to-raw score equivalents for postsmoothing, $S = .01, .05. .10, .20$.

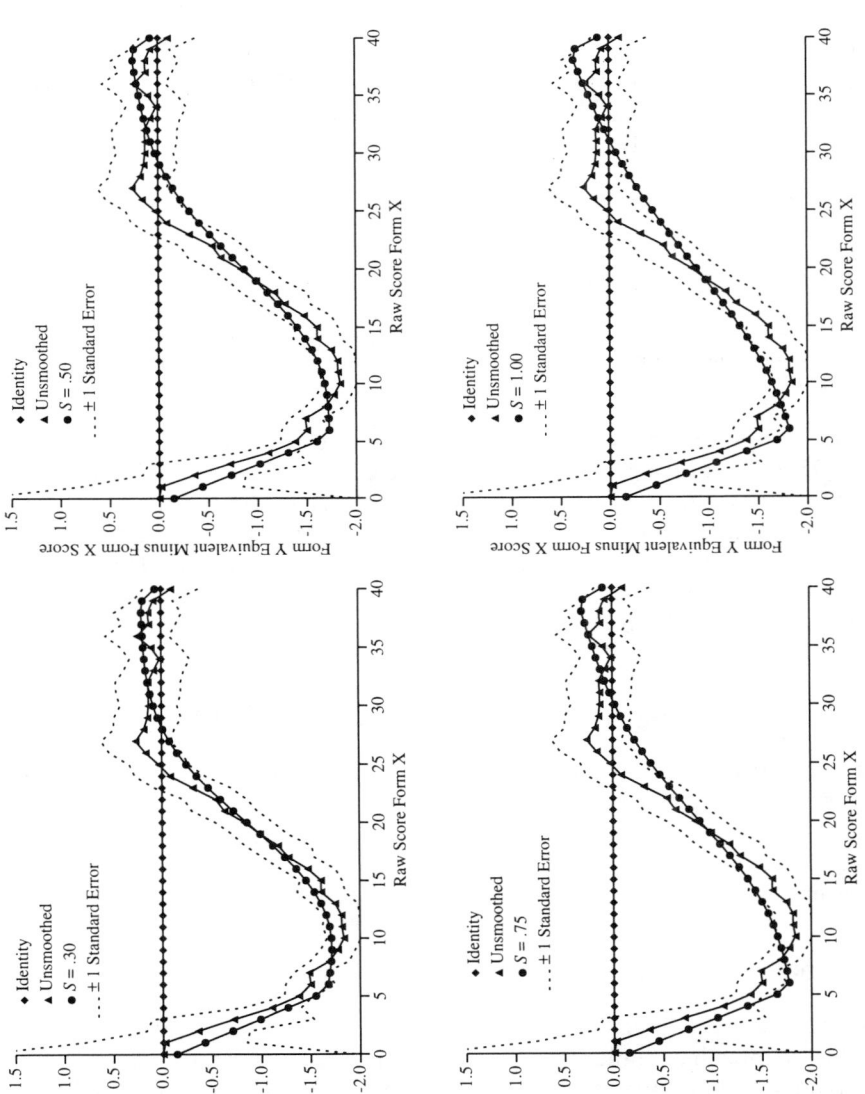

Figure 3.7. Raw-to-raw score equivalents for postsmoothing, $S = .30, .50, .75, 1.00$.

Table 3.8. Raw Score Moments for Postsmoothing.

Test Form	$\hat{\mu}$	$\hat{\sigma}$	$\widehat{sk}$	$\widehat{ku}$
Form Y	18.9798	8.9393	.3527	2.1464
Form X	19.8524	8.2116	.3753	2.3024
Form X Equated to Form Y Scale				
Unsmoothed	18.9799	8.9352	.3545	2.1465
$S = .01$	18.9789*	8.9393*	.3533*	2.1488*
$S = .05$	18.9767	8.9313	.3561	2.1587
$S = .10$	18.9743	8.9172	.3603	2.1738
$S = .20$	18.9717	8.8987	.3644	2.1922
$S = .30$	18.9699	8.8852	.3670	2.2054
$S = .50$	18.9676	8.8643	.3704	2.2258
$S = .75$	18.9656	8.8439	.3733	2.2457
$S = 1.00$	18.9642	8.8271	.3756	2.2624

* Indicates moment closest to Form Y moment among smoothed estimates.

All things considered, the results from these procedures suggest that $S = .05$ is the most appropriate of the values tried. However, this example should not be overgeneralized. The smallest smoothing values do not always appear to produce the most adequate equating. Especially for the rounded conversions, higher values of S often lead to more adequate results. There is no single statistical criterion that can be used. Instead, various values of S need to be tried and the results compared.

Practical Issues in Equipercentile Equating

As was indicated earlier, the purpose of smoothing in equipercentile equating is to reduce equating error. However, there is a danger that smoothing will introduce equating error. Provided next are guidelines to help ensure that smoothing improves the equating process. Guidelines for the sample sizes needed to produce adequate equating are considered subsequently.

Summary of Smoothing Strategies

The strategies for presmoothing and postsmoothing that were illustrated in this chapter have much in common, although the strategies differ. The focus in presmoothing is on finding a method for smoothing score distributions, whereas the focus in postsmoothing is on choosing among degrees of smoothing of the equipercentile relationship. Another difference is that statistical tests can be used with the presmoothing method, whereas no statistical tests exist for the postsmoothing method. The following are

Table 3.9. Scale Score Moments for Postsmoothing.

Test Form	$\hat{\mu}_{SC}$	$\hat{\sigma}_{SC}$	$\widehat{sk}_{SC}$	$\widehat{ku}_{SC}$
Form Y				
unrounded	16.5120	8.3812	−.1344	2.0557
rounded	16.4875	8.3750	−.1025	2.0229
Form X Equated to Form Y Scale for				
Unsmoothed				
unrounded	16.5125	8.3725	−.1300	2.0515
rounded	16.4324	8.3973	−.1212	2.0294
$S = .01$				
unrounded	16.5120*	8.3758*	−.1303*	2.0543*
rounded	16.4823	8.4164	−.1308*	2.0334
$S = .05$				
unrounded	16.5158	8.3638	−.1302	2.0606
rounded	16.5156*	8.3648*	−.1164	2.0262
$S = .10$				
unrounded	16.5236	8.3475	−.1294	2.0737
rounded	16.5366	8.3223	−.1308	2.0597*
$S = .20$				
unrounded	16.5336	8.3284	−.1289	2.0908
rounded	16.5345	8.2576	−.1103	2.0859
$S = .30$				
unrounded	16.5409	8.3152	−.1287	2.1034
rounded	16.5345	8.2576	−.1103	2.0859
$S = .50$				
unrounded	16.5523	8.2956	−.1288	2.1229
rounded	16.5551	8.2288	−.0907	2.1525
$S = .75$				
unrounded	16.5635	8.2770	−.1292	2.1423
rounded	16.5211	8.2165	−.0804	2.1632
$S = 1.00$				
unrounded	16.5731	8.2619	−.1297	2.1586
rounded	16.5211	8.2165	−.0804	2.1632

* Indicates moment closest to unrounded for Form Y among smoothed estimates.

the steps in the smoothing strategies that have been discussed. Step 1 is used only with presmoothing. Differences between presmoothing and post-smoothing strategies are highlighted.

Step 1. Fit the score distributions (presmoothing only). The strategy used for fitting the score distributions involves both graphic inspection and the use of statistical indices. For the log-linear method

Table 3.10. Unrounded Raw-to-Scale Score Conversions for Postsmoothing.

Form X Score	No Smooth				Form Y Equivalent				
		$S = .01$	$S = .05$	$S = .10$	$S = .20$	$S = .30$	$S = .50$	$S = .75$	$S = 1.00$
0	.500	.500	.500	.500	.500	.500	.500	.500	.500
1	.500	.500	.500	.500	.500	.500	.500	.500	.500
2	.500	.500	.500	.500	.500	.500	.500	.500	.500
3	.500	.500	.500	.500	.500	.500	.500	.500	.500
4	.500	.500	.500	.500	.500	.500	.500	.500	.500
5	.500	.500	.500	.500	.500	.500	.500	.500	.500
6	.595	.591	.584	.576	.567	.561	.552	.543	.535
7	1.187	1.118	1.049	1.027	1.002	.985	.958	.931	.908
8	2.110	2.126	2.101	2.095	2.087	2.080	2.069	2.057	2.046
9	3.464	3.455	3.508	3.529	3.550	3.562	3.579	3.595	3.606
10	4.926	4.936	5.004	5.043	5.084	5.110	5.148	5.184	5.213
11	6.368	6.346	6.398	6.447	6.501	6.537	6.591	6.641	6.682
12	7.739	7.752	7.789	7.847	7.914	7.961	8.030	8.095	8.149
13	9.262	9.274	9.284	9.342	9.414	9.465	9.541	9.614	9.674
14	10.846	10.787	10.778	10.827	10.891	10.937	11.006	11.073	11.129
15	12.105	12.136	12.154	12.193	12.246	12.284	12.343	12.401	12.449
16	13.449	13.469	13.484	13.515	13.559	13.591	13.642	13.692	13.734
17	14.874	14.848	14.844	14.866	14.897	14.920	14.959	14.997	15.030
18	16.152	16.177	16.178	16.188	16.202	16.215	16.236	16.259	16.279
19	17.391	17.387	17.393	17.392	17.390	17.391	17.395	17.401	17.408
20	18.496	18.501	18.513	18.501	18.485	18.476	18.465	18.457	18.452
21	19.615	19.588	19.605	19.582	19.552	19.534	19.510	19.489	19.474
22	20.553	20.588	20.610	20.577	20.539	20.515	20.482	20.452	20.429

23	21.479	21.485	21.477	21.439	21.398	21.371	21.333	21.299	21.272
24	22.270	22.254	22.214	22.173	22.131	22.104	22.065	22.028	21.999
25	22.935	22.945	22.885	22.842	22.800	22.771	22.730	22.691	22.659
26	23.617	23.616	23.541	23.498	23.455	23.426	23.382	23.341	23.307
27	24.295	24.264	24.194	24.155	24.114	24.085	24.041	23.998	23.963
28	24.850	24.859	24.828	24.802	24.770	24.746	24.704	24.662	24.627
29	24.354	25.362	25.361	25.347	25.326	25.310	25.282	25.248	25.220
30	25.784	25.787	25.799	25.792	25.777	25.764	25.743	25.719	25.699
31	26.218	26.216	26.234	26.233	26.222	26.211	26.191	26.169	26.151
32	26.728	26.711	26.739	26.746	26.740	26.731	26.715	26.696	26.679
33	27.291	27.289	27.335	27.349	27.350	27.345	27.335	27.322	27.311
34	27.922	27.959	28.015	28.034	28.042	28.044	28.043	28.039	28.034
35	28.800	28.811	28.830	28.856	28.876	28.886	28.899	28.909	28.917
36	30.101	29.988	29.931	29.964	30.001	30.026	30.064	30.100	30.131
37	31.387	31.433	31.380	31.410	31.457	31.494	31.554	31.615	31.667
38	32.890	32.879	32.863	32.889	32.946	32.995	33.076	33.161	33.235
39	34.297	34.311	34.326	34.343	34.391	34.434	34.503	34.571	34.627
40	35.336	35.525	35.533	35.542	35.569	35.592	35.630	35.667	35.698

Table 3.11. Rounded Raw-to-Scale Score Conversions for Postsmoothing.

Form X Score	No Smooth	Form Y Equivalent							
		S = .01	S = .05	S = .10	S = .20	S = .30	S = .50	S = .75	S = 1.00
0	1	1	1	1	1	1	1	1	1
1	1	1	1	1	1	1	1	1	1
2	1	1	1	1	1	1	1	1	1
3	1	1	1	1	1	1	1	1	1
4	1	1	1	1	1	1	1	1	1
5	1	1	1	1	1	1	1	1	1
6	1	1	1	1	1	1	1	1	1
7	1	1	1	1	1	1	1	1	1
8	2	2	2	2	2	2	2	2	2
9	3	3***	4	4	4	4	4	4	4
10	5	5	5	5	5	5	5	5	5
11	6	6	6	6***	7	7	7	7	7
12	8	8	8	8	8	8	8	8	8
13	9	9	9	9	9	9***	10	10	10
14	11	11	11	11	11	11	11	11	11
15	12	12	12	12	12	12	12	12	12
16	13	13	13***	14	14	14	14	14	14
17	15	15	15	15	15	15	15	15	15
18	16	16	16	16	16	16	16	16	16
19	17	17	17	17	17	17	17	17	17
20	18***	19	19	19***	18	18	18	18	18
21	20	20	20	20	20	20	20***	19	19
22	21	21	21	21	21	21***	20	20	20

23	21	21	21	21	21	21	21	21	21
24	22	22	22	22	22	22	22	22	22
25	23	23	23	23	23	23	23	23	23
26	24	24	24***	23	23	23	23	23	23
27	24	24	24	24	24	24	24	24	24
28	25	25	25	25	25	25	25	25	25
29	25	25	25	25	25	25	25	25	25
30	26	26	26	26	26	26	26	26	26
31	26	26	26	26	26	26	26	26	26
32	27	27	27	27	27	27	27	27	27
33	27	27	27	27	27	27	27	27	27
34	28	28	28	28	28	28	28	28	28
35	29	29	29	29	29	29	29	29	29
36	30	30	30	30	30	30	30	30	30
37	31	31	31	31	31	31***	32	32	32
38	33	33	33	33	33	33	33	33	33
39	34	34	34	34	34	34***	35	35	35
40	35***	36	36	36	36	36	36	36	36

*** Indicates a different conversion obtained for adjacent methods.

(a) Examine graphs of the fitted versus the sample distribution. For an adequate fit, the fitted distribution should be smooth without departing more than necessary from the sample distribution.
(b) Examine $\chi^2_C - \chi^2_{C+1}$. Choose a degree among those that are fit which is larger than the largest value of C that has a significant χ^2.
(c) Examine the overall χ^2 fit statistic. The degree chosen should not be significant.

Ideally, a degree of smoothing would be chosen that meets all of these criteria. When fitting the beta4 method, only (a) and (c) are used. Also, when the sample size is very large, minor differences between models might be significant. This issue needs to be considered when setting the overall significance level. This model selection strategy needs to be applied with caution and not followed too rigidly in practice. More than one acceptable set of values for C for the log-linear model can be chosen and evaluated in subsequent steps.

Step 2. Construct the raw-to-raw equivalents. For presmoothing, construct the equipercentile equivalents for the models chosen in Step 1. For postsmoothing, construct the equipercentile equivalents for the degrees of smoothing that are to be evaluated.

(a) Examine the graphs of the raw-to-raw equivalents. For smoothing to be adequate, the relationship should be smooth without departing too much from the unsmoothed equivalents, as indicated by the standard error bands.
(b) Examine the moments of the equated raw scores. The moments of the Form X equated raw scores should be close to those for Form Y.

Models that are judged to produce adequate results are considered further.

Step 3. Construct the raw-to-scale score equivalents. For presmoothing, construct the equivalents for the methods chosen in Step 1. For postsmoothing, construct the equivalents for various degrees of smoothing that are to be considered further.

(a) Examine the moments for the unrounded scale scores. The moments for the Form X scale scores should not be too different from the moments for the Form Y scale scores.
(b) Examine the moments for the rounded Form X scale scores. The moments for the Form X rounded scale scores should be similar to the moments for the unrounded Form Y scales scores.
(c) Consider adjusting the rounded raw-to-scale score equivalents for Form X. If the moments for the Form X rounded scale scores are not close enough to the moments for the unrounded Form Y scale scores, then different adjustments of the conversion should be considered.

The strategy described might result in more than one method or degree of smoothing being adequate, and various subjective judgments could be made. Such judgments are necessarily dependent on the testing program in which the equating is being done. General rules of thumb do not seem possible, because testing programs vary so much in their sample sizes, distribution shapes, numbers of items, and other relevant characteristics. However, rules of thumb for a particular testing program often can be developed after some experience with the program.

Equating Error and Sample Size

Holland *et al.* (1989) developed standard error formulas for equipercentile equating using log-linear presmoothing. Standard error formulas have not been derived for the other smoothing methods, although the bootstrap methods (Efron and Tibshirani, 1993) to be described in Chapter 7 can be used. There is no general analytic procedure for estimating systematic error. Technically, the estimation of both types of error is necessary to thoroughly evaluate the effects of smoothing.

Hanson *et al.* (1994) conducted an empirical comparison of the presmoothing and postsmoothing methods that are presented here. In this study, empirical score distributions were smoothed. The smoothed distributions were assumed to be the population distributions. Random samples of a given size then were drawn from the population distributions. Equipercentile equivalents were estimated from these random samples using both presmoothing and postsmoothing methods. Because the population distributions were known, random and systematic error components could be estimated separately. Note that the use of smoothed distributions as population distributions helps ensure that the distributions are realistic. Still, the Hanson *et al.* (1994) study is but one simulation.

Mean-squared errors for a portion of the Hanson *et al.* (1994) study are presented in Table 3.12 for the enhanced ACT Assessment English and Science Reasoning tests. The values in the table are estimates of the total error of equation (3.8). Larger values indicate more total equating error. The first row in the upper and lower portions of the table is for the identity equating. Note the relatively large value for ACT English compared to that for ACT Science Reasoning. This difference occurs because the two English forms are quite different from one another, whereas the two Science Reasoning forms are very similar. The sample sizes in the table are per form. For the English test with $N = 100$, the identity equating results in less error than some of the smoothing methods. For the Science Reasoning Test with $N = 100$, the identity equating results in the least amount of error of all of the methods. For the English test, one of the smoothed equipercentile methods (postsmoothing $S = .50$) produces the lowest mean-squared error for all sample sizes. For the Science Reasoning test, only at a sample size of 3000 do all of the smoothing methods have mean-squared error values equal to or lower than the value for linear equating.

Table 3.12. Mean-Squared Equating Error from Hanson *et al.* (1994) Study.

Test	Equating Method	$N = 100$	$N = 250$	$N = 500$	$N = 1000$	$N = 3000$
ACT	Identity	5.76	5.76	5.76	5.76	5.76
English	Linear	6.15	3.65	2.80	2.33	2.00
($K = 75$)	Unsmoothed	6.60	2.83	1.50	.75	.25
	Beta4	5.28	2.24	1.22	.63	.24
	Log-linear $C = 3$	5.20	2.30	1.29	.71	.35
	Log-linear $C = 4$	5.66	2.47	1.39	.77	.36
	Log-linear $C = 6$	6.09	2.55	1.33	.67	.23
	Postsmoothing $S = .10$	5.98	2.55	1.33	.67	.22
	Postsmoothing $S = .25$	5.57	2.34	1.23	.62	.21
	Postsmoothing $S = .50$	5.17	2.19	1.17	.59	.21
ACT	Identity	.51	.51	.51	.51	.51
Science	Linear	1.03	.46	.20	.11	.05
Reasoning	Unsmoothed	1.62	.70	.32	.17	.06
($K = 40$)	Beta4	1.28	.55	.24	.12	.04
	Log-linear $C = 3$	1.17	.51	.22	.12	.04
	Log-linear $C = 4$	1.34	.57	.25	.13	.04
	Log-linear $C = 6$	1.52	.63	.28	.14	.05
	Postsmoothing $S = .10$	1.42	.63	.28	.14	.05
	Postsmoothing $S = .25$	1.32	.56	.24	.12	.04
	Postsmoothing $S = .50$	1.26	.51	.22	.11	.04

In comparing the smoothing results to one another, there is no method that appears to be clearly superior to the others. For the English test, the mean-squared error for the best smoothing method is approximately 80% of that of the unsmoothed equipercentile method. For the Science Reasoning test, the mean-squared error for the best smoothing method is approximately 70% of that of the unsmoothed equipercentile method. Thus, smoothed equipercentile equating produces a modest reduction in error compared to unsmoothed equipercentile equating. These results are for equating error averaged over all score points. More detailed results presented by Hanson *et al.* (1994) indicate that the smoothing reduces error, even at extreme scores.

The results from the Hanson *et al.* (1994) study as well as practical experience with these methods suggest the use of the following guidelines:

- Use of the identity equating can be preferable to using one of the other equating methods, especially with sample sizes at or below 100 examinees per test form. The use of equipercentile equating with fewer than 250 examinees per form might even introduce error.
- Smoothing in equipercentile equating can be expected to produce a modest decrease in mean-squared equating error when compared to unsmoothed equipercentile equating.

No clear way exists for choosing whether to use presmoothing versus postsmoothing. One positive characteristic of the presmoothing methods is that there are statistical tests that can be readily used. Such tests do not exist for the postsmoothing method. In addition, the postsmoothing method described here requires averaging two splines, and there is no compelling theoretical reason for doing so other than to produce a symmetric relationship. However, postsmoothing directly smoothes the equipercentile relationship, which is more direct than smoothing the distributions, as is done with the presmoothing methods. The presmoothing and postsmoothing methods have been used in practice in testing programs with good results. Research evidence suggests that both types of methods can produce results which have the potential to improve equating accuracy. Thus, either type of method can function adequately in operational testing programs.

Exercises

3.1. Suppose that, in the population, the Form Y equipercentile equivalent of a Form X score of 26 is 28.3. Also, suppose that the expected (over a large number of random samples) equivalent using a smoothing method is 29.1. Based on a sample, the unsmoothed equivalent is estimated to be 31.1 and the smoothed equipercentile equivalent is estimated to be 31.3. Answer the following questions about finding the Form Y equipercentile equivalent of a Form X score of 26. Indicate if the question cannot be answered from the information given.
   a. What is the systematic error in using the smoothing method?
   b. What is the error in estimating the equipercentile equivalent using the unsmoothed equipercentile method in the sample?
   c. What is the error in estimating the equipercentile equivalent using the smoothed equipercentile method in the sample?
   d. What is the standard error of equating using the unsmoothed equipercentile method?
   e. Which method (smoothed or unsmoothed) was more accurate in the sample?
   f. Which method (smoothed or unsmoothed) would be better over a large number of replications?

3.2. If $C = 3$ in the log-linear method, which of the following would be the same for the observed distribution and smoothed distribution: mean, standard deviation, skewness, kurtosis?

3.3. Suppose a nominal alpha level of .30 had been used. In Table 3.1, what values of C would have been eliminated using the single degree of freedom difference χ^2 statistics for Form X and for Form Y? (The significance level is 1.07.)

3.4. What would be the cubic spline equivalent of a score on x of 28.6 using the data shown in Table 3.6?

3.5. In Table 3.11, which pairs of conversions are identical? Are there any circumstances under which it would matter whether one or the other of the identical conversions was chosen?

3.6. In Figures 3.6 and 3.7, ± 1 standard error bands are presented. If ± 2 standard error bands had been used, which S parameters would have had relationships that fell within the band? How about the relationship for the identity equating?

3.7. In Table 3.12, under what conditions in the studies presented was it better to use the identity equating than to use any of the methods studied? What factor do you think could have led to the identity equating appearing to be relatively better with small samples for the Science Reasoning test than for the English test? Can you think of a situation in which the identity equating would always be better than one of the other equating methods?

CHAPTER 4

Nonequivalent Groups—Linear Methods

Chapter 1 introduced the common-item nonequivalent groups design. For this design, two groups of examinees from different populations are each administered different test forms that have a set of items in common. This design often is used when only one form of a test can be administered on a given test date. As discussed in Chapter 1, the set of common items should be as similar as possible to the full-length forms in both content and statistical characteristics.

There are two special cases of the common-item nonequivalent groups design. The common item set is said to be *internal* when scores on the common items contribute to the total scores for both forms. By contrast, the common items are said to be *external* when their scores do not contribute to total scores. Notationally, denote the new test form and the random variable score on that form as X, the old form and the random variable score on that form as Y, and the common-item set and the random variable score on the common-item set as V. Assume that X and V are taken by a group of examinees from Population 1, and Y and V are taken by a group of examinees from Population 2. If V is an internal set of common items, then X and Y include scores on V. If V is external, then X and Y do not include scores on V. For example, consider an examinee who got 10 common items correct and 40 noncommon items correct. If V is an internal set of common items, then $x = 50$. If V is an external set, then $x = 40$.

In general, the common items are used to adjust for population differences. Doing so requires strong statistical assumptions because each examinee comes from only one population and takes only one form. The various methods for performing equating under the common-item nonequivalent groups design are distinguished in terms of their statistical assumptions.

Even though the design under consideration here involves two populations, an equating function is typically viewed as being defined for a single population. Therefore, Populations 1 and 2 must be combined to obtain a single population for defining an equating relationship. To address this issue Braun and Holland (1982) introduced the concept of a *synthetic population* in which Populations 1 and 2 are weighted by w_1 and w_2, respectively, where $w_1 + w_2 = 1$ and $w_1, w_2 \geq 0$.

The equating methods considered in this chapter are all linear. The first two methods are called *observed score* equating methods because observed scores on X are transformed to observed scores on the scale of Y. The third method is called a *true score* method because it relates true scores on X to the scale of true scores on Y. All of these methods are described by Angoff (1971) and Petersen et al. (1989). However, the presentation here more closely parallels a combination of Kolen and Brennan (1987) and Brennan (1990). Other authors who have provided derivations of one or more of these methods include MacCann (1990) and Woodruff (1986, 1989).

As discussed in Chapter 2, the linear conversion is defined by setting standardized deviation scores (z-scores) equal for the two forms. For the common-item nonequivalent groups design, this results in the following linear equation for equating observed scores on X to the scale of observed scores on Y:

$$l_{Ys}(x) = \frac{\sigma_s(Y)}{\sigma_s(X)}[x - \mu_s(X)] + \mu_s(Y), \tag{4.1}$$

where s indicates the synthetic population. The four synthetic population parameters in equation (4.1) can be expressed in terms of parameters for Populations 1 and 2 as follows:

$$\mu_s(X) = w_1\mu_1(X) + w_2\mu_2(X), \tag{4.2}$$

$$\mu_s(Y) = w_1\mu_1(Y) + w_2\mu_2(Y), \tag{4.3}$$

$$\sigma_s^2(X) = w_1\sigma_1^2(X) + w_2\sigma_2^2(X) + w_1w_2[\mu_1(X) - \mu_2(X)]^2, \tag{4.4}$$

and

$$\sigma_s^2(Y) = w_1\sigma_1^2(Y) + w_2\sigma_2^2(Y) + w_1w_2[\mu_1(Y) - \mu_2(Y)]^2, \tag{4.5}$$

where the subscripts 1 and 2 refer to Populations 1 and 2, respectively.

For the common-item nonequivalent groups design, X is not administered to examinees in Population 2, and Y is not administered to examinees in Population 1. Therefore, $\mu_2(X), \sigma_2^2(X), \mu_1(Y)$, and $\sigma_1^2(Y)$ in equations (4.2)–(4.5) cannot be estimated directly. The Tucker and Levine observed score methods considered next make different statistical assumptions in order to express these four parameters as functions of directly estimable parameters. Throughout this chapter, all results are reported in terms of parameters, some of which are directly estimable [e.g., $\mu_1(X)$],

Parameters Estimated From Data

Form X Administered in Population 1:	Form Y Administered in Population 2:
$\mu_1(X)$ and $\sigma_1^2(X)$	$\mu_2(Y)$ and $\sigma_2^2(Y)$

Parameters Estimated From Assumptions

Form X Moments in Population 2:	Form Y Moments in Population 1:
$\mu_2(X)$ and $\sigma_2^2(X)$	$\mu_1(Y)$ and $\sigma_1^2(Y)$

Parameters for Synthetic Population

$$\mu_s(X) = w_1\mu_1(X) + w_2\,\mu_2(X)$$

$$\mu_s(Y) = w_1\,\mu_1(Y) + w_2\,\mu_2(Y)$$

$$\sigma_s^2(X) = w_1\,\sigma_1^2(X) + w_2\,\sigma_2^2(X) + w_1\,w_2\,[\mu_1(X) - \mu_2(X)]^2$$

$$\sigma_s^2(Y) = w_1\,\sigma_1^2(Y) + w_2\,\sigma_2^2(Y) + w_1\,w_2\,[\mu_1(Y) - \mu_2(Y)]^2$$

Figure 4.1. Linear equating parameters for the common-item nonequivalent groups design.

while others are not [e.g., $\mu_2(X)$]. In practice, of course, the results are used by replacing all parameters with estimates. The parameters estimated from the data and from assumptions are distinguished in Figure 4.1.

Tucker Method

The Tucker method was described by Gulliksen (1950, pp. 299–301), who attributed it to Ledyard Tucker. This method makes two types of assumptions in order to estimate the parameters in equations (4.2)–(4.5) that cannot be estimated directly. The first type of assumption concerns the regressions of total scores on common-item scores. The second type of assumption concerns the conditional variances of total scores given common-item scores. Basically, these are the assumptions of univariate selection theory (see Gulliksen, 1950, pp. 131–132).

Linear Regression Assumptions

First, the regression of X on V is assumed to be the same linear function for both Populations 1 and 2. A similar assumption is made for Y on V. Letting α represent a regression slope and β a regression intercept,

$$\alpha_1(X|V) = \sigma_1(X, V)/\sigma_1^2(V) \tag{4.6}$$

and

$$\beta_1(X|V) = \mu_1(X) - \alpha_1(X|V)\mu_1(V) \tag{4.7}$$

are the slope and intercept, respectively, for the regression of X on V in Population 1. These two quantities are directly observed. In Population 2, the slope and intercept are

$$\alpha_2(X|V) = \sigma_2(X, V)/\sigma_2^2(V) \tag{4.8}$$

and

$$\beta_2(X|V) = \mu_2(X) - \alpha_2(X|V)\mu_2(V). \tag{4.9}$$

These two quantities are not directly observed. For X and V, then, the regression assumption is

$$\alpha_2(X|V) = \alpha_1(X|V) \tag{4.10}$$

and

$$\beta_2(X|V) = \beta_1(X|V), \tag{4.11}$$

where the quantities to the left of the equal sign are not directly observable. Similarly, for Y and V, the regression assumption is

$$\alpha_1(Y|V) = \alpha_2(Y|V)$$

and

$$\beta_1(Y|V) = \beta_2(Y|V).$$

Conditional Variance Assumptions

Also, for the Tucker method, the conditional variance of X given V is assumed to be the same for Populations 1 and 2. A similar statement holds for Y given V. Stated explicitly, these assumptions are

$$\sigma_2^2(X)[1 - \rho_2^2(X, V)] = \sigma_1^2(X)[1 - \rho_1^2(X, V)] \tag{4.12}$$

and

$$\sigma_1^2(Y)[1 - \rho_1^2(Y, V)] = \sigma_2^2(Y)[1 - \rho_2^2(Y, V)],$$

where ρ is a correlation and the quantities that are not directly observable are to the left of the equalities.

Intermediate Results

The above assumptions are sufficient to solve for $\mu_2(X), \sigma_2(X), \mu_1(Y)$, and $\sigma_1(Y)$ in terms of observable quantities. Consider, for example, $\mu_2(X)$. Because the regression of X on V is assumed to be linear,

$$\mu_2(X) = \beta_2(X|V) + \alpha_2(X|V)\mu_2(V).$$

Using equations (4.10) and (4.11),

$$\mu_2(X) = \beta_1(X|V) + \alpha_1(X|V)\mu_2(V).$$

Now, using equation (4.7),

$$\mu_2(X) = [\mu_1(X) - \alpha_1(X|V)\mu_1(V)] + \alpha_1(X|V)\mu_2(V)$$
$$= \mu_1(X) - \alpha_1(X|V)[\mu_1(V) - \mu_2(V)]. \tag{4.13}$$

Following a similar approach,

$$\mu_1(Y) = \mu_2(Y) + \alpha_2(Y|V)[\mu_1(V) - \mu_2(V)]. \tag{4.14}$$

To obtain $\sigma_2^2(X)$, begin by noting that

$$\rho_1(X, V) = \sigma_1(X, V)/[\sigma_1(X)\sigma_1(V)],$$

where $\sigma_1(X, V)$ is a covariance. Rearranging terms in equation (4.6),

$$\sigma_1(X, V) = \alpha_1(X|V)\sigma_1^2(V).$$

Therefore,

$$\rho_1(X, V) = \alpha_1(X|V)\sigma_1(V)/\sigma_1(X),$$

and, with a little bit of algebra,

$$\sigma_1^2(X)[1 - \rho_1^2(X, V)] = \sigma_1^2(X) - \alpha_1^2(X|V)\sigma_1^2(V).$$

Similarly,

$$\sigma_2^2(X)[1 - \rho_2^2(X, V)] = \sigma_2^2(X) - \alpha_2^2(X|V)\sigma_2^2(V).$$

Now, using equation (4.12),

$$\sigma_2^2(X) - \alpha_2^2(X|V)\sigma_2^2(V) = \sigma_1^2(X) - \alpha_1^2(X|V)\sigma_1^2(V).$$

Because $\alpha_2(X|V) = \alpha_1(X|V)$ by assumption,

$$\sigma_2^2(X) = \sigma_1^2(X) - \alpha_1^2(X|V)[\sigma_1^2(V) - \sigma_2^2(V)]. \tag{4.15}$$

A similar derivation gives,

$$\sigma_1^2(Y) = \sigma_2^2(Y) + \alpha_2^2(Y|V)[\sigma_1^2(V) - \sigma_2^2(V)]. \tag{4.16}$$

Final Results

Given the results in equations (4.13)–(4.16), the synthetic population means and variances in equations (4.2)–(4.5) can be shown to be

$$\mu_s(X) = \mu_1(X) - w_2\gamma_1[\mu_1(V) - \mu_2(V)], \tag{4.17}$$

$$\mu_s(Y) = \mu_2(Y) + w_1\gamma_2[\mu_1(V) - \mu_2(V)], \tag{4.18}$$

$$\sigma_s^2(X) = \sigma_1^2(X) - w_2\gamma_1^2[\sigma_1^2(V) - \sigma_2^2(V)] + w_1w_2\gamma_1^2[\mu_1(V) - \mu_2(V)]^2, \tag{4.19}$$

and

$$\sigma_s^2(Y) = \sigma_2^2(Y) + w_1\gamma_2^2[\sigma_1^2(V) - \sigma_2^2(V)] + w_1w_2\gamma_2^2[\mu_1(V) - \mu_2(V)]^2, \tag{4.20}$$

where the γ-terms are the regression slopes

$$\gamma_1 = \alpha_1(X|V) = \sigma_1(X, V)/\sigma_1^2(V) \tag{4.21}$$

and

$$\gamma_2 = \alpha_2(Y|V) = \sigma_2(Y, V)/\sigma_2^2(V), \tag{4.22}$$

and the parameters to the right of the equal signs can be estimated directly from the data. The Tucker linear equating function is obtained by using the results from equations (4.17)–(4.22) in equation (4.1).

It is evident from the form of equations (4.17)–(4.20) that the synthetic population means and variances for X and Y can be viewed as adjustments to directly observable quantities. The adjustments are functions of differences in means and variances for the common items. If $\mu_1(V) = \mu_2(V)$ and $\sigma_1^2(V) = \sigma_2^2(V)$ then the synthetic population parameters would equal observable means and variances.

Note that the foregoing derivation does not require specifying whether the common-item set is an internal scored set of items or an external unscored set. Consequently, the results apply to both possibilities, provided, of course, that X is correctly specified as the total set of items that directly contribute to an examinee's score. That is, scores on X include scores on V if V is an internal common-item set, and scores on X do not include scores on V if V is an external common-item set.

Special Cases

Equations (4.17)–(4.22) apply for any set of nonnegative weights, w_1 and w_2, provided that $w_1 + w_2 = 1$. At least three special cases are sometimes considered. First, Gulliksen's (1950, pp. 299–301) initial presentation of the Tucker method can be obtained by setting $w_1 = 1$ and $w_2 = 0$, in which case the synthetic population is the population that took the new form. Second, Angoff (1971, p. 580) provides formulas for the Tucker method based on weights that are proportional to sample sizes—i.e., $w_1 = N_1/(N_1 + N_2)$ and $w_2 = N_2/(N_1 + N_2)$, where N_1 and N_2 are the sample sizes from Populations 1 and 2, respectively. Third, the weights are sometimes set equal (i.e., $w_1 = w_2 = .5$), reflecting an a priori judgment that both Populations 1 and 2 are equally relevant for the investigator's conception of the synthetic population.

Levine Observed Score Method

The assumptions of the Tucker method involve only observable quantities. No reference is made to true scores. Yet, it would seem that for equating to be sensible, true scores must be functionally related. Otherwise, it would not be sensible to talk about scores being interchangeable. This argument per se does not render the Tucker method inappropriate, but it does suggest that there may be merit in deriving equating results based on assumptions about true scores. One such method is discussed in this section.

The Levine observed score method was originally developed by Levine (1955), although he did not explicitly consider the concept of a synthetic population. Consequently, the development here is more general than Levine's (1955). This method is an observed score equating method in the sense that it uses equation (4.1) to relate *observed* scores on X to the scale of *observed* scores on Y. However, the assumptions for this method pertain to true scores T_X, T_Y, and T_V which are assumed to be related to observed scores according to the classical test theory model (see Feldt and Brennan, 1989):

$$X = T_X + E_X, \tag{4.23}$$

$$Y = T_Y + E_Y, \tag{4.24}$$

and

$$V = T_V + E_V, \tag{4.25}$$

where E_X, E_Y, and E_V are errors that have zero expectations and are uncorrelated with true scores.

Correlational Assumptions

The Levine method assumes that X, Y, and V are all measuring the same thing in the sense that T_X and T_V as well as T_Y and T_V correlate perfectly in both Populations 1 and 2:

$$\rho_1(T_X, T_V) = \rho_2(T_X, T_V) = 1 \qquad (4.26)$$

and

$$\rho_1(T_Y, T_V) = \rho_2(T_Y, T_V) = 1. \qquad (4.27)$$

Note that equations (4.26) and (4.27) imply that T_X and T_Y are functionally related in both populations.

Linear Regression Assumptions

Also for the Levine method, the regression of T_X on T_V is assumed to be the same linear function for both Populations 1 and 2, and a similar assumption is made for the regression of T_Y on T_V.

By definition, the slope of T_X on T_V is $\alpha_1(T_X|T_V) = \rho_1(T_X, T_V)\sigma_1(T_X)/\sigma_1(T_V)$. Since $\rho_1(T_X, T_V) = 1$ from the correlational assumption in equation (4.26), $\alpha_1(T_X|T_V) = \sigma_1(T_X)/\sigma_1(T_V)$. Similarly, $\alpha_2(T_X|T_V) = \sigma_2(T_X)/\sigma_2(T_V)$. Consequently, the assumption of equal true score regression slopes for T_X on T_V in Populations 1 and 2 is effectively

$$\frac{\sigma_2(T_X)}{\sigma_2(T_V)} = \frac{\sigma_1(T_X)}{\sigma_1(T_V)}. \qquad (4.28)$$

By an analogous derivation,

$$\frac{\sigma_1(T_Y)}{\sigma_1(T_V)} = \frac{\sigma_2(T_Y)}{\sigma_2(T_V)}. \qquad (4.29)$$

For each of the classical test theory model equations (4.23)–(4.25), the mean of observed scores equals the mean of true scores. Consequently, the assumption of equal true score regression intercepts for T_X on T_V in Populations 1 and 2 is

$$\mu_2(X) - \frac{\sigma_2(T_X)}{\sigma_2(T_V)}\mu_2(V) = \mu_1(X) - \frac{\sigma_1(T_X)}{\sigma_1(T_V)}\mu_1(V). \qquad (4.30)$$

Similarly, for the intercepts of T_Y on T_V,

$$\mu_1(Y) - \frac{\sigma_1(T_Y)}{\sigma_1(T_V)}\mu_1(V) = \mu_2(Y) - \frac{\sigma_2(T_Y)}{\sigma_2(T_V)}\mu_2(V). \qquad (4.31)$$

Error Variance Assumptions

The Levine method also assumes that the measurement error variance for X is the same for Populations 1 and 2. A similar assumption is made for Y and V. Because true scores and errors are uncorrelated under the classical test theory model, error variance is the difference between observed score variance and true score variance. Therefore, the error variance assumptions are

$$\sigma_2^2(X) - \sigma_2^2(T_X) = \sigma_1^2(X) - \sigma_1^2(T_X), \tag{4.32}$$

$$\sigma_1^2(Y) - \sigma_1^2(T_Y) = \sigma_2^2(Y) - \sigma_2^2(T_Y),$$

and

$$\sigma_1^2(V) - \sigma_1^2(T_V) = \sigma_2^2(V) - \sigma_2^2(T_V) \tag{4.33}$$

Intermediate Results

Recall that expressions for $\mu_2(X), \sigma_2(X), \mu_1(Y)$, and $\sigma_1(Y)$ are needed in order to obtain the synthetic population means and variances in equations (4.2)–(4.5).

By rearranging terms in equation (4.30) and then using equation (4.28),

$$\mu_2(X) = \mu_1(X) - \frac{\sigma_1(T_X)}{\sigma_1(T_V)}[\mu_1(V) - \mu_2(V)]. \tag{4.34}$$

Similarly, using equations (4.31) and (4.29),

$$\mu_1(Y) = \mu_2(Y) + \frac{\sigma_2(T_Y)}{\sigma_2(T_V)}[\mu_1(V) - \mu_2(V)]. \tag{4.35}$$

From equation (4.32) an expression for $\sigma_2^2(X)$ is

$$\sigma_2^2(X) = \sigma_1^2(X) - \sigma_1^2(T_X) + \sigma_2^2(T_X).$$

From equation (4.28), $\sigma_2(T_X) = \sigma_1(T_X)\sigma_2(T_V)/\sigma_1(T_V)$. It follows that

$$\sigma_2^2(X) = \sigma_1^2(X) - \sigma_1^2(T_X)[1 - \sigma_2^2(T_V)/\sigma_1^2(T_V)]$$

$$= \sigma_1^2(X) - \frac{\sigma_1^2(T_X)}{\sigma_1^2(T_V)}[\sigma_1^2(T_V) - \sigma_2^2(T_V)].$$

Using equation (4.33),

$$\sigma_2^2(X) = \sigma_1^2(X) - \frac{\sigma_1^2(T_X)}{\sigma_1^2(T_V)}[\sigma_1^2(V) - \sigma_2^2(V)]. \tag{4.36}$$

Similarly,

$$\sigma_1^2(Y) = \sigma_2^2(Y) + \frac{\sigma_2^2(T_Y)}{\sigma_2^2(T_V)}[\sigma_1^2(V) - \sigma_2^2(V)]. \tag{4.37}$$

General Results

Given the results in equations (4.34)–(4.37), it can be shown algebraically that the synthetic population means and variances in equations (4.2)–(4.5) are given by equations (4.17)–(4.20) with

$$\gamma_1 = \sigma_1(T_X)/\sigma_1(T_V) \tag{4.38}$$

and

$$\gamma_2 = \sigma_2(T_Y)/\sigma_2(T_V). \tag{4.39}$$

That is, under the Levine assumptions, the γ-terms are ratios of true score standard deviations. Note that the derivation of these results did not require specifying whether V was an internal or external set of common items.

The expressions for the γ-terms in equations (4.38) and (4.39) are not immediately usable because they are ratios of true score standard deviations, which are not directly observed. Given the assumptions of classical test theory, and letting $\rho(X, X') = \sigma^2(T_X)/\sigma^2(X)$ denote the reliability of X, it follows that $\sigma(T_X) = \sigma(X)\sqrt{\rho(X, X')}$. Similarly, $\sigma(T_Y) = \sigma(Y)\sqrt{\rho(Y, Y')}$ and $\sigma(T_V) = \sigma(V)\sqrt{\rho(V, V')}$. Consequently, the γ-terms can be expressed as:

$$\gamma_1 = \frac{\sigma_1(X)\sqrt{\rho_1(X, X')}}{\sigma_1(V)\sqrt{\rho_1(V, V')}} \tag{4.40}$$

and

$$\gamma_2 = \frac{\sigma_2(Y)\sqrt{\rho_2(Y, Y')}}{\sigma_2(V)\sqrt{\rho_2(V, V')}}. \tag{4.41}$$

In principal, any defensible estimates of the reliabilities in equations (4.40) and (4.41) could be used to estimate γ_1 and γ_2. In practice, the most frequently used equations for the Levine method can be shown to result from applying what will be called the "classical congeneric" test theory model (see Feldt and Brennan, 1989, pp. 111–112). [Note that Levine's 1955 derivation effectively stopped with equations (4.40) and (4.41).]

Classical Congeneric Model Results

In this section, unless otherwise noted, the classical congeneric model is assumed for X and V, and for a single population. It is straightforward to extend the results presented here to Y and V, and to Populations 1 and 2.

Recall from equations (4.23) and (4.25) that for the classical model $X = T_X + E_X$ and $V = T_V + E_V$, where E_X and T_X, as well as E_V and T_V, are assumed to be uncorrelated. The congeneric model goes one step further in specifying that T_X and T_V are linearly related, which is consistent with the assumption in equation (4.26) that T_X and T_V are perfectly correlated.

For our present purposes, a convenient way to represent that T_X and T_V are linearly related is to set $T_X = \lambda_X T + \delta_X$ and $T_V = \lambda_V T + \delta_V$, where the λ's are slopes and the δ's are constant intercepts (see Feldt and Brennan,

1989, pp. 110–111). This implies that $T_X = (\lambda_X/\lambda_V)T_V + [\delta_X - (\lambda_X/\lambda_V)\delta_V]$, although this expression is not required in the subsequent derivation. Under the congeneric model, then, the equations for X and V can be expressed as:

$$X = T_X + E_X = (\lambda_X T + \delta_X) + E_X \tag{4.42}$$

and

$$V = T_V + E_V = (\lambda_V T + \delta_V) + E_V. \tag{4.43}$$

The classical congeneric model adds the assumptions that

$$\sigma^2(E_X) = \lambda_X \sigma^2(E) \tag{4.44}$$

and

$$\sigma^2(E_V) = \lambda_V \sigma^2(E). \tag{4.45}$$

In classical test theory, error variances are proportional to test length. Here, error variances are proportional to λ_X and λ_V which are called "effective" test lengths. Note also that the ratio $\sigma^2(E_X)/\sigma^2(E_V)$ is simply λ_X/λ_V.

Given equations (4.42)–(4.45), the following can be shown relatively easily:

$$\sigma^2(X) = \lambda_X^2 \sigma^2(T) + \lambda_X \sigma^2(E), \tag{4.46}$$

$$\sigma^2(V) = \lambda_V^2 \sigma^2(T) + \lambda_V \sigma^2(E), \tag{4.47}$$

and

$$\sigma(X, V) = \lambda_X \lambda_V \sigma^2(T) + \sigma(E_X, E_V). \tag{4.48}$$

Here, we make use of the classical congeneric model to obtain an expression for $\sigma(T_X)/\sigma(T_V)$, which is the γ-term in equation (4.38). From equations (4.42) and (4.43),

$$\gamma = \frac{\sigma(T_X)}{\sigma(T_V)} = \frac{\lambda_X \sigma(T)}{\lambda_V \sigma(T)} = \frac{\lambda_X}{\lambda_V}, \tag{4.49}$$

which means that γ can be interpreted as the ratio of effective test lengths for X and V, respectively. Two cases need to be considered: (a) an internal anchor in which all items in V are included in X, and (b) an external anchor in which V and X consist of entirely different sets of items. These two cases can be distinguished in terms of the error covariance $\sigma(E_X, E_V)$ in equation (4.48).

Internal Anchor. When V is included in X, the full-length test is X. Now, let A be the noncommon part of X such that $X = A + V$. Under the congeneric model, the covariance between the errors for A and V is assumed to be 0 because these two parts of X consist of entirely different items. Consequently,

$$\sigma(E_X, E_V) = \sigma(E_{A+V}, E_V) = \sigma(E_V, E_V) = \sigma^2(E_V) = \lambda_V \sigma^2(E). \qquad (4.50)$$

That is, the covariance between E_X and E_V is simply the variance of E_V.

Using equation (4.50) in (4.48) gives

$$\sigma(X, V) = \lambda_X \lambda_V \sigma^2(T) + \lambda_V \sigma^2(E)$$
$$= \lambda_V [\lambda_X \sigma^2(T) + \sigma^2(E)]. \qquad (4.51)$$

After rewriting equation (4.46) as

$$\sigma^2(X) = \lambda_X [\lambda_X \sigma^2(T) + \sigma^2(E)],$$

it is evident from equation (4.51) and the above expression for $\sigma^2(X)$ that γ in equation (4.49) is

$$\gamma = \lambda_X / \lambda_V = \sigma^2(X)/\sigma(X, V) = 1/\alpha(V|X). \qquad (4.52)$$

Therefore, for the internal anchor case, the results for Levine's observed score method under the classical congeneric model are obtained by using

$$\gamma_1 = 1/\alpha_1(V|X) = \sigma_1^2(X)/\sigma_1(X, V) \qquad (4.53)$$

and

$$\gamma_2 = 1/\alpha_2(V|Y) = \sigma_2^2(Y)/\sigma_2(Y, V). \qquad (4.54)$$

That is, with an internal anchor, the γ-terms in equations (4.17)–(4.20) under the classical congeneric model are the inverses of the regression slopes of V on X and V on Y.

External Anchor. When X and V contain no items in common, under the congeneric model,

$$\sigma(E_X, E_V) = 0. \qquad (4.55)$$

Using equation (4.55) in (4.48) gives

$$\sigma(X, V) = \lambda_X \lambda_V \sigma^2(T). \qquad (4.56)$$

From equations (4.46) and (4.56),

$$\sigma^2(X) + \sigma(X, V) = \lambda_X [(\lambda_X + \lambda_V)\sigma^2(T) + \sigma^2(E)].$$

Similarly, using equations (4.47) and (4.56),

$$\sigma^2(V) + \sigma(X, V) = \lambda_V [(\lambda_X + \lambda_V)\sigma^2(T) + \sigma^2(E)].$$

It follows that γ in equation (4.49) is

$$\gamma = \frac{\lambda_X}{\lambda_V} = \frac{\sigma^2(X) + \sigma(X, V)}{\sigma^2(V) + \sigma(X, V)}. \qquad (4.57)$$

Therefore, for the external anchor case, the results for Levine's observed score method under the classical congeneric model are obtained by using

$$\gamma_1 = \frac{\sigma_1^2(X) + \sigma_1(X, V)}{\sigma_1^2(V) + \sigma_1(X, V)} \qquad (4.58)$$

and

$$\gamma_2 = \frac{\sigma_2^2(Y) + \sigma_2(Y, V)}{\sigma_2^2(V) + \sigma_2(Y, V)} \qquad (4.59)$$

in equations (4.17)–(4.20).

Comments. Under the assumption that $w_1 = N_1/(N_1 + N_2)$ and $w_2 = N_2/(N_1 + N_2)$, the results for Levine's observed score method and a classical congeneric model are identical to those reported by Angoff (1971), although the derivation is different. Angoff's (1971) results are sometimes called the Levine-Angoff method, or described as "Levine's method using Angoff error variances." The error variances are those in Angoff (1953), which are also reported by Petersen *et al.* (1989, p. 254). Brennan (1990) has shown that Angoff's error variances are derivable from the classical congeneric model. Table 4.1 reports these error variances along with other results for the classical congeneric model that can be used to express the quantities illustrated in Figure 4.1.

Levine True Score Method

Levine (1955) also derived results for a true score equating method using the same assumptions about true scores discussed in the previous section. The principal difference between the observed score and true score methods is that the observed score method uses equation (4.1) to equate observed scores on X to the scale of observed scores on Y, whereas the true score method equates true scores. Specifically, the following equation is used to equate true scores on X to the scale of true scores on Y:

$$l_{Ys}(t_X) = \frac{\sigma_s(T_Y)}{\sigma_s(T_X)}[t_X - \mu_s(T_X)] + \mu_s(T_Y).$$

In classical theory, observed score means equal true score means. Therefore,

$$l_{Ys}(t_X) = \frac{\sigma_s(T_Y)}{\sigma_s(T_X)}[t_X - \mu_s(X)] + \mu_s(Y). \qquad (4.60)$$

Table 4.1. Classical Congeneric Model Results.

Quantity	Anchor	
	Internal	External
$\gamma = \dfrac{\lambda_X}{\lambda_V}$	$\dfrac{1}{\alpha(V\|X)} = \dfrac{\sigma^2(X)}{\sigma(X,V)}$	$\dfrac{\sigma^2(X) + \sigma(X,V)}{\sigma^2(V) + \sigma(X,V)}$
$\sigma^2(T_X)$	$\dfrac{\gamma^2[\sigma(X,V) - \sigma^2(V)]}{\gamma - 1}$	$\gamma\sigma(X,V)$
$\sigma^2(T_V)$	$\dfrac{\sigma(X,V) - \sigma^2(V)}{\gamma - 1}$	$\dfrac{\sigma(X,V)}{\gamma}$
$\sigma^2(E_X)$	$\dfrac{\gamma^2\sigma^2(V) - \gamma\sigma(X,V)}{\gamma - 1}$	$\sigma^2(X) - \gamma\sigma(X,V)$
$\sigma^2(E_V)$	$\dfrac{\gamma\sigma^2(V) - \sigma(X,V)}{\gamma - 1}$	$\sigma^2(V) - \dfrac{\sigma(X,V)}{\gamma}$
$\rho(X, X')$	$\dfrac{\gamma^2[\sigma(X,V) - \sigma^2(V)]}{(\gamma - 1)\sigma^2(X)}$	$\dfrac{\gamma\sigma(X,V)}{\sigma^2(X)}$
$\rho(V, V')$	$\dfrac{\sigma(X,V) - \sigma^2(V)}{(\gamma - 1)\sigma^2(V)}$	$\dfrac{\sigma(X,V)}{\gamma\sigma^2(V)}$

Note: Here, the population subscript "1" has been suppressed.

Results

Equations (4.2) and (4.3) are still appropriate for $\mu_s(X)$ and $\mu_s(Y)$, respectively. Also, under Levine's assumptions, equations (4.34) and (4.35) still apply for $\mu_2(X)$ and $\mu_1(Y)$, respectively. Consequently, equations (4.17) and (4.18) for $\mu_s(X)$ and $\mu_s(Y)$ are valid for both the Levine observed score and the Levine true score methods, with the γ-terms given by equations (4.38) and (4.39). For ease of reference, these results are repeated below:

$$\mu_s(X) = \mu_1(X) - w_2\gamma_1[\mu_1(V) - \mu_2(V)], \qquad (4.17)$$

and

$$\mu_s(Y) = \mu_2(Y) + w_1\gamma_2[\mu_1(V) - \mu_2(V)], \qquad (4.18)$$

where

$$\gamma_1 = \sigma_1(T_X)/\sigma_1(T_V) \qquad (4.38)$$

and

$$\gamma_2 = \sigma_2(T_Y)/\sigma_2(T_V). \tag{4.39}$$

Using Levine's true score assumptions, the derivation of expressions for the variance of T_X and T_Y for the synthetic population is tedious (see Appendix), although the results are simple:

$$\sigma_s^2(T_X) = \gamma_1^2 \sigma_s^2(T_V) \tag{4.61}$$

and

$$\sigma_s^2(T_Y) = \gamma_2^2 \sigma_s^2(T_V), \tag{4.62}$$

where

$$\sigma_s^2(T_V) = w_1 \sigma_1^2(T_V) + w_2 \sigma_2^2(T_V) + w_1 w_2 [\mu_1(V) - \mu_2(V)]^2.$$

From equations (4.61) and (4.62), the slope of the equating relationship $l_{Y_s}(t_X)$ in equation (4.60) is

$$\sigma_s(T_Y)/\sigma_s(T_X) = \gamma_2/\gamma_1, \tag{4.63}$$

where the γ-terms are given by equations (4.38) and (4.39).

These results are quite general, but they are not directly usable without expressions for the true score standard deviations $\sigma_1(T_X)$, $\sigma_2(T_Y)$, $\sigma_1(T_V)$, and $\sigma_2(T_V)$, which are incorporated in γ_1 and γ_2. As with the Levine observed score method, $\sigma_1(X)\sqrt{\rho_1(X, X')}$ can be used for $\sigma_1(T_X)$, and corresponding expressions can be used for the other true score standard deviations. Then, given estimates of the required reliabilities, the linear equating relationship $l_{Y_s}(t_X)$ in equation (4.60) can be determined.

One counterintuitive property of the Levine true score method is that the slope and intercept do not depend on the synthetic population weights w_1 and w_2. Clearly, this is true for the slope in equation (4.63). From equations (4.60) and (4.63), the intercept is

$$\mu_s(Y) - (\gamma_2/\gamma_1)\mu_s(X),$$

and, using equations (4.17) and (4.18), it can be expressed as:

$$\mu_2(Y) + w_1\gamma_2[\mu_1(V) - \mu_2(V)] - (\gamma_2/\gamma_1)\{\mu_1(X) - w_2\gamma_1[\mu_1(V) - \mu_2(V)]\}$$
$$= \mu_2(Y) - (\gamma_2/\gamma_1)\mu_1(X) + \gamma_2(w_1 + w_2)[\mu_1(V) - \mu_2(V)]$$
$$= \mu_2(Y) - (\gamma_2/\gamma_1)\mu_1(X) + \gamma_2[\mu_1(V) - \mu_2(V)], \tag{4.64}$$

which does not depend on the weights w_1 and w_2.

Given the slope and intercept in equations (4.63) and (4.64), respectively, the linear equating relationship for Levine's true score method can be expressed as:

$$l_Y(t_X) = (\gamma_2/\gamma_1)[t_X - \mu_1(X)] + \mu_2(Y) + \gamma_2[\mu_1(V) - \mu_2(V)], \tag{4.65}$$

which gives the same Form Y equivalents as equation (4.60). Note, however, that s does not appear as a subscript of l in equation (4.65) because this expression for Levine's true score method does not involve a synthetic population. In short, Levine's true score method does not require the conceptual framework of a synthetic population and is invariant with respect to the weights w_1 and w_2.

Classical Congeneric Model. Results for Levine true score equating under the classical congeneric model with an internal anchor are obtained simply by using equations (4.53) and (4.54) for γ_1 and γ_2, respectively. For an external anchor, equations (4.58) and (4.59) are used.

Using Levine's True Score Method with Observed Scores. Equations (4.60) and (4.65) were derived for true scores, not observed scores. Even so, in practice, observed scores are used in place of true scores. That is, observed scores on X are assumed to be related to the scale of observed scores on Y by the equation:

$$l_Y(x) = (\gamma_2/\gamma_1)[x - \mu_1(X)] + \mu_2(Y) + \gamma_2[\mu_1(V) - \mu_2(V)]. \qquad (4.66)$$

Although replacing true scores with observed scores may appear sensible, there is no seemingly compelling logical basis for doing so. Note, in particular, that the transformed observed scores on X [i.e., $l_Y(x)$] typically do not have the same standard deviation as either the true scores on Y or the observed scores on Y. However, as will be discussed next, Levine's true score method applied to observed scores has an interesting property.

First-Order Equity

Although the logic of using observed scores in Levine's true score equating function appears somewhat less than compelling, Hanson (1991a) has shown that using observed scores in Levine's true score equating function for the common-item nonequivalent groups design results in first-order equity (see Chapter 1) of the equated test scores under the classical congeneric model. Hanson's (1991a) result gives Levine's true score equating method applied to observed scores a well-grounded theoretical justification. In general, his result means that, for the population of persons with a particular true score on Y, the expected value of the linearly transformed scores on X [equation (4.66)] equals the expected value of the scores on Y, and this statement holds for all true scores on Y. In formal terms, first-order equity means that

$$\mathbf{E}[l_Y(X)|\psi(T_X) = \tau] = \mathbf{E}[Y|T_Y = \tau] \qquad \text{for all } \tau, \qquad (4.67)$$

where ψ is a function that relates true scores on X to true scores on Y, and

X is capitalized in $l_Y(X)$ to emphasize that interest is focused here on the variable X rather than on a realization x.

Before treating the specific case of the common-item nonequivalent groups design, it is shown next that first-order equity holds whenever there exists a population such that Forms X and Y are congeneric and true scores are replaced by observed scores. As was discussed previously, for the congeneric model,

$$X = T_X + E_X = (\lambda_X T + \delta_X) + E_X \quad \text{and} \quad Y = T_Y + E_Y = (\lambda_Y T + \delta_Y) + E_Y.$$

To convert true scores on X to the scale of true scores on Y, it can be shown that

$$T_Y = \Psi(T_X) = \frac{\lambda_Y}{\lambda_X}(T_X - \delta_X) + \delta_Y.$$

Substituting X for T_X gives

$$l_Y(X) = \frac{\lambda_Y}{\lambda_X}(X - \delta_X) + \delta_Y. \tag{4.68}$$

In congeneric theory, the expected value of errors is 0. Thus,

$$\mathbf{E}(X|T = \tau) = \mathbf{E}[\lambda_X T + \delta_X + E_X] = \lambda_X T + \delta_X \quad \text{and}$$

$$\mathbf{E}(Y|T = \tau) = \mathbf{E}[\lambda_Y T + \delta_Y + E_Y] = \lambda_Y T + \delta_Y.$$

First-order equity holds for $l_Y(X)$ because the expected value of $l_Y(X)$ given $\Psi(T_X) = \tau$ equals the expected value of Y given $T_Y = \tau$:

$$\mathbf{E}\left[\frac{\lambda_Y}{\lambda_X}(X - \delta_X) + \delta_Y | \Psi(T_X) = \tau\right]$$

$$= \mathbf{E}\left[\frac{\lambda_Y}{\lambda_X}(\lambda_X T + \delta_X + E_X - \delta_X) + \delta_Y | T_Y = \tau\right]$$

$$= \lambda_Y T + \delta_Y = \mathbf{E}[Y|T_Y = \tau],$$

as was previously indicated.

For the common-item nonequivalent groups design, one parameterization of the classical congeneric model is

$$\left.\begin{array}{ll}
X_1 = (\lambda_X T_1 + \delta_X) + E_{X_1}, & \sigma_1^2(E_X) = \lambda_X \sigma_1^2(E), \\
Y_2 = (\lambda_Y T_2 + \delta_Y) + E_{Y_2}, & \sigma_2^2(E_Y) = \lambda_Y \sigma_2^2(E), \\
V_1 = (\lambda_V T_1 + \delta_V) + E_{V_1}, & \sigma_1^2(E_V) = \lambda_V \sigma_1^2(E), \\
V_2 = (\lambda_V T_2 + \delta_V) + E_{V_2}, & \sigma_2^2(E_V) = \lambda_V \sigma_2^2(E),
\end{array}\right\} \tag{4.69}$$

where the subscripts 1 and 2 designate the populations. This parameterization is different from that in Hanson (1991a), but it is consistent with the parameterization introduced previously.

Given the parameterization in equation set (4.69),

$$\left.\begin{aligned}
\mu_1(X) &= \lambda_X \mu_1(T) + \delta_X, & \mu_2(Y) &= \lambda_Y \mu_2(T) + \delta_Y, \\
\mu_1(V) &= \lambda_V \mu_1(T) + \delta_V, & \mu_2(V) &= \lambda_V \mu_2(T) + \delta_V, \\
\sigma_1^2(X) &= \lambda_X^2 \sigma_1^2(T) + \lambda_X \sigma_1^2(E), & \sigma_2^2(Y) &= \lambda_Y^2 \sigma_2^2(T) + \lambda_Y \sigma_2^2(E), \\
\sigma_1^2(V) &= \lambda_V^2 \sigma_1^2(T) + \lambda_V \sigma_1^2(E), & \sigma_2^2(V) &= \lambda_V^2 \sigma_2^2(T) + \lambda_V \sigma_2^2(E), \\
\sigma_1(X, V) &= \lambda_X \lambda_V \sigma_1^2(T) & \sigma_2(Y, V) &= \lambda_Y \lambda_V \sigma_2^2(T) \\
&\quad + \sigma_1(E_X, E_V), & &\quad + \sigma_2(E_Y, E_V).
\end{aligned}\right\} \quad (4.70)$$

From equation (4.50), for the internal case, $\sigma_1(E_X, E_V) = \lambda_V \sigma_1^2(E)$; similarly, $\sigma_2(E_Y, E_V) = \lambda_V \sigma_2^2(E)$. From equation (4.55), for the external case, $\sigma_1(E_X, E_V) = 0$; similarly, $\sigma_2(E_Y, E_V) = 0$.

To prove that first-order equity holds for Levine's true score method applied to observed scores, it is sufficient to show that the slope and intercept in the Levine equation (4.66) equal the slope and intercept, respectively, in equation (4.68).

To prove the equality of slopes, it is necessary to show that

$$\gamma_2/\gamma_1 = \lambda_Y/\lambda_X.$$

For the internal case, from equation (4.53),

$$\begin{aligned}
\gamma_1 &= \sigma_1^2(X)/\sigma_1(X, V) \\
&= \frac{\lambda_X^2 \sigma_1^2(T) + \lambda_X \sigma_1^2(E)}{\lambda_X \lambda_V \sigma_1^2(T) + \lambda_V \sigma_1^2(E)} \\
&= \lambda_X/\lambda_V.
\end{aligned}$$

Similarly,

$$\gamma_2 = \lambda_Y/\lambda_V \tag{4.71}$$

and, consequently,

$$\gamma_2/\gamma_1 = \lambda_Y/\lambda_X. \tag{4.72}$$

The external case is left as an exercise for the reader.

To prove the equality of intercepts, it is necessary to show that

$$\mu_2(Y) - (\gamma_2/\gamma_1)\mu_1(X) + \gamma_2[\mu_1(V) - \mu_2(V)] = \delta_Y - (\lambda_Y/\lambda_X)\delta_X.$$

For the internal case, from equations (4.71) and (4.72), the intercept is

$$\begin{aligned}
&\mu_2(Y) - (\lambda_Y/\lambda_X)\mu_1(X) + (\lambda_Y/\lambda_V)[\mu_1(V) - \mu_2(V)] \\
&= [\lambda_Y \mu_2(T) + \delta_Y] - (\lambda_Y/\lambda_X)[\lambda_X \mu_1(T) + \delta_X] \\
&\quad + (\lambda_Y/\lambda_V)[\lambda_V \mu_1(T) + \delta_V - \lambda_V \mu_2(T) - \delta_V] \\
&= \lambda_Y[\mu_2(T) - \mu_1(T)] + [\delta_Y - (\lambda_Y/\lambda_X)\delta_X] + \lambda_Y[\mu_1(T) - \mu_2(T)] \\
&= \delta_Y - (\lambda_Y/\lambda_X)\delta_X.
\end{aligned}$$

The external case is left as an exercise for the reader.

Table 4.2. Computational Formulas and Equations for Linear Equating Methods with the Common-Item Nonequivalent Groups Design.

Tucker and Levine Observed Score Methods

$$l_{Ys}(x) = [\sigma_s(Y)/\sigma_s(X)][x - \mu_s(X)] + \mu_s(Y) \tag{4.1}$$

Levine True Score Method Applied to Observed Scores

$$l_Y(x) = (\gamma_2/\gamma_1)[x - \mu_1(X)] + \mu_2(Y) + \gamma_2[\mu_1(V) - \mu_2(V)] \tag{4.66}$$

$$\mu_s(X) = \mu_1(X) \qquad - w_2\gamma_1[\mu_1(V) - \mu_2(V)] \tag{4.17}$$

$$\mu_s(Y) = \mu_2(Y) + w_1\gamma_2[\mu_1(V) - \mu_2(V)] \tag{4.18}$$

$$\sigma_s^2(X) = \sigma_1^2(X) \qquad - w_2\gamma_1^2[\sigma_1^2(V) - \sigma_2^2(V)] + w_1w_2\gamma_1^2[\mu_1(V) - \mu_2(V)]^2 \tag{4.19}$$

$$\sigma_s^2(Y) = \sigma_2^2(Y) + w_1\gamma_2^2[\sigma_1^2(V) - \sigma_2^2(V)] \qquad + w_1w_2\gamma_2^2[\mu_1(V) - \mu_2(V)]^2 \tag{4.20}$$

Tucker Observed Score Method

$$\gamma_1 = \alpha_1(X|V) = \sigma_1(X, V)/\sigma_1^2(V) \quad \text{internal anchor} \tag{4.21}$$

$$\text{and}$$

$$\gamma_2 = \alpha_2(Y|V) = \sigma_2(Y, V)/\sigma_2^2(V) \quad \text{external anchor} \tag{4.22}$$

Levine Methods Under a Classical Congeneric Model

$$\gamma_1 = 1/\alpha_1(V|X) = \sigma_1^2(X)/\sigma_1(X, V) \tag{4.53}$$

$$\text{internal anchor}$$

$$\gamma_2 = 1/\alpha_2(V|Y) = \sigma_2^2(Y)/\sigma_2(Y, V) \tag{4.54}$$

$$\gamma_1 = \frac{\sigma_1^2(X) + \sigma_1(X, V)}{\sigma_1^2(V) + \sigma_1(X, V)} \tag{4.58}$$

$$\text{external anchor}$$

$$\gamma_2 = \frac{\sigma_2^2(Y) + \sigma_2(Y, V)}{\sigma_2^2(V) + \sigma_2(Y, V)} \tag{4.59}$$

Note: For the Levine methods, without assuming a classical congeneric model, for both the internal and external anchor cases [see equations (4.40) and (4.41)],

$$\gamma_1 = \frac{\sigma_1(X)\sqrt{\rho_1(X, X')}}{\sigma_1(V)\sqrt{\rho_1(V, V')}} \quad \text{and} \quad \gamma_2 = \frac{\sigma_2(Y)\sqrt{\rho_2(Y, Y')}}{\sigma_2(V)\sqrt{\rho_2(V, V')}}.$$

Illustrative Example and Other Topics

Table 4.2 provides the principal computational equations for the three linear equating methods that have been developed in this chapter. Note that terms containing w_2 in equations (4.17)–(4.20) in Table 4.2 are slightly separated from the other terms. Doing so more clearly reveals the simplifications in synthetic population means and variances when $w_1 = 1$ and

Table 4.3. Directly Observable Statistics for an Illustrative Example of Equating Forms X and Y Using the Common-Item Nonequivalent Groups Design.

Group	Score	$\hat{\mu}$	$\hat{\sigma}$	Covariance	Correlation
1	X	15.8205	6.5278	13.4088	.8645
1	V	5.1063	2.3760		
2	Y	18.6728	6.8784	14.7603	.8753
2	V	5.8626	2.4515		

Note: $N_1 = 1655$ and $N_2 = 1638$.

$w_2 = 0$. In this section, all references to Levine methods (except for parts of Table 4.2) assume the classical congeneric model.

Illustrative Example

Table 4.3 provides statistics for a real data example that employs two 36-item forms, Form X and Form Y, in which every third item in both forms is a common item. Therefore, items 3, 6, 9,..., 36 constitute the 12-item common set V. Scores on V are contained in X, so V is an internal set of items. Form X was administered to 1655 examinees, and Form Y was administered to 1638 examinees. Method of moments estimates of directly observable parameters are presented in Table 4.3. The Tucker and Levine observed score analyses were conducted using the *CIPE* Macintosh computer program described in Appendix B.

To simplify computations, let $w_1 = 1$ and $w_2 = 1 - w_1 = 0$ for the Tucker and Levine observed score methods. For this synthetic population, using equations (4.17) and (4.19),

$$\hat{\mu}_s(X) = \hat{\mu}_1(X) = 15.8205$$

and

$$\hat{\sigma}_s(X) = \hat{\sigma}_1(X) = 6.5278.$$

Now, for the Tucker method, using equation (4.22),

$$\hat{\gamma}_2 = \hat{\sigma}_2(Y, V)/\hat{\sigma}_2^2(V) = 14.7603/2.4515^2 = 2.4560.$$

Using this value in equations (4.18) and (4.20) gives

$$\hat{\mu}_s(Y) = 18.6728 + 2.4560 \, (5.1063 - 5.8626) = 16.8153$$

and

$$\hat{\sigma}_s(Y) = \sqrt{6.8784^2 + 2.4560^2(2.3760^2 - 2.4515^2)} = 6.7167.$$

Applying these results in equation (4.1) gives

$$\hat{l}_{Ys}(x) = (6.7167/6.5278)(x - 15.8205) + 16.8153$$

$$= .5370 + 1.0289x. \tag{4.73}$$

For the Levine observed score method, with $w_1 = 1$, $\hat{\mu}_s(X) = 15.8205$ and $\hat{\sigma}_s(X) = 6.5278$, as for the Tucker method. However, for the Levine method under the classical congeneric model, using equation (4.54),

$$\hat{\gamma}_2 = \hat{\sigma}_2^2(Y)/\hat{\sigma}_2(Y, V) = 6.8784^2/14.7603 = 3.2054. \tag{4.74}$$

Then, using equations (4.18) and (4.20),

$$\hat{\mu}_s(Y) = 18.6728 + 3.2054 \, (5.1063 - 5.8626) = 16.2486,$$

and

$$\hat{\sigma}_s(Y) = \sqrt{6.8784^2 + 3.2054^2(2.3760^2 - 2.4515^2)} = 6.6006.$$

Applying these results in equation (4.1) gives

$$\hat{l}_{Ys}(x) = (6.6006/6.5278)(x - 15.8205) + 16.2486$$

$$= .2517 + 1.0112x. \tag{4.75}$$

For the Levine true score method applied to observed scores, $\hat{\gamma}_2 = 3.2054$ in equation (4.74) still applies and, using equation (4.53),

$$\hat{\gamma}_1 = \hat{\sigma}_1^2(X)/\hat{\sigma}_1(X, V) = 6.5278^2/13.4088 = 3.1779.$$

Therefore, equation (4.66) gives

$$\hat{l}_Y(x) = (3.2054/3.1779)(x - 15.8205) + 18.6728$$

$$+ 3.2054 \, (5.1063 - 5.8626)$$

$$= .2912 + 1.0087x. \tag{4.76}$$

These results are summarized in Table 4.4. The slight discrepancies in slopes and intercepts in equations (4.73), (4.75), and (4.76) compared to those in Table 4.4 are due to rounding error; the results in Table 4.4 are more accurate. In practice, it is generally advisable to perform computations with more decimal digits than presented here for illustrative purposes, especially for accurate estimates of intercepts.

The similarity of slopes and intercepts for the three methods suggests that the Form Y equivalents will be about the same for all three methods. This finding is illustrated by the results provided in Table 4.5. The Form Y equivalents for the three methods are very similar, although there is a greater difference between the equivalents for the Tucker method and either Levine method than between the equivalents for the two Levine methods.

The new Form X is more difficult than the old Form Y for very high achieving examinees as suggested in Table 4.5, where, for all three meth-

Table 4.4. Linear Equating Results for the Illustrative Example in Table 4.3 Using the Classical Congeneric Model with Levine's Methods.

w_1	Method	$\hat{\gamma}_1$	$\hat{\gamma}_2$	$\hat{\mu}_s(X)$	$\hat{\mu}_s(Y)$	$\hat{\sigma}_s(X)$	$\hat{\sigma}_s(Y)$	$\widehat{int}$	$\widehat{slope}$
1	Tucker	(a)	2.4560	15.8205	16.8153	6.5278	6.7168	.5368	1.0289
	Levine Obs. Sc.	(a)	3.2054	15.8205	16.2485	6.5278	6.6007	.2513	1.0112
.5	Tucker	2.3751	2.4560	16.7187	17.7440	6.6668	6.8612	.5378	1.0292
	Levine Obs. Sc.	3.1779	3.2054	17.0223	17.4607	6.7747	6.8491	.2514	1.0110
.5026[c]	Tucker	2.3751	2.4560	16.7141	17.7392	6.6664	6.8608	.5378	1.0292
	Levine Obs. Sc.	3.1779	3.2054	17.0161	17.4544	6.7740	6.8484	.2514	1.0110
—	Levine True Sc.	3.1779	3.2054	15.8205	16.2485	(b)	(b)	.2914	1.0086

[a] Not required when $w_1 = 1$.

[b] Not required for Levine true score equating.

[c] Proportional to sample size [i.e, $w_1 = N_1/(N_1 + N_2) = .5026$].

Table 4.5. Selected Form Y Equivalents for Illustrative
Example with $w_1 = 1$.

x	Tucker	Levine Observed Score	Levine True Score
0	.54	.25	.29
10	10.83	10.36	10.38
20	21.12	20.47	20.46
30	31.41	30.59	30.55
36	37.58	36.65	36.60

ods, the Form Y equivalent of $x = 36$ is a score greater than the maximum possible score of 36. As was discussed in Chapter 2, raw score equivalents that are out of the range of possible scores can be problematic. Sometimes, equivalents greater than the maximum observable raw score are set to this maximum score. In other cases, this problem is handled through the transformation to scale scores. In most cases, doing so has little practical importance, but this issue could be consequential when various test forms are used for scholarship decisions. The occasional need to truncate Form Y equivalents is a limitation of linear equating procedures. This issue will be discussed further in Chapter 8.

Synthetic Population Weights

As noted previously, the synthetic population weights (w_1 and $w_2 = 1 - w_1$) have no bearing on Levine's true score method. That is why the results for this method appear on a separate line in Table 4.4. For the Tucker and Levine observed score methods, however, the weights do matter, in the sense that they are required to derive the results. From a practical perspective, however, the weights seldom make much difference in the Form Y equivalents. This observation is illustrated in Table 4.4 by the fact that the intercepts and slopes for Tucker equating are almost identical under very different weighting schemes (e.g., $w_1 = 1$ and $w_1 = .5$), and the same is true for Levine observed score equating.

Although the choice of weights makes little practical difference in the vast majority of real equating contexts, many equations are simplified considerably by choosing $w_1 = 1$ and $w_2 = 0$. This observation is evident from examining equations (4.17)–(4.20) in Table 4.2. Furthermore, setting $w_1 = 1$ means that the synthetic group is simply the new population, which is often the only population that will take the new form under the nonequivalent groups design. Therefore, using $w_1 = 1$ often results in some conceptual simplifications. For these reasons, setting $w_1 = 1$ appears to have merit. However, the choice of synthetic population weights ultimately is a judgment that should be based on an investigator's conceptualization of the synthetic population. It is not the authors' intent to suggest that $w_1 = 1$ be used routinely or thoughtlessly. (See Angoff, 1987; Kolen and Brennan, 1987; and Brennan and Kolen, 1987a, for further discussion and debate about choosing w_1 and w_2.)

Mean Equating

If sample sizes are quite small (say less than 100), the standard errors of linear equating (as will be discussed in Chapter 7) may be unacceptably large. In such cases, mean equating might be considered. Form Y equivalents for mean equating under the Tucker and Levine observed score methods are obtained by setting $\sigma_s(Y)/\sigma_s(X) = 1$ in equation (4.1), which gives

$$m_{Ys}(x) = [x - \mu_s(X)] + \mu_s(Y), \tag{4.77}$$

where $\mu_s(X)$ and $\mu_s(Y)$ are given by equations (4.17) and (4.18). Effectively, the Form Y equivalent of a Form X score is obtained by adding the same constant, $\mu_s(Y) - \mu_s(X)$, to all scores on Form X.

Form Y equivalents under Levine's true score method are obtained by setting $\gamma_2/\gamma_1 = 1$ in equation (4.66), which gives

$$m_Y(x) = [x - \mu_1(X)] + \{\mu_2(Y) + \gamma_2[\mu_1(V) - \mu_2(V)]\}. \tag{4.78}$$

Note that, if $w_1 = 1$, equations (4.77) and (4.78) are identical because $\mu_s(X) = \mu_1(X)$ and $\mu_s(Y)$ is given by the term in braces in equation (4.78). Since γ_2 is the same for both of Levine's methods, this implies that, when $w_1 = 1$, mean equating results are identical for Levine's observed score and true score methods.

Decomposing Observed Differences in Means and Variances

In the common-item nonequivalent groups design, differences in the observable means $\mu_1(X) - \mu_2(Y)$ and observable variances $\sigma_1^2(X) - \sigma_2^2(Y)$ are due to the confounded effects of group and form differences. Since estimates of these parameters are directly observed, a natural question is, "How much of the observed difference in means (or variances) is attributable to group differences, and how much is attributable to form differences?" An answer to this question is of some consequence to both test developers and psychometricians responsible for equating. There is nothing a test developer can do about group differences; but in principle, if form differences are known to be relatively large, test developers can take steps to create more similar forms in the future. Furthermore, if a psychometrician notices that group differences or form differences are very large, this should alert him or her to the possibility that equating results may be suspect.

One way to answer the question posed in the previous paragraph is discussed by Kolen and Brennan (1987). Their treatment is briefly summarized here.

Decomposing Differences in Means.   Begin with the tautology

$$\mu_1(X) - \mu_2(Y) = \mu_s(X) - \mu_s(Y) + \{[\mu_1(X) - \mu_s(X)] - [\mu_2(Y) - \mu_s(Y)]\}.$$
$$(4.79)$$

Note that $\mu_s(X) - \mu_s(Y)$ is the mean difference for the two forms for the synthetic population. Since the synthetic population is constant, the difference is entirely attributable to forms and will be called the *form difference factor*. The remaining terms in braces will be called the *population difference factor*. [Note that, since equation (4.79) involves a synthetic population, it does not apply to Levine's true score procedure.]

After replacing equations (4.2) and (4.3) in equation (4.79), it can be shown that

$$\mu_1(X) - \mu_2(Y) = w_1\{\mu_1(X) - \mu_1(Y)\} \quad \text{Form difference for Population 1}$$
$$+ w_2\{\mu_2(X) - \mu_2(Y)\} \quad \text{Form difference for Population 2}$$
$$+ w_2\{\mu_1(X) - \mu_2(X)\} \quad \text{Population difference on } X \text{ scale}$$
$$+ w_1\{\mu_1(Y) - \mu_2(Y)\} \quad \text{Population difference on } Y \text{ scale,}$$
$$(4.80)$$

where the descriptions on the right describe the mathematical terms in braces (i.e., excluding the w_1 and w_2 weights). This expression states that $\mu_1(X) - \mu_2(Y)$ is a function of two weighted form difference factors (one for each population) and two weighted population difference factors (one for each scale). Since this result is rather complicated, it is probably of little practical value in most circumstances.

Equation (4.80) simplifies considerably, however, if $w_1 = 1$. Then

$$\mu_1(X) - \mu_2(Y) = \{\mu_1(X) - \mu_1(Y)\} \quad \text{Form difference for Population 1}$$
$$+ \{\mu_1(Y) - \mu_2(Y)\} \quad \text{Population difference on } Y \text{ scale.}$$
$$(4.81)$$

When $w_1 = 1$ in equation (4.18),

$$\mu_s(Y) = \mu_1(Y) = \mu_2(Y) + \gamma_2[\mu_1(V) - \mu_2(V)].$$

Therefore, equation (4.81) results in

$$\mu_1(X) - \mu_2(Y) = \{\mu_1(X) - \mu_2(Y)$$

$$- \gamma_2[\mu_1(V) - \mu_2(V)]\} \quad \text{Form difference for Population 1}$$

$$+ \{\gamma_2[\mu_1(V) - \mu_2(V)]\} \quad \text{Population difference on Y scale.}$$

$$(4.82)$$

Note that equation (4.82) applies to the Tucker method as well as both the Levine observed score and the Levine true score methods. As was discussed previously, the choice of synthetic population weights generally has little effect on Form Y equivalents. Consequently, equation (4.82) should be adequate for practical use in partitioning $\mu_1(X) - \mu_2(Y)$ into parts attributable to group and form differences.

Refer again to the example in Table 4.3 and the associated results in Table 4.4. For the Tucker method, equation (4.82) gives

$$15.8205 - 18.6728 = \{15.8205 - 18.6728 - 2.4560\ (5.1063 - 5.8626)\}$$

$$+ \{2.4560\ (5.1063 - 5.8626)\},$$

which simplifies to

$$-2.85 = -.99 - 1.86. \tag{4.83}$$

This result means that, on average: (a) the new group (Population 1) is lower achieving than the old group (Population 2) by 1.86 units on the Form Y scale; and (b) for the new group, the new Form X is more difficult than the old Form Y by .99 unit.

The corresponding result for both of the Levine methods under the classical congeneric model is obtained by using $\gamma_2 = 3.2054$ (rather than $\gamma_2 = 2.4560$) in equation (4.82), which gives

$$-2.85 = -.43 - 2.42. \tag{4.84}$$

Under the Levine assumptions, population mean differences on the Form Y scale are greater than under the Tucker assumptions by $2.42 - 1.86 = .56$ unit.

Decomposing Differences in Variances. As has been shown by Kolen and Brennan (1987), decomposing $\sigma_1^2(X) - \sigma_2^2(Y)$ is considerably more complicated, in general. However, for all three equating methods discussed in this chapter, when $w_1 = 1$ the result is quite simple:

$$\sigma_1^2(X) - \sigma_2^2(Y) = \{\sigma_1^2(X) - \sigma_2^2(Y)$$

$$- \gamma_2^2[\sigma_1^2(V) - \sigma_2^2(V)]\} \quad \text{Form difference for Population 1}$$

$$+ \{\gamma_2^2[\sigma_1^2(V) - \sigma_2^2(V)]\}. \quad \text{Population difference on Y scale.}$$

$$(4.85)$$

The form of equation (4.85) parallels that of equation (4.82) for decomposing the difference in means.

For the example in Tables 4.3 and 4.4, under Tucker assumptions, using equation (4.85),

$$6.5278^2 - 6.8784^2 = \{6.5278^2 - 6.8784^2 - [2.4560^2(2.3760^2 - 2.4515^2)]\}$$
$$+ \{2.4560^2(2.3760^2 - 2.4515^2)\},$$

which gives approximately

$$-4.70 = -2.50 - 2.20,$$

where -2.50 is the form difference factor, and -2.20 is the population difference factor. This result means that, on average: (a) on the old Form Y scale, the new group (Population 1) has smaller variance than the old group by 2.20 units; and (b) for the new group, the new Form X has smaller variance than the old Form Y by 2.50 units.

The reader can verify that the corresponding result for both Levine methods under the classical congeneric model is

$$-4.70 = -.96 - 3.74.$$

Under the Levine assumptions, population differences in variances on the Form Y scale are greater than under the Tucker assumptions by $3.74 - 2.20 = 1.54$ units.

Relationships Among Tucker and Levine Equating Methods

Kolen and Brennan (1987) show that, for both an internal and external anchor, if $\sigma_1(X, V) > 0$, then γ_1 for the Levine methods under a classical congeneric model is larger than γ_1 for the Tucker method. A similar result holds for the γ_2-terms. As is evident from equations (4.17)–(4.20) for the observed score methods and from equation (4.66) for the Levine true score method, the γ-terms are "expansion factors" in the sense that they multiply the group differences $\mu_1(V) - \mu_2(V)$ and $\sigma_1^2(V) - \sigma_2^2(V)$. This relationship is also evident from the form of equations (4.82) and (4.85). Therefore, because the Levine γ's are larger than the Tucker γ's population differences under the Levine assumptions are greater than under the Tucker assumptions, as is illustrated by the example results for the decompositions of observed means and variances. This may be one reason why the Levine methods sometimes are said to be more appropriate than the Tucker method when groups are rather dissimilar. Although this ad hoc reasoning is by no means definitive, it does suggest that an investigator might choose one of the Levine methods when it is known, or strongly suspected, that populations differ substantially. This logic is especially compelling if there is also reason to believe that the forms are quite similar, because in that case the true score assumptions of the Levine methods are plausible. However, if the populations are too dissimilar, any equating is suspect.

Along a similar line of reasoning, if the forms are known or suspected to be dissimilar, the Levine true score assumptions are likely violated, which may lead an investigator to choose the Tucker method. Of course, it should be kept in mind that if forms are too dissimilar, any equating is suspect. It is virtually impossible to provide strict and all-inclusive guidelines about what characterizes forms that are "too dissimilar." However, forms that do not share common content and statistical specifications certainly are "too dissimilar" to justify a claim that their scores can be equated, as the term is used in this book, no matter what method is chosen.

The Tucker and Levine methods make linearity assumptions that are, in some cases, amenable to direct examination. For example, the regression of X on V in Population 1 can be examined directly. If it is not linear, then at least one of the assumptions of the Tucker method is false, and an alternative procedure (the Braun-Holland method) discussed in Chapter 5 might be considered.

Since the Levine γ-terms are larger than the Tucker γ-terms, under some circumstances it is possible to predict whether the mean for Form Y equivalents under Levine equating will be larger or smaller than the mean under Tucker equating. For example, when $w_1 = 1$ and $x = \mu_1(X)$, both equations (4.1) and (4.66) reduce to

$$l_Y[\mu_1(X)] = \mu_1(Y) = \mu_2(Y) + \gamma_2[\mu_1(V) - \mu_2(V)],$$

which is the Form Y equivalent of the mean score for X in Population 1 when $w_1 = 1$. Clearly, in this case, if the new group is higher achieving than the old group on the common items [i.e., $\mu_1(V) > \mu_2(V)$], then $l_Y[\mu_1(X)]$ is greater under both of the Levine methods than under the Tucker method. Of course, when the common item means are quite similar in the two groups, there will be little difference in the Form Y equivalents of $\mu_1(X)$ under the three methods.

When Levine (1955) developed his methods, he referred to the observed score method as a method for use with "equally reliable" tests, and he referred to the true score method as a method for "unequally reliable" tests. This terminology, which is also found in Angoff (1971) and other publications, is not used here for two reasons. First, as is shown in this chapter, the derivations of Levine's methods do not require any assumptions about the equality or inequality of reliabilities. (It is possible to derive Levine's methods using such assumptions, but it is not necessary to do so.) Second, this terminology seems to suggest that the two methods should give the same results if Forms X and Y are equally reliable. This conclusion does not necessarily follow, however, because it fails to explicitly take into account the facts that reliabilities are population dependent, Levine's observed score method involves a synthetic population, and Levine's true score method does not. For example, suppose that $\rho_1(X, X') = \rho_2(Y, Y')$, which means that Forms X and Y are equally reliable for Populations 1 and 2, respectively. It does not necessarily follow that $\rho_s(X, X') = \rho_s(Y, Y')$ for the particular synthetic population used in Levine's observed score method. This is evident from the fact that, even if $\sigma_1^2(X) = \sigma_2^2(Y)$, the synthetic group variances for Forms X and Y are not likely to be equal [see equations (4.19) and (4.20)]. Thus, it is quite possible for forms to be equally reliable, in some sense, without having the two Levine methods give the same Form Y equivalents.

Even though the derivations of the methods described in this chapter do not directly require assumptions about reliability, if Forms X and Y are not approximately equal in reliability then the equating will be suspect, at best. For example, suppose that Form X is very short relative to Form Y. Under these circumstances, even after "equating," it will not be a matter of indifference to examinees which form they take. Because Form X has more measurement error than Form Y, well-prepared examinees are likely to be more advantaged by taking Form Y, and poorly prepared examinees are likely to be more advantaged by taking Form X. Probably the most favorable characteristic that such an equating might possess is first-order equity.

Scale Scores

In most testing programs, equated raw scores (e.g., Form Y equivalents) are not reported to examinees and users of scores. Rather, scale scores are reported, where the scale is defined as a transformation of the raw scores for the initial form of the test, as was discussed in Chapter 1. In principle, the scale scores could be either a linear or nonlinear transformation of the raw scores. This section extends the discussion of linear conversions in Chapter 2.

Let S represent scale scores. If Form Y is the initial test form and the raw-to-scale score transformation is linear,

$$sc(y) = B_{Y|sc} + A_{Y|sc}(y). \tag{4.86}$$

The linear equation for equating raw scores on Form X to the raw score scale of Form Y can be represented as

$$l_Y(x) = y = B_{X|Y} + A_{X|Y}(x). \tag{4.87}$$

Therefore, to obtain scale scores associated with the Form X raw scores, y in equation (4.87) is replaced in equation (4.86) giving

$$sc(x) = B_{Y|sc} + A_{Y|sc}[B_{X|Y} + A_{X|Y}(x)]$$

$$= (B_{Y|sc} + A_{Y|sc}B_{X|Y}) + A_{Y|sc}A_{X|Y}(x) \tag{4.88}$$

$$= B_{X|sc} + A_{X|sc}(x), \tag{4.89}$$

where the intercept and slope are, respectively,

$$B_{X|sc} = B_{Y|sc} + A_{Y|sc}B_{X|Y} \quad \text{and} \quad A_{X|sc} = A_{Y|sc}A_{X|Y}.$$

Suppose that $A_{Y|sc} = 2$ and $B_{Y|sc} = 100$. Then, for the illustrative example, assuming Tucker equating with $w_1 = .5$ (see Table 4.4), equation (4.88) gives:

$$sc(x) = [100 + 2(.5378)] + 2(1.0291)(x)$$
$$= 101.08 + 2.06(x).$$

For example, if $x = 25$,

$$sc(x = 25) = 101.08 + 2.06(25) = 152.58.$$

[Alternatively, the Form Y equivalent of $x = 25$ could be obtained first and then used as y in equation (4.86).]

The same process can be used for obtaining scale scores for scores on a subsequent form, say Z, that is equated to Form X. The transformation has the same form as equations (4.88) and (4.89):

$$sc(z) = (B_{X|sc} + A_{X|sc}B_{Z|X}) + A_{X|sc}A_{Z|X}(z)$$
$$= B_{Z|sc} + A_{Z|sc}(z).$$

If the transformation of raw scores on the initial form to scale scores is nonlinear, then equation (4.86) is not valid and the process described in this section will not work. In that case, the scale score intercepts and slopes for each form [e.g., equation (4.89)] are replaced by a conversion table that maps the raw score on each form to a scale score, as was discussed in Chapter 1 and illustrated in Chapter 2.

Appendix
Proof that $\sigma_s^2(T_X) = \gamma_1^2\sigma_s^2(T_V)$ Under the Classical Congeneric Model

The true score analogue of equation (4.4) (see also Exercise 4.1) is

$$\sigma_s^2(T_X) = w_1\sigma_1^2(T_X) + w_2\sigma_2^2(T_X) + w_1w_2[\mu_1(T_X) - \mu_2(T_X)]^2.$$

For the classical congeneric model, $\mu_1(T_X) = \mu_1(X), \mu_2(T_X) = \mu_2(X)$ and, from equation (4.34),

$$\mu_2(X) = \mu_1(X) - [\sigma_1(T_X)/\sigma_1(T_V)][\mu_1(V) - \mu_2(V)].$$

It follows that

$$\sigma_s^2(T_X) = w_1\sigma_1^2(T_X) + w_2\sigma_2^2(T_X) + w_1w_2[\sigma_1^2(T_X)/\sigma_1^2(T_V)][\mu_1(V) - \mu_2(V)]^2$$

$$= \frac{\sigma_1^2(T_X)}{\sigma_1^2(T_V)}\left\{ w_1\sigma_1^2(T_V) + w_2\frac{\sigma_1^2(T_V)}{\sigma_1^2(T_X)}\sigma_2^2(T_X) + w_1w_2[\mu_1(V) - \mu_2(V)]^2\right\}.$$

Under the Levine assumptions, the slope of the linear regression of T_X on T_V in both Populations 1 and 2 is given by equation (4.28):

$$\sigma_1(T_X)/\sigma_1(T_V) = \sigma_2(T_X)/\sigma_2(T_V).$$

Applying this equation to the second term in braces in the previous equation gives

$$\sigma_s^2(T_X) = \frac{\sigma_1^2(T_X)}{\sigma_1^2(T_V)}\{w_1\sigma_1^2(T_V) + w_2\sigma_2^2(T_V) + w_1w_2[\mu_1(V) - \mu_2(V)]^2\}.$$

The term in braces is $\sigma_s^2(T_V)$ and, by equation (4.38), $\sigma_1^2(T_X)/\sigma_1^2(T_V) = \gamma_1^2$. Thus,

$$\sigma_s^2(T_X) = \gamma_1^2\sigma_s^2(T_V),$$

as was to be proved.

Exercises

4.1. Prove equation (4.4). [Hint:

$$\sigma_s^2(X) = w_1\underset{1}{E}[X - \mu_s(X)]^2 + w_2\underset{2}{E}[X - \mu_s(X)]^2,$$

where $\underset{i}{E}$ means the expected value in Population $i(i = 1$ or $2)$].

4.2. Using the notation of this chapter, Angoff (1971, p. 580) provides the following equations for the synthetic group means and variances under Tucker assumptions:

$$\mu_s(X) = \mu_1(X) + \alpha_1(X|V)[\mu_s(V) - \mu_1(V)],$$
$$\mu_s(Y) = \mu_2(Y) + \alpha_2(Y|V)[\mu_s(V) - \mu_2(V)],$$
$$\sigma_s^2(X) = \sigma_1^2(X) + \alpha_1^2(X|V)[\sigma_s^2(V) - \sigma_1^2(V)],$$

and

$$\sigma_s^2(Y) = \sigma_2^2(Y) + \alpha_2^2(Y|V)[\sigma_s^2(V) - \sigma_2^2(V)].$$

Show that Angoff's equations give results identical to equations (4.17)–(4.20), using equations (4.21) for γ_1 and (4.22) for γ_2. (Strictly speaking, Angoff refers to a "total" group rather than a synthetic group with the notion of a total group being all examinees used for equating, which implies that Angoff's weights are proportional to sample sizes for the two groups.)

4.3. Verify the results in Table 4.4 when $w_1 = .5$ and $w_1 = .5026$.

4.4. Suppose the data in Table 4.3 were for an external anchor of 12 items, and X and Y both contain 36 items. If $w_1 = .5$, what are the linear equations for the Tucker and Levine observed score methods?

4.5. Under the classical congeneric model, what are the reliabilities $\rho_1(X, X')$ and $\rho_2(Y, Y')$ for the illustrative example?

4.6. Suppose the Levine assumptions are invoked and X, Y, and V are assumed to satisfy the classical test theory model assumptions for both populations, such that $\sigma_1(T_X) = (K_X/K_V)\sigma_1(T_V)$ and $\sigma_2(T_Y) = (K_Y/K_V)\sigma_2(T_V)$.
   a. Under these circumstances, what are the γ's given by equations (4.38) and (4.39)?
   b. Provide a brief verbal interpretation of these γ's as contrasted with the γ's under the classical congeneric model.

4.7. If $w_1 = 1$ and the common-item means for the two groups are identical, how much of the difference $\mu_1(X) - \mu_2(Y)$ is attributable to forms?

4.8. Jessica is a test development specialist for a program in which test forms are equated. She has been taught in an introductory measurement course that good items are highly discriminating items. Therefore, in developing a new form of a test, she satisfies the content requirements using more highly discriminating items than were used in constructing previous forms. From an equating perspective, is this good practice? Why? [Hint: If p_i is the difficulty level for item i and r_i is the point-biserial discrimination index for item i, then the standard deviation of total test scores is $\sum_i r_i \sqrt{p_i(1 - p_i)}$.]

4.9. Given equation set (4.70), show that the external anchor γ_2 given by equation (4.59) is λ_Y/λ_V.

4.10. Let V be an internal anchor such that $X = A + V$ and assume that $0 < \rho_1(X, V) < 1$. Show that
   a. $\sigma_1^2(V) < \sigma_1(X, V) < \sigma_1^2(X)$ and
   b. $1 < \gamma_{1T} < \gamma_{1L}$, where T stands for Tucker equating and L stands for Levine observed score equating under the classical congeneric model.
   c. Name one condition under which the result in (a) would not hold if V is an external anchor.

CHAPTER 5

Nonequivalent Groups—Equipercentile Methods

Equipercentile equating methods have been developed for the common-item nonequivalent groups design. These methods are similar to the equipercentile methods for random groups described in Chapter 2. Equipercentile methods with nonequivalent groups consider the distributions of total score and scores on the common items, rather than only the means, standard deviations, and covariances that were considered in Chapter 4. As has been indicated previously, equipercentile equating is an observed score equating procedure that is developed from the perspective of the observed score equating property described in Chapter 1. Thus, equipercentile equating with the common-item nonequivalent groups design requires that a synthetic population, as defined in Chapter 4, be considered. In this chapter, we present an equipercentile method that we show to be closely allied to the Tucker linear method of Chapter 4. We also describe how smoothing methods, such as those described in Chapter 3, can be used when conducting equipercentile equating with nonequivalent groups. The methods described in this chapter are illustrated using the same data that were used in Chapter 4, and the results are compared to the linear results from Chapter 4.

Frequency Estimation Equipercentile Equating

The *frequency estimation method* described by Angoff (1971) and Braun and Holland (1982) provides a method for estimating the cumulative distributions of scores on Form X and Form Y for a synthetic population from data that are collected using the common-item nonequivalent groups design. Percentile ranks then are obtained from the cumulative distributions and the forms equated by equipercentile methods, as was done in Chapter 2.

Conditional Distributions

Conditional score distributions are required for using these statistical methods. Two identities are particularly useful, and they are presented here. The use of these identities is illustrated later, in connection with the frequency estimation method.

Define $f(x, v)$ as the joint distribution of total score and common-item score, so that $f(x, v)$ represents the probability of earning a score of x on Form X and a score of v on the common items. Specifically, $f(x, v)$ is the probability that $X = x$ and $V = v$. Define $f(x)$ as the *marginal distribution* of scores on Form X, so that $f(x)$ represents the probability of earning a score of x on Form X. That is, $f(x)$ represents the probability that $X = x$. Also define $h(v)$ as the marginal distribution of scores on the common items, so that $h(v)$ represents the probability that $V = v$, and define $f(x|v)$ as the conditional distribution of scores on Form X for examinees earning a particular score on the common items. Thus, $f(x|v)$ represents the probability that $X = x$ *given that* $V = v$. Using standard results from conditional expectations, it can be shown that

$$f(x|v) = \frac{f(x, v)}{h(v)}. \tag{5.1}$$

From equation (5.1), it follows that

$$f(x, v) = f(x|v)h(v). \tag{5.2}$$

These identities are used to develop the frequency estimation method.

Frequency Estimation Method

To conduct *frequency estimation equipercentile equating*, it is necessary to express the distributions for the synthetic population. These distributions are considered to be a weighted combination of the distributions for each population. Specifically, for Form X and Form Y,

$$f_s(x) = w_1 f_1(x) + w_2 f_2(x) \tag{5.3}$$

and

$$g_s(y) = w_1 g_1(y) + w_2 g_2(y),$$

where the subscript s refers to the synthetic population, the subscript 1 refers to the population administered Form X, and the subscript 2 refers to the population administered Form Y. As before, f and g refer to distributions for Form X and Form Y, respectively, and w_1 and $w_2(w_1 + w_2 = 1)$ are used to weight Populations 1 and 2 to form the synthetic population.

From the data that are collected in the common-item nonequivalent groups design, direct estimates of $f_1(x)$ and $g_2(y)$ may be obtained. Because Form X is not administered to examinees from Population 2, a direct estimate of $f_2(x)$ is unavailable. Also, because Form Y is not administered to examinees from Population 1, a direct estimate of $g_1(y)$ is unavailable. Statistical assumptions need to be invoked to obtain expressions for these functions using quantities for which direct estimates are available from data that are collected.

The assumption made in the frequency estimation method is that, for both Form X and Form Y, the conditional distribution of total score given each common-item score, $V = v$, is the same in both populations. The same assumption is made whether the common items are internal or external. This assumption is stated as follows:

$$f_1(x|v) = f_2(x|v), \quad \text{for all } v \quad \text{and} \quad g_1(y|v) = g_2(y|v), \quad \text{for all } v. \tag{5.4}$$

For example, $f_1(x|v)$ represents the probability that total score $X = x$ given that $V = v$ in Population 1. The other conditional distributions are interpreted similarly. Equation (5.2) can be used to obtain expressions for these functions using quantities for which direct estimates are available from data that are collected.

The following discussion describes how the assumptions presented in equation (5.4) can be used to find expressions for $f_2(x)$ and $g_1(y)$ using quantities for which direct estimates are available.

From equation (5.2), the following equalities hold:

$$f_2(x, v) = f_2(x|v)h_2(v) \quad \text{and} \quad g_1(y, v) = g_1(y|v)h_1(v). \tag{5.5}$$

For Population 2, $f_2(x, v)$ represents the joint distribution of total scores and common-item scores. Specifically, $f_2(x, v)$ represents the probability that $X = x$ and $V = v$ in Population 2. For Population 2, $h_2(v)$ represents the distribution of scores on the common items. Thus, $h_2(v)$ represents the probability that $V = v$ in Population 2. The expressions $g_1(y, v)$ and $h_1(v)$ are similarly defined for Population 1.

Combining the equalities in equation (5.5) with the assumptions in equation (5.4), $f_2(x, v)$ and $g_1(y, v)$ can be expressed using quantities for which direct estimates are available from data that are collected as follows:

$$f_2(x, v) = f_1(x|v)h_2(v) \quad \text{and} \quad g_1(y, v) = g_2(y|v)h_1(v). \tag{5.6}$$

For the first equality, $f_1(x|v)$ can be estimated directly from the Population 1 examinees who take Form X. The quantity $h_2(v)$ can be estimated directly from the Population 2 examinees who take Form Y. For the second equality, $g_2(y|v)$ can be estimated directly from the Population 2 examinees who take Form Y, and $h_1(v)$ can be estimated directly from the Population 1 examinees who take Form X.

The associated marginal distributions can be found by summing over common-item scores as follows:

$$f_2(x) = \sum_v f_2(x, v) = \sum_v f_1(x|v)h_2(v) \qquad \text{and}$$

$$g_1(y) = \sum_v g_1(y, v) = \sum_v g_2(y|v)h_1(v). \tag{5.7}$$

In this equation, $f_2(x)$ represents the probability that $X = x$ in Population 2, and $g_1(y)$ represents the probability that $Y = y$ in Population 1.

All of the terms in equation (5.7) use quantities for which direct estimates are available from data. The expressions in equation (5.7) can be substituted into equation (5.3) to provide expressions for the synthetic population as follows:

$$f_s(x) = w_1 f_1(x) + w_2 \sum_v f_1(x|v)h_2(v) \qquad \text{and}$$

$$g_s(y) = w_1 \sum_v g_2(y|v)h_1(v) + w_2 g_2(y). \tag{5.8}$$

Equation (5.8) uses quantities for which direct estimates are available from data.

For the synthetic population, $f_s(x)$ can be cumulated over values of x to produce the cumulative distribution $F_s(x)$. The cumulative distribution $G_s(y)$ is similarly derived. Define P_s as the percentile rank function for Form X and Q_s as the percentile rank function for Form Y, using the definitions for percentile ranks that were developed in Chapter 2. Similarly, P_s^{-1} and Q_s^{-1} are the percentile functions.

The equipercentile function for the synthetic population is

$$e_{Y_s}(x) = Q_s^{-1}[P_s(x)], \tag{5.9}$$

which is analogous to the equipercentile relationship for random groups equipercentile equating in equation (2.17).

Evaluating the Frequency Estimation Assumption

The frequency estimation assumption of equation (5.4) cannot be tested using data collected using the common-item nonequivalent groups design. To test this assumption, a representative group of examinees from Population 1 would need to take Form Y, and a representative group of examinees from Population 2 would need to take Form X. Unfortunately, these data are not available in practice. If Populations 1 and 2 were identical, then the assumption in equation (5.4) would be met. Logically, then, the more similar Populations 1 and 2 are to one another, the more likely it is that this assumption will hold. Thus, frequency estimation equating should be conducted only when the two populations are reasonably similar to one another. How similar is "reasonably similar" depends on the context of the equating and on empirical evidence of the degree of similarity required. When the populations differ considerably, methods based on true score models, such as the Levine linear method described in Chapter 4 or item response theory methods described in Chapter 6, should be considered, although adequate equating might not be possible when populations differ considerably. This problem will be considered further in Chapter 8.

Numerical Example

A numerical example based on synthetic data is used here to aid in the understanding of this method. In this example, Form X has 5 items, Form Y has 5 items, and there are 3 common items. Assume that the common items are external.

Table 5.1 presents the data for Population 1 for the hypothetical example. The values in the body of the table represent the joint distribution, $f_1(x, v)$. For example, the upper left-hand value is .04. This value represents the probability that an examinee from Population 1 would earn a score of 0 on Form X *and* a score of 0 on the common items. The values in the body of Table 5.1 sum to 1. The values at the bottom of the table are for the marginal distribution on the common items for Population 1, $h_1(v)$. For example, the table indicates that the probability of earning a common-item score of 0 is .20 over all examinees in Population 1. The values listed

Table 5.1. Form X and Common-Item Distributions for Population 1 in a Hypothetical Example.

	v					
x	0	1	2	3	$f_1(x)$	$F_1(x)$
0	.04	.04	.02	.00	.10	.10
1	.04	.08	.02	.01	.15	.25
2	.06	.12	.05	.02	.25	.50
3	.03	.12	.05	.05	.25	.75
4	.02	.03	.04	.06	.15	.90
5	.01	.01	.02	.06	.10	1.00
$h_1(v)$	.20	.40	.20	.20		

Note: Values shown in the body of table are for $f_1(x, v)$.

Table 5.2. Form Y and Common-Item Distributions for Population 2 in a Hypothetical Example.

y	0	1	2	3	$g_2(y)$	$G_2(y)$
0	.04	.03	.01	.00	.08	.08
1	.07	.05	.07	.01	.20	.28
2	.03	.05	.12	.02	.22	.50
3	.03	.04	.13	.05	.25	.75
4	.02	.02	.05	.06	.15	.90
5	.01	.01	.02	.06	.10	1.00
$h_2(v)$	.20	.20	.40	.20		

The column group 0, 1, 2, 3 is headed by v.

Note: Values shown in the body of table are for $g_2(y,v)$.

under the column labeled $f_1(x)$ represent the marginal distribution for total score on Form X. The sum of the values in each row in the body of the table equals the value for the marginal shown for $f_1(x)$ and the sum of the marginal distribution values for $f_1(x)$ equals 1. The rightmost column is the cumulative distribution for Form X scores, $F_1(x)$. The values in this column are obtained by cumulating the probabilities shown in the $f_1(x)$ column. Table 5.2 presents the joint and marginal distributions for Form Y and common-item scores in Population 2.

Estimates of the distributions presented in Tables 5.1 and 5.2 would be available from the common-item nonequivalent groups design. Estimates of the distribution for Form X in Population 2 would be unavailable, because Form X is not administered in Population 2. Similarly, estimates of the distribution for Form Y in Population 1 would be unavailable. However, equating still can proceed by making the frequency estimation assumption in equation (5.4).

To simplify the example, assume that $w_1 = 1$, which results in the following simplification of equation (5.8):

$$f_s(x) = f_1(x) \quad \text{and} \quad g_s(y) = \sum_v g_2(y|v)h_1(v). \quad (5.10)$$

The first of equations (5.10) indicates that the distribution of Form X scores for the synthetic population is the same as the distribution in Population 1. Thus, the rightmost column in Table 5.1 labeled $F_1(x)$ also gives $F_s(x)$ for $w_1 = 1$.

The synthetic group is Population 1, because $w_1 = 1$ in the example. Thus, the second of the equations (5.10) provides an expression for the cumulative distribution of Form Y scores for examinees in Population 1. Because Form Y was not administered in Population 1, it is necessary to use the conditional distribution of Form Y scores given common-item

Table 5.3. Conditional Distributions of Form Y
Given Common-Item Scores for Population 2 in a
Hypothetical Example.

			v	
y	0	1	2	3
0	.20	.15	.025	.00
1	.35	.25	.175	.05
2	.15	.25	.30	.10
3	.15	.20	.325	.25
4	.10	.10	.125	.30
5	.05	.05	.05	.30
$h_2(v)$	.20	.20	.40	.20

Note: Values in the body of the table are for $g_2(y|v) = \dfrac{g_2(y, v)}{h_2(v)}$.

scores in Population 2 and assume that this conditional distribution also
would hold in Population 1 at all common-item scores [see equation (5.4)].

Table 5.3 presents the Form Y conditional distribution for Population 2.
To calculate the values in the table, take the joint probability in Table 5.2
and divide it by its associated marginal probability on the common items.
Specifically,

$$g_2(y|v) = \frac{g_2(y, v)}{h_2(v)}, \qquad (5.11)$$

which follows from equation (5.1). For example, the .20 value in the upper
left cell of Table 5.3 equals .04 from the upper left cell of Table 5.2 divided
by .20, which is the probability of earning a score of $V = 0$ as shown in
Table 5.2. Note that the conditional probabilities in each column of the
body of Table 5.3 sum to 1.

To find the values to substitute into equation (5.10), at each v the condi-
tional distribution in Population 2, $g_2(y|v)$, is multiplied by the marginal
distribution for common items for Population 1, $h_1(v)$. The result is the
joint distribution in Population 1 under the frequency estimation assumption of equation (5.4). The results are shown in Table 5.4.

Table 5.5 presents the cumulative distributions, percentile ranks, and
equipercentile equivalents. These values can be verified using the computational procedures described in Chapter 2.

Refer to Table 5.4 to gain a conceptual understanding of what was done.
In this table, the joint distribution of Form Y total scores and common-
item scores was calculated for Population 1. As was indicated earlier, Population 1 did not even take Form Y. The way that the values in this table
could be calculated was by making the statistical assumptions associated

Table 5.4. Calculation of Distribution of Form Y and Common-Item Scores for Population 1 Using Frequency Estimation Assumptions in a Hypothetical Example.

y	v 0	1	2	3	$g_1(y)$	$G_1(y)$
0	.20(.20) = .04	.15(.40) = .06	.025(.20) = .005	.00(.20) = .00	.105	.105
1	.35(.20) = .07	.25(.40) = .10	.175(.20) = .035	.05(.20) = .01	.215	.320
2	.15(.20) = .03	.25(.40) = .10	.30(.20) = .06	.10(.20) = .02	.210	.530
3	.15(.20) = .03	.20(.40) = .08	.325(.20) = .065	.25(.20) = .05	.225	.755
4	.10(.20) = .02	.10(.40) = .04	.125(.20) = .025	.30(.20) = .06	.145	.900
5	.05(.20) = .01	.05(.40) = .02	.05(.20) = .01	.30(.20) = .06	.100	1.000
$h_1(v)$	.20	.40	.20	.20		

Note: Values in the body of the table are for $g_1(y,v) = g_2(y|v)h_1(v)$.

Table 5.5. Cumulative Distributions and Finding Equipercentile Equivalents for $w_1 = 1$.

x	$F_1(x)$	$P_1(x)$	y	$G_1(y)$	$Q_1(y)$	x	$e_{Ys}(x)$
0	.100	5.0	0	.105	5.25	0	−.02
1	.250	17.5	1	.320	21.25	1	.83
2	.500	37.5	2	.530	42.50	2	1.76
3	.750	62.5	3	.755	64.25	3	2.92
4	.900	82.5	4	.900	82.75	4	3.98
5	1.000	95.0	5	1.000	95.00	5	5.00

with frequency estimation. To estimate this joint distribution, the conditional distribution observed in Population 2 was assumed to hold for Population 1 at all common-item scores. The Population 2 conditional distribution was multiplied by the Population 1 common-item marginal distributions to form the joint probabilities shown in Table 5.4. The Population 1 marginal distribution on the common items can be viewed as providing weights that are multiplied by the Population 2 conditional distribution at each score on the common items.

Estimating the Distributions

Estimates of distributions can be used in place of the parameters when using frequency estimation in practice. However, a problem occurs when no examinees earn a particular common-item score in one of the groups but some examinees earn that score in the other group. When estimating the Form Y distribution in Population 1, the assumption is made in equation (5.4) that $g_1(y|v) = g_2(y|v)$, for all v. If no Population 2 examinees earn a particular score on v in a sample, then no estimate of $g_1(y|v)$ exists at that v. However, such an estimate would be needed to conduct the equating if some examinees in Population 1 earned that v. Jarjoura and Kolen (1985) recommended using the conditional distribution at a score close to that v (e.g., at $v + 1$) as an estimate for what the conditional distribution would be at v. On logical grounds, they argued that this substitution would cause insignificant bias in practice in those cases where very few examinees in one population earn a score that has a frequency of 0 in the other population. A practical solution is to use the conditional distribution for the v with nonzero frequency that is closest to the v in question as we move toward the median of the distribution of v.

Smoothing methods also can be used with the frequency estimation method. An extension of the log-linear presmoothing method was described by Holland and Thayer (1987, 1989) and Rosenbaum and Thayer (1987) in the context of frequency estimation. In this extension, the joint distributions of scores on the items that are not common and scores on the common items are fit using a log-linear model. The resulting smoothed joint distributions then are used to equate forms using the frequency estimation method described in this chapter. Model fitting using this method requires the fitting of a joint distribution, which makes the moment preservation property for this method more complicated than with the random groups design. To fit the joint distribution, the number of moments for each fitted marginal distribution that are the same as those for the observed distribution need to be specified. In addition, the cross-product moments for the fitted joint distribution that are the same as those for the observed distribution need to be specified. For example, a model might be specified so that the first four moments of each marginal distribution and the covariance for the fitted and observed distributions are equal. The fit of this model could be compared to other more and other less complicated models.

Lord's (1965) beta4 method that was described in Chapter 3 also can be used to fit the joint distributions of total scores and common-item scores. In this application, the assumption is made that true score on the common items and true score on the total tests are functionally related. That is, the total test and common items are measuring precisely the same construct. Hanson (1991c) described an implementation of this method. Empirical research conducted by Hanson (1991c) and Livingston and Feryok (1987) indicates that the bivariate smoothing techniques can improve equating precision when using the frequency estimation method.

The cubic spline postsmoothing method described by Kolen and Jarjoura (1987) is a straightforward extension of the random groups method described in Chapter 3. In this method, unsmoothed equipercentile equivalents are estimated using frequency estimation as described in this chapter. The cubic spline method described in Chapter 3 then is implemented. The only difference in methodology is that standard errors of frequency estimation equating developed by Jarjoura and Kolen (1985) are used in place of the random groups standard errors. Kolen and Jarjoura (1987) reported that the cubic spline method used with frequency estimation increased equating precision.

Braun-Holland Linear Method

Braun and Holland (1982) presented a linear method that uses the mean and standard deviation which arise from using the frequency estimation assumptions to conduct linear equating. This method is closely related to the Tucker linear method presented in Chapter 4. Under the frequency estimation assumptions in equation (5.4), the mean and standard deviation of scores on Form X for the synthetic population can be expressed as

$$\mu_s(X) = \sum_x x f_s(x), \tag{5.12}$$

$$\sigma_s^2(X) = \sum_x [x - \mu_s(X)]^2 f_s(x), \tag{5.13}$$

where $f_s(x)$ is taken from equation (5.8). The synthetic population mean and standard deviation for Form Y are expressed similarly. The resulting means and standard deviations then can be substituted into the general form of a linear equating relationship for the common-item nonequivalent groups design that was described in Chapter 4. The resulting equation is referred to here as the Braun-Holland linear method.

Braun and Holland (1982) showed that an equating which results from using the Braun-Holland linear method is identical to the Tucker linear method described in Chapter 4 when the following conditions hold:

(1) The regressions of X on V and Y on V are linear.
(2) The regressions of X on V and Y on V are homoscedastistic. This property means that the variance of X given v is the same for all v, and the variance of Y given v is the same for all v.

Thus, the Braun-Holland method can be viewed as a generalization of the Tucker method when the regressions of total test on common items are nonlinear. Braun and Holland (1982) suggested that the regression of X on V for Population 1 and Y on V for Population 2 be examined for nonlinearity. The Braun-Holland method is more complicated computationally than the Tucker method, and it also has been used much less in practice. Still, the Braun-Holland method should be considered when nonlinear regressions are suspected.

Table 5.6. Computation of Equating Relationship for
Braun-Holland Method in a Hypothetical Example.

From Table 5.1		From Table 5.4	
x	$f_1(x)$	y	$g_1(y)$
0	.100	0	.105
1	.150	1	.215
2	.250	2	.210
3	.250	3	.225
4	.150	4	.145
5	.100	5	.100
$\mu_1(X)$	2.5000	$\mu_1(Y)$	2.3900
$\sigma_1(X)$	1.4318	$\sigma_1(Y)$	1.4792

$$slope = \frac{1.4792}{1.4318} = 1.0331$$

$intercept = 2.3900 - 1.0331(2.5000) = -.1927$

$l_{Ys}(x = 0) = -.1927, l_{Ys}(x = 1) = .8404, l_{Ys}(x = 2) = 1.8735,$

$l_{Ys}(x = 3) = 2.9066, l_{Ys}(x = 4) = 3.9397, l_{Ys}(x = 5) = 4.9728$

The results of using the Braun-Holland method with the hypothetical
data in the frequency estimation example with $w_1 = 1$ are presented in
Table 5.6. In this table, the distribution for Form X was taken from Table
5.1. The distribution for Form Y, which was calculated using the frequency
estimation assumption, was taken from Table 5.4. Means and standard deviations were calculated using equations (5.12) and (5.13). The slope and
intercept were calculated from the means and standard deviations. The linear equivalents were calculated using this slope and intercept. Note that
the linear equivalents differ somewhat from the equipercentile equivalents
shown in Table 5.5, indicating that the equating relationship is not linear
when frequency estimation assumptions are used.

Chained Equipercentile Equating

Angoff (1971) described an alternative equipercentile method that Marco *et
al.* (1983) referred to as the *direct equipercentile method*. Dorans (1990a) referred to this method as *chained equipercentile equating*. In this method,
Form X scores are converted to scores on the common items using examinees from Population 1. Then scores on the common items are equated to
Form Y scores using examinees from Population 2. These two conversions
are chained together to produce a conversion of Form X scores to Form Y
scores.

More specifically, the steps are as follows:

1. Find the equipercentile equating relationship for converting scores on Form X to the common items based on examinees from Population 1 using the equipercentile method described in Chapter 2. (Note that the same examinees take Form X and the common items.) This equipercentile function is referred to as $e_{V1}(x)$.
2. Find the equipercentile equating relationship for converting scores on the common items to scores on Form Y based on examinees from Population 2. Refer to the resulting function as $e_{Y2}(v)$.
3. To equate a Form X score to a score on Form Y, first convert the Form X score to a common-item score using $e_{V1}(x)$. Then equate the resulting common-item score to Form Y using $e_{Y2}(v)$.

Specifically, to find the Form Y equipercentile equivalent of a Form X score, take

$$e_{Y(chain)} = e_{Y2}[e_{V1}(x)]. \tag{5.14}$$

This method is referred to as *chained equipercentile equating* because it involves a chain of two equipercentile equatings, one in Population 1 and another in Population 2. This chaining process is similar to the chaining process described in Chapter 2, where scores on the new form were converted to scale scores by a chain involving the old form.

Livingston *et al.* (1990) suggested that the chained equipercentile method sometimes can produce accurate and stable results in practice, and they suggested that smoothing methods might be used to improve the stability of the method. Livingston (1993) suggested that the log-linear smoothing method which was described previously for the frequency estimation be used for smoothing the joint distribution of total test scores and common-item scores. The resulting smoothed marginal distributions are all that would be used in chained equipercentile equating. Alternatively, only the marginal distributions of total scores and common items could be smoothed using the log-linear method described in Chapter 3 and then applying equipercentile equating. As still another alternative, the cubic spline postsmoothing method could be used to smooth estimates of $e_{V1}(x)$ and $e_{Y2}(v)$. The only modification of the cubic spline method described in Chapter 3 would be to use standard errors of single group equating rather than standard errors of random groups equating in implementing the spline method. These smoothed relationships could be used in place of the population relationships in equation (5.14).

Chained equipercentile equating does not require consideration of the joint distribution of total score and common-item score, so computationally it is much less intensive than frequency estimation equipercentile equating. However, chained equipercentile equating has theoretical shortcomings. First, this method involves equating a long test (total test) to a short test (common items). Tests of considerably unequal lengths cannot be equated in the sense that scores on the long and short tests can be used interchangeably. Second, this method does not directly incorporate a synthetic population, so it is unclear for what population the relationship holds or is intended to hold. Braun and Holland (1982, p. 42) demonstrated that chained equipercentile and frequency estimation equipercentile do not, in general, produce the same results, even when the assumptions for frequency estimation equipercentile equating hold. Harris and Kolen (1990) demonstrated that these methods can produce equating relationships which differ from a practical perspective. However, the chained equipercentile method does not explicitly require that the two populations be very similar, so that this method might be useful in situations where the two groups differ. For example, results presented by Marco et al. (1983) and Livingston et al. (1990) suggest that chained equipercentile equating should be considered when groups differ considerably.

Illustrative Example

The real data example from Chapter 4 is used to illustrate some aspects of frequency estimation equating. As was indicated in that chapter, the test used in this example is a 36-item multiple-choice test. Two forms of the test, Form X and Form Y, were used. Every third item on the test forms is a common item, and the common items are in the same position on each form. Thus, items 3, 6, 9,..., 36 on each form represent the 12 common items. Form X was administered to 1655 examinees and Form Y to 1638 examinees.

Illustrative Results

Summary statistics for this example are shown in Table 5.7 ($\widehat{sk}$ refers to estimated skewness and $\widehat{ku}$ to estimated kurtosis). The examinees who were administered Form X had a number-correct score mean of 5.1063 and a

Table 5.7. Moments for Equating Form X and Form Y in the Common-Item Nonequivalent Groups Design.

Group	Score	$\hat{\mu}$	$\hat{\sigma}$	$\widehat{sk}$	$\widehat{ku}$	Correlation
1	X	15.8205	6.5278	.5799	2.7217	$\hat{\rho}_1(X,V) =$
1	V	5.1063	2.3760	.4117	2.7683	.8645
2	Y	18.6728	6.8784	.2051	2.3028	$\hat{\rho}_2(Y,V) =$
2	V	5.8626	2.4515	.1072	2.5104	.8753

standard deviation of 2.3760 on the common items. The examinees who were administered Form Y had a number-correct score mean of 5.8626 and a standard deviation of 2.4515 on the common items. Thus, based on the common-item statistics, the group taking Form Y appears to be higher achieving than the group taking Form X. The statistics shown in this table were also used to calculate the Tucker and Levine equating functions described in Chapter 4. Some of the statistics shown in Table 5.7 were also presented in Table 4.3. The analyses were conducted using the *CIPE* Macintosh computer program described in Appendix B.

For frequency estimation equating, the joint distributions of total score and common-item score also need to be considered. As was indicated earlier in this chapter, the assumptions in frequency estimation equating require that the distribution of total score given common-item score be the same for both populations. However, from the data that are collected, no data are available to directly address this assumption. The linearity of the regressions of total test on common items can be addressed, however. If the regression is nonlinear, then the use of the Tucker method might be questionable, and the Braun-Holland method might be preferred.

Statistics relevant to the regression of X on V for Group 1 are shown in Table 5.8. The first column lists the possible scores on the common items. The second column lists the number of examinees in Group 1 earning each score on the common items. The third column lists the mean total score given common-item score. For example, the mean total score on Form X

Table 5.8. Analysis of Residuals from the Linear Regression of Total Score on Common-Item Score For Group 1.

v	Number of Examinees	Mean X Given v	Standard Deviation X Given v	Mean X Given v, Linear Regression	Residual Mean
0	14	6.2143	2.2097	3.6923	2.5220
1	54	7.5741	2.2657	6.0674	1.5067
2	142	9.1901	2.6429	8.4425	.7476
3	249	10.8032	2.9243	10.8177	−.0145
4	274	12.7628	3.1701	13.1928	−.4300
5	247	15.1377	3.3302	15.5680	−.4303
6	232	16.9957	3.6982	17.9431	−.9474
7	173	20.5260	3.5654	20.3182	.2078
8	118	23.1610	3.5150	22.6934	.4676
9	75	25.6533	2.8542	25.0685	.5848
10	42	28.5000	3.4658	27.4436	1.0564
11	27	31.1852	2.1780	29.8188	1.3664
12	8	33.2500	1.6394	32.1939	1.0561

for the 14 examinees earning a common-item score of zero is 6.2143. Note that, as expected, the means increase as v increases. The fourth column presents the standard deviation, and the fifth column is based on estimating the mean on Form X given v using standard linear regression. The slope and intercept of the regression equation can be estimated directly from the data in Table 5.7 as follows:

$$regression\ slope = \hat{\rho}_1(X, V)\frac{\hat{\sigma}_1(X)}{\hat{\sigma}_1(V)} = .8645\frac{6.5278}{2.3760} = 2.3751.$$

$$regression\ intercept = \hat{\mu}_1(X) - (regression\ slope)\hat{\mu}_1(V)$$

$$= 15.8205 - (2.3751)5.1063 = 3.6923,$$

apart from the effects of rounding. The slope and intercept can be used to produce the values in the fifth column. The residual mean equals the third column minus the fifth column. The residual mean indicates the extent to which the mean predicted using linear regression differs from the mean that was observed. The mean residuals for Form X are plotted in Figure 5.1. If the regression was truly linear, then the mean residuals would vary randomly around 0. However, the residual means are positive for low and high scores on v and are negative for scores from 3 through 6. This pattern suggests that the regression is not linear. More sophisticated methods for testing hypotheses about the linearity of regression could also be used (e.g., see Draper and Smith, 1981). The regression of Y on V for Group 2 is

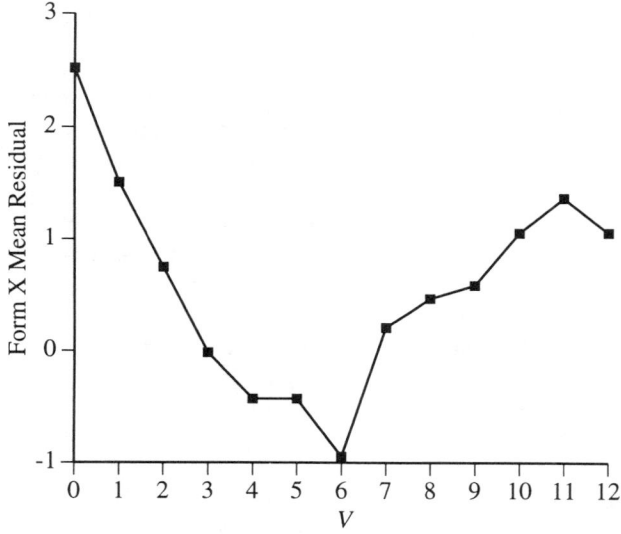

Figure 5.1. Form X mean residual plot.

Table 5.9. Analysis of Residuals from the Linear Regression of Total Score on Common-Item Score For Group 2.

v	Number of Examinees	Mean Y Given v	Standard Deviation Y Given v	Mean Y Given v, Linear Regression	Residual Mean
0	11	6.2727	2.1780	4.2740	1.9988
1	36	8.0000	2.2361	6.7300	1.2700
2	88	9.6023	3.0359	9.1860	.4162
3	159	12.1195	3.2435	11.6421	.4774
4	213	13.9202	3.3829	14.0981	−.1779
5	240	16.0750	3.4234	16.5541	−.4791
6	232	18.3147	3.5623	19.0101	−.6955
7	246	21.2073	3.4854	21.4662	−.2588
8	161	24.1801	3.3731	23.9222	.2579
9	120	27.3333	2.9533	26.3782	.9551
10	85	29.1294	2.8811	28.8343	.2952
11	34	31.8235	1.8386	31.2903	.5332
12	13	33.6154	1.7338	33.7463	−.1309

shown in Table 5.9, and the mean residuals are plotted in Figure 5.2. This regression also appears to be somewhat nonlinear. These nonlinear regressions suggest that the Braun-Holland method might be preferable to the Tucker method.

Comparison Among Methods

The Tucker and Braun-Holland linear methods and frequency estimation equipercentile equating with cubic spline smoothing were all applied to these data. The Levine observed score method under a congeneric model was also applied. The resulting moments are shown in Table 5.10, and the equating relationships are shown in Figure 5.3. First, refer to Figure 5.3. The Levine relationship seems to differ from the others. As was indicated in Chapter 4, the Levine method is based on assumptions about true scores, whereas the other methods make assumptions about observed scores. The differences in assumptions are likely the reason for the discrepancy. Unfortunately, data are not available that allow a judgment about whether the Levine method assumptions (other than possibly linearity of regression) are more or less preferable than the assumptions for the other methods in this example.

The Tucker, Braun-Holland, and frequency estimation methods all require assumptions about characteristics of the observed relationship between total score and score on the common items being the same for the

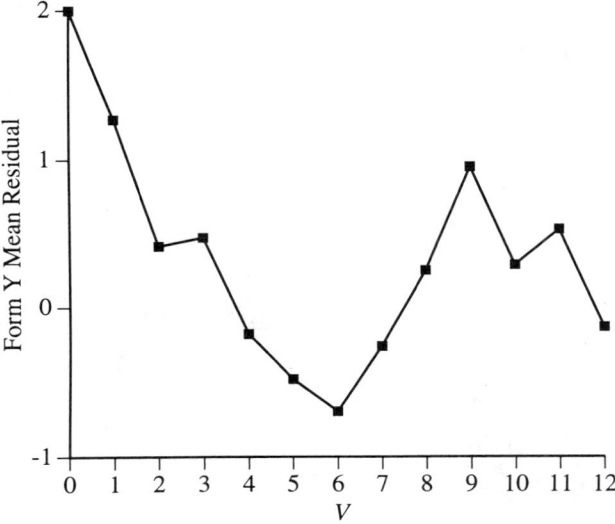

Figure 5.2. Form Y mean residual plot.

Table 5.10. Moments of Form X Scores Converted to Form Y Scores Using Various Methods for Examinees from Population 1.

Method	$\hat{\mu}$	$\hat{\sigma}$	$\widehat{sk}$	$\widehat{ku}$
Tucker linear	16.8153	6.7168	.5799	2.7217
Levine linear	16.2485	6.6007	.5799	2.7217
Braun-Holland linear	16.8329	6.6017	.5799	2.7217
Equipercentile				
Unsmoothed	16.8329	6.6017	.4622	2.6229
$S = .10$	16.8334	6.5983	.4617	2.6234
$S = .25$	16.8333	6.5947	.4674	2.6249
$S = .50$	16.8192	6.5904	.4983	2.6255
$S = .75$	16.8033	6.5858	.5286	2.6508
$S = 1.00$	16.7928	6.5821	.5501	2.6745

two populations. These methods differ with respect to which characteristics of the relationship are assumed to be the same.

First consider the Tucker and Braun-Holland methods. The major difference between these methods is in the assumption of linearity of regression. Thus, the relatively small differences between the two methods in the example are due to the differences in assumptions. The Braun-Holland method might be preferred, because the regression was judged to be nonlinear.

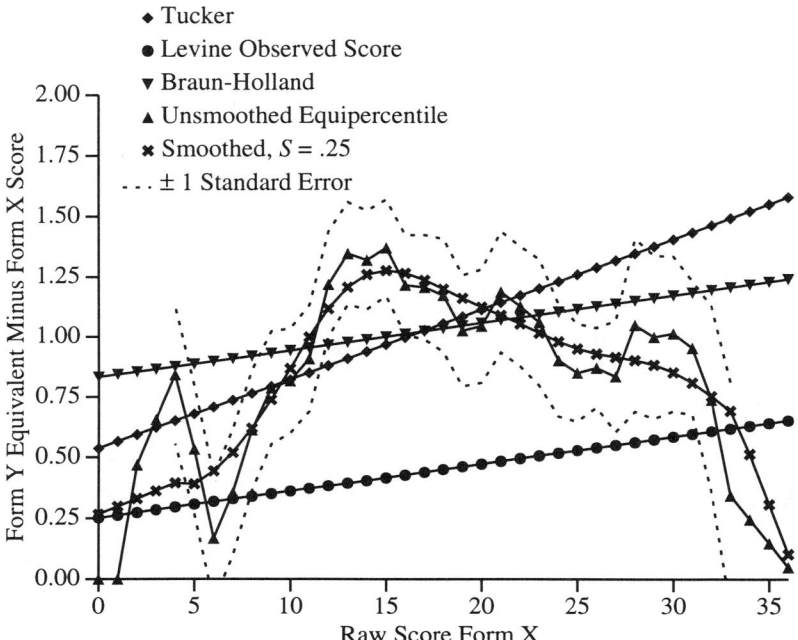

Figure 5.3. Equating relationships for frequency estimation equipercentile equating and linear methods.

Next compare the Braun-Holland and frequency estimation method, referred to as unsmoothed, in Table 5.10 and Figure 5.3. The relationship appears to be nonlinear. The Braun-Holland relationship falls outside the standard error band for the frequency estimation method over parts of the score range. Thus, the frequency estimation method (labeled unsmoothed) appears to more accurately reflect the equipercentile relationship between the forms than does the Braun-Holland method in this example.

Table 5.10 presents the results for various degrees of cubic spline smoothing. The moments for values of S that are greater than .25 seem to differ more than would be desired from those for the unsmoothed equating. For this reason, the relationship for $S = .25$ is plotted in Figure 5.3. This relationship stays within the standard error bands and seems to be smooth without deviating too far from the unsmoothed values.

Practical Issues in Equipercentile Equating with Common Items

A series of additional practical issues should be considered when deciding on which method to use when equating is conducted in practice. First, scale score moments and conversions should be considered, as was done in

Chapter 2. Second, the reasonableness of the frequency estimation assumptions should be evaluated. Third, practical considerations might dictate that a linear method would need to be used with a particular testing program. For example, suppose that the major focus of the testing program was on deciding whether examinees were above or below a cutting score that was near the mean. Then a linear equating method (or even a mean equating method) might be considered adequate, because the results for the Tucker and Braun-Holland linear methods are typically similar to those for frequency estimation equipercentile equating near the mean, and linear methods are less complicated computationally. Practical issues in choosing among methods are considered further in Chapter 8.

Sometimes it is possible to equate forms that have items in common when using the random groups design. Such a design is referred to as the *common item random groups design*. In this design, the use of the common items can lead to greater precision than would be attained using the random groups design without considering the common items. Computationally, equipercentile equating would proceed in the same way as it would for frequency estimation equating. The linear methods described in Chapter 4 also could be applied in this design. The increase in equating precision that is achieved by using common items is discussed briefly in Chapter 7.

Exercises

5.1. Using the data in Table 5.1, find the conditional distribution of X given v, and display the results in a format similar to Table 5.3.

5.2. Using frequency estimation assumptions, find the joint distribution of X and V in Population 2 and display the results in a format similar to Table 5.4. Also display the marginal distributions.

5.3. Using the data in Tables 5.1 and 5.4, the results shown in Table 5.4, the results from Exercise 5.2, and assuming that $w_1 = w_2 = .5$, find the Form Y equipercentile equivalents of Form X integer scores 0, 1, 2, 3, 4, and 5.

5.4. Find the Braun-Holland and Tucker linear equations for the equating relationship for the data in the example associated with Table 5.1 and 5.2 for $w_1 = w_2 = .5$.

5.5. Do the relationships between X and V and Y and V in Tables 5.1 and 5.2 appear to be linear? How can you tell? How would you explain the difference in results for the Braun-Holland and Tucker methods in Exercise 5.4?

5.6. Use chained equipercentile equating to find the Form Y equivalents of Form X integer scores 1 and 3 using the data in Tables 5.1 and 5.2.

CHAPTER 6

Item Response Theory Methods

The use of item response theory (IRT) in testing applications has grown considerably over the past 15 years. This growth has been reinforced by the many publications in the area (e.g., Baker, 1992a; Hambleton and Swaminathan, 1985; Hambleton *et al.*, 1991; Lord, 1980; Wright and Stone, 1979). Applications of IRT include test development, item banking, differential item functioning, adaptive testing, test equating, and test scaling. A major appeal of IRT is that it provides an integrated psychometric framework for developing and scoring tests. Much of the power of IRT results from it explicitly modeling examinee responses at the item level, whereas, for example, the focus of classical test models and strong true score models is on responses at the level of test scores.

Unidimensional IRT models have been developed for tests that are intended to measure a single dimension, and *multidimensional IRT models* have been developed for tests that are intended to measure simultaneously along multiple dimensions. IRT models have been developed for tests whose items are scored dichotomously (0/1) as well as for tests whose items are scored polytomously (e.g., a short answer test in which examinees can earn a score of 0, 1, or 2 on each item).

Many testing programs use unidimensional IRT models to assemble tests. In these testing programs, the use of IRT equating methods often seems natural. Also, IRT methods can be used for equating in some situations in which traditional methods typically are not used, such as equating to an item pool. Thus, IRT methods are an important component of equating methodology. However, IRT models gain their flexibility by making strong statistical assumptions, which likely do not hold precisely in real testing situations. For this reason, studying the robustness of the models to violations of the assumptions, as well as studying the fit of the IRT model, is a crucial aspect of IRT applications. See Hambleton and Swaminathan (1985) and Hambleton *et al.* (1991) for general discussions of testing model fit.

This chapter is concerned primarily with equating dichotomously (0/1) scored tests forms using the unidimensional IRT model referred to as the *three-parameter logistic model* (Lord, 1980). This model, which is described more fully later in this chapter, is the most general unidimensional model for dichotomously scored tests that is in widespread use. The *Rasch model* (Rasch, 1960; Wright and Stone, 1979) also is discussed briefly. See Thissen and Steinberg (1986) for a taxonomy of item response models, including IRT models, that can be used with other scoring schemes. Equating polytomously scored tests is discussed briefly in Chapter 8. IRT also has been generalized to multidimensional models. In this chapter, after an introduction to IRT, methods of transforming IRT scales are discussed. Then IRT true score equating and IRT observed score equating are treated. The methods are illustrated using the same data that were used in Chapters 4 and 5. Equating using IRT-based item pools also is discussed. Issues in equating computer administered and computer adaptive tests are considered in Chapter 8.

As will be described more fully later in this chapter, equating using IRT typically is a three-step process. First, item parameters are estimated using a computer program such as BILOG 3 (Mislevy and Bock, 1990) or LOGIST (Wingersky *et al.*, 1982). Second, parameter estimates are scaled to a base IRT scale using a linear transformation. Third, if number-correct scoring is used, number-correct scores on the new form are converted to the number-correct scale on an old form and then to scale scores.

Some Necessary IRT Concepts

A description of some necessary concepts in IRT is presented here to provide a base for understanding unidimensional IRT equating of dichotomously scored tests. References cited earlier provide a much more complete presentation of IRT. Instructional modules on IRT by Harris (1989) and on IRT equating by Cook and Eignor (1991) can be used as supplements to the material presented here.

Unidimensionality and Local Independence Assumptions

Unidimensional item response theory (IRT) models for dichotomously (0/1) scored tests assume that *examinee ability* is described by a single latent variable, referred to as θ, defined so that $-\infty < \theta < \infty$. The use of a single latent variable implies that the construct being measured by the test is *uni-*

dimensional. In practical terms, the unidimensionality assumption in IRT requires that tests measure only one ability. For example, a mathematics test that contains some items that are strictly computational and other items that involve verbal material likely is not unidimensional.

The *item characteristic curve* for each item relates the probability of correctly answering the item to examinee ability. The item characteristic curve for item j is symbolized by $p_j(\theta)$, which represents the probability of correctly answering item j for examinees with ability θ. For example, if 50% of the examinees with ability $\theta = 1.5$ can be expected to correctly answer item 1, then the probability can be symbolized as $p_1(\theta = 1.5) = .5$. Note that p_j is written as a function of the variable θ. IRT models typically assume a specified functional form for the item characteristic curve, which is what distinguishes IRT models from one another.

An assumption of *local independence* is made in applying IRT models. Local independence means that, after taking into account examinee ability, examinee responses to the items are statistically independent. Under local independence, the probability that examinees of ability θ correctly answer *both* item 1 *and* item 2 equals the product of the probability of correctly answering item 1 and the probability of correctly answering item 2. For example, if examinees of ability $\theta = 1.5$ have a .5 probability of answering item 1 correctly and a .6 probability of answering item 2 correctly, for such examinees the probability of correctly answering *both* items correctly under local independence is $.30 = .50(.60)$.

The local independence assumption implies that there are no dependencies among items other than those that are attributable to latent ability. One example where local independence likely would not hold is when tests are composed of sets of items that are based on common stimuli, such as reading passages or charts. In this case, local independence probably would be violated because items associated with one stimulus are likely to be more related to one another than to items associated with another stimulus.

Although the IRT unidimensionality and local independence assumptions might not hold strictly, they might hold closely enough for IRT to be used advantageously in many practical situations. In using IRT equating, it is important to choose an equating design that minimizes the effects of violations of model assumptions.

IRT Models

Various IRT models are in use that differ in the functional form of the item characteristic curve. Among unidimensional models, the three-parameter logistic model is the most general of the forms in widespread use. In this model, the functional form for an item characteristic curve is characterized by three item parameters. Under the three-parameter logistic model, the probability that persons of ability equal to the ability of person i correctly answer item j is defined as

$$p_{ij} = p_{ij}(\theta_i; a_j, b_j, c_j) = c_j + (1 - c_j)\frac{\exp[Da_j(\theta_i - b_j)]}{1 + \exp[Da_j(\theta_i - b_j)]}. \quad (6.1)$$

In this equation, θ_i is the ability parameter for person i. Ability, θ, is defined over the range $-\infty < \theta < \infty$ and often is scaled to be normally distributed with a mean of 0 and standard deviation of 1. In this case, nearly all of the persons have θ values in the range -3 to $+3$. The expression "exp" in equation (6.1) stands for the natural logarithm "exponential." That is, the quantity in brackets after exp is the exponent of $e = 2.72828 \ldots$. The constant D typically is set to 1.7 so that the logistic item response curve and the normal ogive differ by no more than .01 for all values of θ.

The item parameters a_j, b_j, and c_j are associated with item j. The meanings of these parameters are illustrated in the portion of Table 6.1 labeled "Item Parameters" and in Figure 6.1. For now, consider only the item parameters for the three items listed below the labeled portion "Scale I" on the left-hand side of the table. Also ignore the I subscript for the present.

Table 6.1. Item and Person Parameters on Two Scales for a Hypothetical Test.

	Scale I			Scale J		
Item Parameters						
Item	a_{Ij}	b_{Ij}	c_{Ij}	a_{Jj}	b_{Jj}	c_{Jj}
$j = 1$	1.30	-1.30	.10	2.60	-1.15	.10
$j = 2$	.60	$-.10$	.17	1.20	$-.55$	.17
$j = 3$	1.70	.90	.18	3.40	$-.05$	.18
Person Abilities						
Person	θ_{Ii}			θ_{Ji}		
$i = 1$	-2.00			-1.50		
$i = 2$	1.00			.00		

Scale Transformation Constants

$A = .5 \quad B = -.5$

Probability of Correctly Answering Items

	$p_{ij}(\theta_{Ii}; a_{Ij}, b_{Ij}, c_{Ij})$		$p_{ij}(\theta_{Ji}; a_{Jj}, b_{Jj}, c_{Jj})$	
	Person		Person	
Item	$i = 1$	$i = 2$	$i = 1$	$i = 2$
$j = 1$	.26	.99	.26	.99
$j = 2$	.27	.80	.27	.80
$j = 3$	.18	.65	.18	.65

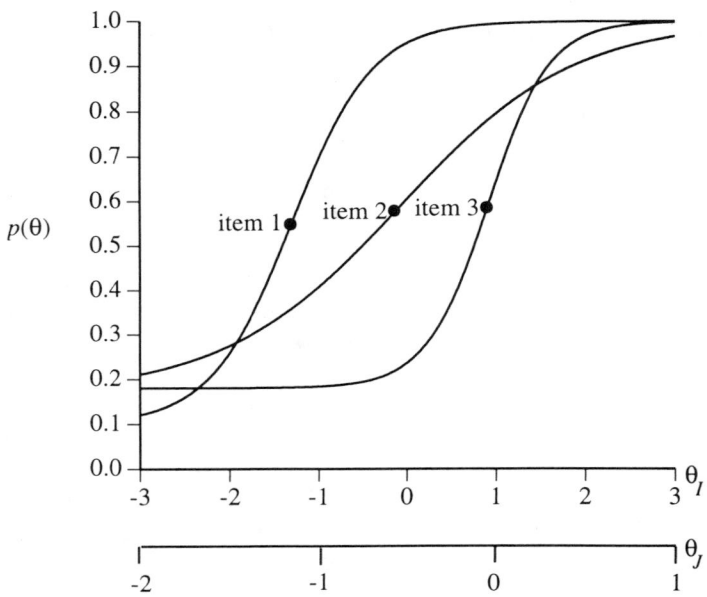

Figure 6.1. Hypothetical example of scale transformations.

The item parameter c_j is the *lower asymptote* or *pseudochance level pa-rameter* for item j. The parameter c_j represents the probability that an ex-aminee with very low ability (actually, $\theta = -\infty$) correctly answers the item. For example, for low ability examinees, the curve for item 3 in Figure 6.1 appears to be leveling off (have a lower asymptote) at a probability of .18, which corresponds to the c-parameter for this item listed in Table 6.1. If the horizontal axis in Figure 6.1 were extended beyond $\theta = -3$, items 1 and 2 would appear to have the lower asymptotes of .10 and .17 shown in Table 6.1. The c-parameter for an item must be in the range 0 to 1. Typi-cally, the c-parameter for an item is somewhere in the range of 0 to the probability of correctly answering an item by random guessing (1 divided by the number of options).

The item parameter b_j is referred to as the *difficulty* or *location parame-ter* for item j. The logistic curve has an inflexion point at $\theta = b$. When $c = 0, b$ is the level of ability where the probability of a correct answer is .5. Otherwise, b is the ability level where the probability of a correct response is halfway between c and 1.0. The inflexion point of each curve is indicated by the circular symbol on each item characteristic curve in Figure 6.1. Typically, b is in the range -3 to $+3$. Higher values of b are associated with more difficult items. As an illustration, item 3 has the highest b-parameter in Table 6.1. Of the three items in Figure 6.1, the item charac-teristic curve for item 3 tends to be shifted the farthest to the right.

The item parameter a_j is referred to as the *discrimination parameter* for item j. The a-parameter is proportional to the slope of the item character-istic curve at the inflexion point. As can be seen in Table 6.1, item 3 has the highest a-parameter (1.7) and item 3 also has the steepest item character-istic curve in Figure 6.1.

The abilities for two persons are shown in the middle of Table 6.1 under the heading "Person Abilities." The probabilities of correctly answering each of the three items for examinees of ability $\theta = -2.00$ and $\theta = 1.00$ are shown at the bottom of Table 6.1 under the heading "Probability of Cor-rectly Answering Items." For example, the probability of person $i = 1$ with ability $\theta_{1i} = -2.00$ correctly answering the first item can be calculated as follows using equation (6.1):

$$p_{ij} = .10 + (1 - .10) \frac{\exp\{1.7(1.30)[-2.00 - (-1.30)]\}}{1 + \exp\{1.7(1.30)[-2.00 - (-1.30)]\}} = .26.$$

The reader should verify the computation of the other probabilities by substituting the abilities and item parameters into equation (6.1).

Various simplifications of the three-parameter logistic model have been used. One variation can be obtained by setting c_j equal to a constant other than 0. The *two-parameter logistic model* is obtained from equation (6.1) by setting c_j equal to 0. This model does not explicitly accommodate ex-aminee guessing. The Rasch model is obtained from equation (6.1) by setting c_j equal to 0, a_j equal to 1, and D equal to 1. The Rasch model, therefore, requires all items to be equally discriminating, and it does not explicitly accommodate guessing. Other models exist that use a normal ogive to model p_{ij}.

The three-parameter logistic model is the only one of the three models presented that explicitly accommodates items which vary in difficulty, which vary in discrimination, and for which there is a nonzero probability of obtaining the correct answer by guessing. Because of its generality, the three-parameter model is the focus of this chapter. However, the assumed form of the relationship between ability and the probability of a correct re-sponse (e.g., the three-parameter logistic curve) is chosen primarily for rea-sons of mathematical tractability. No reason exists for this relationship to hold, precisely, for actual test items.

IRT Parameter Estimation

IRT parameters need to be estimated when using IRT methods in practice. LOGIST (Wingersky et al., 1982) and BILOG 3 (Mislevy and Bock, 1990) are two computer programs for the three-parameter logistic model that often are used to estimate ability and item parameters. An IRT parameter estimation computer program, such as BILOG or LOGIST, is necessary if the concepts and methodologies described in this chapter are to be applied to equating tests other than those used in the examples. Mislevy and

Stocking (1989) compared these two programs, and Baker (1992a) wrote extensively about IRT estimation. The basic input for these programs is 0/1 (wrong or right) scores of examinees at the item level. These programs use different estimation methods, have various options for estimating parameters, and produce somewhat different results. Unless the user chooses to do otherwise, the scaling of ability in LOGIST removes persons with extreme ability estimates before the ability estimates are scaled to have a mean of 0 and a standard deviation of 1. Extreme ability estimates are not included in the computation of the mean and standard deviation. These estimates are considered to be poor because they are likely to have considerable estimation error. Unless the user chooses otherwise, the ability scaling in BILOG also results in abilities with means of approximately 0 and standard deviations of approximately 1.

One important characteristic of ability estimation that arises from either of these programs is that the ability estimates depend on the pattern of item responses, rather than just on the number of items an examinee answers correctly (except for the Rasch model). That is, examinees who earn the same number-correct score would likely earn different estimated θ's if some of the items that they correctly answered were different. The use of such *pattern scoring* in IRT increases the precision of the IRT ability estimates over using the number-correct score if the IRT model holds. However, for many practical reasons, including equating, number-correct scoring often is used.

Transformations of IRT Scales

When conducting equating with nonequivalent groups, the parameters from different test forms need to be on the same IRT scale. However, the parameter estimates that result from IRT parameter estimation procedures are often on different IRT scales. For example, assume that the parameters for the IRT model are estimated for Form X based on a sample of examinees from Population 1 and separately for Form Y based on a sample of examinees from Population 2, where the two populations are not equivalent. As was already indicated, computer programs often define the θ-scale as having a mean of 0 and a standard deviation of 1 for the set of data being analyzed. In this case, the abilities for each group would be scaled to have a mean of 0 and a standard deviation 1, even though the groups differed in ability. Thus, a transformation of IRT scales is needed.

As will be demonstrated later in this section, if an IRT model fits a set of data, then any linear transformation of the θ-scale also fits the set of data, provided that the item parameters also are transformed. When the IRT model holds, the parameter estimates from different computer runs are on linearly related θ-scales. Thus, a linear equation can be used to convert IRT parameter estimates to the same scale. After conversion, the means and standard deviations of the abilities for the two groups on the common scale would be expected to differ. The resulting transformed parameter estimates, which sometimes are referred to as being *calibrated*, then can be used to establish score equivalents between number-correct scores on Form X and Form Y, and then to scale scores.

Transformation Equations

Define Scale I and Scale J as three-parameter logistic IRT scales that differ by a linear transformation. Then the θ-values for the two scales are related as follows:

$$\theta_{Ji} = A\theta_{Ii} + B, \qquad (6.2)$$

where A and B are constants in the linear equation and θ_{Ji} and θ_{Ii} are values of θ for individual i on Scale J and Scale I. The item parameters on the two scales are related as follows:

$$a_{Jj} = \frac{a_{Ij}}{A}, \qquad (6.3)$$

$$b_{Jj} = Ab_{Ij} + B, \qquad (6.4)$$

and

$$c_{Jj} = c_{Ij}, \qquad (6.5)$$

where a_{Jj}, b_{Jj}, and c_{Jj} are the item parameters for item j on Scale J and a_{Ij}, b_{Ij}, and c_{Ij} are the item parameters for item j on Scale I. The lower asymptote parameter is independent of the scale transformation, as is indicated by equation (6.5).

Demonstrating the Appropriateness of Scale Transformations

To demonstrate that there is an A and a B which result in the scale transformation that correctly transforms parameters from Scale I to Scale J, note that the right-hand side of equation (6.1) for Scale J equals

$$c_{Jj} + (1 - c_{Jj}) \frac{\exp[Da_{Jj}(\theta_{Ji} - b_{Jj})]}{1 + \exp[Da_{Jj}(\theta_{Ji} - b_{Jj})]}.$$

Now replace $\theta_{Ji}, a_{Jj}, b_{Jj}, c_{Jj}$ with the expressions from equations (6.2)–(6.5) as follows:

$$c_{Ij} + (1 - c_{Ij}) \frac{\exp\left\{D\frac{a_{Ij}}{A}[A\theta_{Ii} + B - (Ab_{Ij} + B)]\right\}}{1 + \exp\left\{D\frac{a_{Ij}}{A}(A\theta_{Ii} + B - (Ab_{Ij} + B)]\right\}}$$

$$= c_{Ij} + (1 - c_{Ij}) \frac{\exp[Da_{Ij}(\theta_{Ii} - b_{Ij})]}{1 + \exp[Da_{Ij}(\theta_{Ii} - b_{Ij})]}$$

This resulting expression is the right-hand portion of equation (6.1) for Scale I, which demonstrates that A and B in equations (6.2)–(6.5) provide the scale transformation.

Expressing A and B Constants

One way to express the constants A and B is as follows. For any two individuals, i and i^*, or any two items, j and j^*, A and B in equations (6.2)–(6.5) can be expressed as

$$A = \frac{\theta_{Ji} - \theta_{Ji^*}}{\theta_{Ii} - \theta_{Ii^*}} = \frac{b_{Jj} - b_{Jj^*}}{b_{Ij} - b_{Ij^*}} = \frac{a_{Ij}}{a_{Jj}} \qquad (6.6)$$

and

$$B = b_{Jj} - Ab_{Ij} = \theta_{Ji} - A\theta_{Ii}. \qquad (6.7)$$

To illustrate these equalities, refer back to Table 6.1 and Figure 6.1 for a hypothetical example of scale transformations. Parameters for three items are presented in the portion of Table 6.1 labeled "Item Parameters." Parameters for these items are given for Scale I and for Scale J. The item characteristic curves for these three items are presented in Figure 6.1. Note that horizontal scales are presented in this figure for Scale I and Scale J, and these are labeled θ_I and θ_J. As is evident from this figure, the item characteristic curves are the same shape on either scale. To calculate A from equation (6.6) using the difficulty parameters for items 1 and 2 ($j = 1$ and $j^* = 2$), take

$$A = \frac{(-1.15) - (-.55)}{(-1.30) - (-.10)} = \frac{-.6}{-1.2} = .5.$$

Alternatively, using the slope parameters for item 1,

$$A = \frac{1.3}{2.6} = .5.$$

Using equation (6.7) with the difficulty parameters for item 1,

$$B = (-1.15) - (.5)(-1.30) = -.5.$$

These values agree with those in the section labeled "Scale Transformation Constants" in Table 6.1. Equations (6.6) and (6.7) also can be used to calculate A and B using the θ-values for Persons 1 and 2. These A and B values can be used to transform parameters from Scale I to Scale J using equations (6.2)–(6.5). For example, to transform the ability of Person 1 from Scale I to Scale J using equation (6.2), take

$$\theta_{J1} = A\theta_{I1} + B = .5(-2.00) + (-.5) = -1.5,$$

which is the value for Person 1 shown under "Person Abilities" in Table 6.1. To convert the parameters for item 3 from Scale I to Scale J using equations (6.3)–(6.5), take

$$a_{J3} = \frac{a_{I3}}{A} = \frac{1.7}{.5} = 3.4,$$

$$b_{J3} = Ab_{I3} + B = .5(.90) - .5 = -.05,$$

and

$$c_{J3} = c_{I3} = .18.$$

These values agree with the Scale J values in the portion of Table 6.1 labeled "Item Parameters."

The p_{ij} values based on equation (6.1) are presented in the portion of Table 6.1 labeled "Probability of Correctly Answering Items." These values can be calculated from the item and person parameters presented in Table 6.1; they are the same for Scales I and J, and the p_{ij} values will be identical for any linearly related scales. This property often is referred to as *indeterminacy of scale location and spread.*

Expressing A and B Constants in Terms of Groups of Items and/or Persons

So far, the relationships between scales have been expressed by two abilities and two items. Often, it is more useful to express the relationships in terms of groups of items or people. From equations (6.6) and (6.7) it follows that (see Exercise 6.3)

$$A = \frac{\sigma(b_J)}{\sigma(b_I)}, \tag{6.8a}$$

$$= \frac{\mu(a_I)}{\mu(a_J)}, \tag{6.8b}$$

$$= \frac{\sigma(\theta_J)}{\sigma(\theta_I)}, \tag{6.8c}$$

$$B = \mu(b_J) - A\mu(b_I), \quad \text{and} \tag{6.9a}$$

$$= \mu(\theta_J) - A\mu(\theta_I). \tag{6.9b}$$

The means $\mu(b_J), \mu(b_I), \mu(a_I)$, and $\mu(a_J)$ in these equations are defined over one or more items with parameters that are expressed on both Scale I and Scale J. The standard deviations $\sigma(b_J)$ and $\sigma(b_I)$ are defined over two or more items with parameters that are expressed on both Scale I and Scale J. The means $\mu(\theta_J)$ and $\mu(\theta_I)$ are defined over one or more examinees with ability parameters that are expressed on both Scale I and Scale J. The standard deviations $\sigma(\theta_J)$ and $\sigma(\theta_I)$ are defined over two or more examinees with parameters that are expressed on both Scale I and Scale J.

To illustrate the use of equations (6.8a), (6.8b), and (6.9a), the following quantities can be calculated for the three items from the example in Table 6.1: $\mu(b_I) = -.1667$, $\sigma(b_I) = .8994$, $\mu(a_I) = 1.2$, $\mu(b_J) = -.5833$, $\sigma(b_J) = .4497$, and $\mu(a_J) = 2.4$. From equations (6.8) and (6.9),

$$A = \frac{\sigma(b_J)}{\sigma(b_I)} = \frac{\mu(a_I)}{\mu(a_J)} = \frac{.4497}{.8994} = \frac{1.2000}{2.4000} = .5000,$$

and

$$B = \mu(b_J) - A\mu(b_I) = -.5833 - .5000(-.1667) = -.5000.$$

Similar calculations can be made using the mean and standard deviations for the two ability scales in Table 6.1.

In equating with nonequivalent groups, parameter estimates for the common items would be available for examinees in the two groups. The parameter estimates on the common items could be used to find the scaling constants by substituting parameters for these estimates in the preceding equations.

Consider a situation in which the mean and standard deviation of the abilities on Scale I are known for one group of examinees. Also, the mean and standard deviation of the abilities are known for a different group of examinees on Scale J. Is there any way equations (6.8c) and (6.9b) can be used to transform Scale I to Scale J? No! These equations can be used only if the parameters for the *same* group of examinees are expressed on *both* scales.

Consider a different situation, in which the mean and standard deviation of abilities on Scale I are 0 and 1, respectively. For the *same* group of examinees, the mean and standard deviation of abilities are 50 and 10, respectively, on Scale J. Can equations (6.8c) and (6.9b) be used to transform parameters from Scale I to Scale J? Yes. The resulting scaling constants calculated using equations (6.8c) and (6.9b) are as follows:

$$A = \frac{\sigma(\theta_J)}{\sigma(\theta_I)} = \frac{10}{1} = 10 \quad \text{and} \quad B = \mu(\theta_J) - A\mu(\theta_I) = 50 - 10(0) = 50.$$

These equations might be used to transform IRT parameters to a different scale when the means and standard deviations of the abilities are known.

Transforming IRT Scales When Parameters Are Estimated

The estimation of item parameters complicates the problem of transforming IRT scales. The process that needs to be followed depends on the design used for data collection. As was indicated earlier, when either LOGIST or BILOG are used for parameter estimation, unless the user chooses to do otherwise, ability estimates are scaled to have a mean of 0 and a standard deviation of 1.

Designs

In the *random groups equating design*, the IRT parameters for Form X can be estimated separately from the parameters for Form Y. If the same scaling convention (e.g., mean of 0 and standard deviation of 1) for ability is used in the separate estimations, then the parameter estimates for the two forms are assumed to be on the same scale without further transformation. No further transformation is assumed to be required because the groups are randomly equivalent, and the abilities are scaled to have the same mean and standard deviation in both groups. If, for some reason, different scaling conventions were used for the two forms, then estimates of the mean and standard deviations of the θ-estimates could be used in place of the mean and standard deviations of the θ-parameters in equations (6.8c) and (6.9b).

In the *single group design with counterbalancing*, the parameters for all examinees on both forms can be estimated together. Because the parameters for the two forms are estimated together on the same examinees, the parameter estimates are assumed to be on the same scale. If the parameters for the two forms are estimated separately using the same scaling conventions, the parameter estimates can be assumed to be on the same scale following the logic discussed previously for the random groups design.

In the *common-item nonequivalent groups equating design*, the Form Y item and ability parameters typically are estimated at the time Form Y is first administered. Consequently, only the Form X parameters need to be estimated when Form X is equated to Form Y. Because the examinees who took Form X are not considered to be equivalent to the examinees who took Form Y, parameter estimates for the two estimations are not on the same scale. However, there is a set of items that is common to the two forms. (See Chapter 8 for rules of thumb on the number of common items to use and a discussion of characteristics of common items.) The estimates of the item parameters for these common items can be used to estimate the scale transformation.

As an alternative, the parameters for Form X and Form Y can be estimated together. For example, a single run of LOGIST can be conducted using the item level data for Form X and Form Y on the two examinee groups, and indicating that the items which an examinee did not take are "not reached." [Consult the LOGIST manual (Wingersky *et al.*, 1982) for a description of this procedure.] When the estimation is conducted in this manner, the resulting estimates are all on the same scale.

Parameter estimates must be on the same scale to proceed with equating number-correct scores on alternate forms and converting them to scale scores. Methods for equating number-correct scores are described later in this chapter.

Mean/Sigma and Mean/Mean Transformation Methods

The most straightforward way to transform the scales in the common-item nonequivalent groups design is to substitute the means and standard deviations of the item parameter estimates of the common items for the parameters in equations (6.8) and (6.9). After transformation, the item parameter estimates are often referred to as being *calibrated*. One procedure, described by Marco (1977) and referred to here as the *mean/sigma method*, uses the means and standard deviations of the b-parameter estimates from the common items in place of the parameters in equations (6.8a) and (6.9a). In another method, described by Loyd and Hoover (1980) and referred to here as the *mean/mean* method, the mean of the a-parameter estimates for the common items is used in place of the parameters in equation (6.8b) to estimate the A-constant. Then, the mean of the b-parameter estimates of the common items is used in place of the parameters in equation (6.9a) to estimate the B-constant. The values of A and B then can be substituted into equations (6.2)–(6.5) to obtain the rescaled parameter estimates.

When estimates are used in place of the parameters, or when the IRT model does not hold precisely, the equalities shown in equations (6.8) and (6.9) do not necessarily hold. So, the mean/sigma and the mean/mean methods typically produce different results. One reason that the mean/sigma method is sometimes preferred over the mean/mean method is that estimates of b-parameters are more stable than estimates of the a-parameters. However, Baker and Al-Karni (1991) pointed out that the mean/mean method might be preferable because means are typically more stable than standard deviations, and the mean/mean method uses only means. Empirical research comparing these two methods is lacking, so the approach suggested here is to consider both procedures, and compare the raw-to-scale score conversions that result from the application of both methods when equating is conducted.

Mislevy and Bock (1990) recommended a further variation that uses the means of the b-parameters and the geometric means of the a-parameters. Stocking and Lord (1983) also discussed procedures for using robust estimates of the means and standard deviations of estimates of the b-parameters, although they were not satisfied with the performance of these robust methods.

Characteristic Curve Transformation Methods

One potential problem with the methods considered so far arises when various combinations of a-, b-, and c-parameter estimates produce almost identical item characteristic curves over the range of ability at which most examinees score. For example, in two estimations an item with very different b-parameter estimates could have very similar item characteristic curves. In this case, the mean/sigma method could be overly influenced by the difference between the b-parameter estimates, even though the item characteristic curves for the items on the two estimations were very similar. This problem arises because the scale conversion methods described so far do not consider all of the item parameter estimates simultaneously. As a response to this problem, Haebara (1980) presented a method that does consider all of the item parameters simultaneously, and Stocking and Lord (1983) developed a method similar to Haebara's. Stocking and Lord (1983) referred to both their method and the Haebara method as *characteristic curve methods*. To develop these methods, note that the indeterminacy of scale location and spread property which was described earlier implies that, for ability Scales I and J,

$$p_{ij}(\theta_{Ji}; a_{Jj}, b_{Jj}, c_{Jj}) = p_{ij}\left(A\theta_{Ii} + B; \frac{a_{Ij}}{A}, Ab_{Ij} + B, c_{Ij}\right), \qquad (6.10)$$

for examinee i and item j. Equation (6.10) states that the probability that examinees of a given ability will answer a particular item correctly is the same regardless of the scale that is used to report the scores.

If estimates are used in place of the parameters in equation (6.10), then there is no guarantee that the equality will hold over all items and examinees for any A and B. This lack of equality is exploited by the characteristic curve methods.

Haebara Approach. The function used by Haebara (1980) to express the difference between the item characteristic curves is the sum of the squared difference between the item characteristic curves for each item for examinees of a particular ability. For a given θ_i, the sum, over items, of the squared difference can be displayed as

$$Hdiff(\theta_i) = \sum_{j:V}\left[p_{ij}(\theta_{Ji}; \hat{a}_{Jj}, \hat{b}_{Jj}, \hat{c}_{Jj}) - p_{ij}\left(\theta_{Ji}; \frac{\hat{a}_{Ij}}{A}, A\hat{b}_{Ij} + B, \hat{c}_{Ij}\right)\right]^2. \quad (6.11)$$

The summation is over the common items ($j:V$). In this equation, the difference between each item characteristic curve on the two scales is squared and summed.

Hdiff then is cumulated over examinees. The estimation process proceeds by finding A and B that minimize the following criterion:

$$Hcrit = \sum_i Hdiff(\theta_i). \quad (6.12)$$

The summation in equation (6.12) is over examinees.

Stocking and Lord Approach. In contrast to the Haebara approach, Stocking and Lord (1983) used the sum, over items, of the squared difference,

$$SLdiff(\theta_i) = \left[\sum_{j:V} p_{ij}(\theta_{Ji}; \hat{a}_{Jj}, \hat{b}_{Jj}, \hat{c}_{Jj}) - \sum_{j:V} p_{ij}\left(\theta_{Ji}; \frac{\hat{a}_{Ij}}{A}, A\hat{b}_{Ij} + B, \hat{c}_{Ij}\right)\right]^2.$$

$$(6.13)$$

In the Stocking and Lord (1983) approach, the summation is taken over items for each set of parameter estimates before squaring. Note that in IRT, the function

$$\tau(\theta_i) = \sum_j p_{ij}(\theta_i) \quad (6.14)$$

is referred to as the *test characteristic curve*. So, the expression $SLdiff(\theta_i)$ is the squared difference between the test characteristic curves for a given θ_i. In contrast, the expression $Hdiff(\theta_i)$ is the sum of the squared difference between the item characteristic curves for a given θ_i. *SLdiff* then is cumulated over examinees. The estimation proceeds by finding the combination of A and B that minimizes the following criterion:

$$SLcrit = \sum_i SLdiff(\theta_i). \quad (6.15)$$

The summation in equation (6.15) is over examinees. The approach to solving for A and B in equations (6.12) and (6.15) is a computationally intensive iterative approach.

Specifying the Summation Over Examinees. In addition to differences in the function used to express the difference between the characteristic curves described in equations (6.11) and (6.13), these methods differ in how they cumulate the differences between the characteristic curves. Various ways to specify the examinees have been used in the summations in equations (6.12) and (6.15). Some of these ways are as follows:

1. Sum over estimated abilities of examinees who were administered the old form. (Stocking and Lord, 1983, used a random sample of 200 ability estimates.)
2. Sum over estimated abilities of examinees who were administered the new form and sum over estimated abilities of examinees who were administered the old form [used by Haebara (1980)].
3. Sum over estimated abilities that are grouped into intervals and then weight the differences by the proportion of examinees in each interval [used by Haebara (1980)].
4. Sum over a set of equally spaced values of ability [implemented by Baker and Al-Karni (1991)].
5. If a continuous distribution of ability is known or estimated, the summation over examinees could be replaced by integration over the ability distribution [see Zeng and Kolen (1994)].

A decision needs to be made about which of these options (or others) are used when implementing the characteristic curve procedures. The Macintosh computer program *ST* that is listed in Appendix B can be used to implement these schemes for summation over examinees.

Hypothetical Example. A hypothetical example is presented in Table 6.2 that illustrates part of the process of scaling item parameter estimates. Assume that the three items listed are common items in a common-item nonequivalent groups equating design, and that the resulting estimates are on different linearly related ability scales. Estimates of A and B based on these parameter estimates for the mean/sigma and mean/mean methods are presented in the top portion of Table 6.2. The Scale I parameter estimates are converted to Scale J in the middle portion of the table. The results for the two methods differ somewhat. These differences likely would cause some differences in raw-to-scale score conversions, which could be studied if equating relationships subsequently were estimated.

The probability of a correct response, using equation (6.1), is shown in the bottom portion of Table 6.2 for examinees with ability $\theta_i = 0$. In this example, the mean/sigma and mean/mean methods are compared using *Hdiff* and *SLdiff* as criteria. The criteria can be calculated at $\theta_i = 0$ using the estimated probabilities at the bottom of Table 6.2. To calculate *Hdiff*(θ_i) using equation (6.11), sum, over items, the squared difference between the estimated probabilities for the original Scale J and for the transformed scale that results from the application of one of the methods. For example, for the mean/sigma method,

$$Hdiff(\theta_i = 0) = (.8034 - .7556)^2 + (.3634 - .3359)^2 + (.1291 - .1202)^2$$
$$= .003120.$$

Table 6.2. Hypothetical Example for Characteristic Curve Methods Using Estimated Parameters.

Item	Scale I			Scale J		
	$\hat{a}$	$\hat{b}$	$\hat{c}$	$\hat{a}$	$\hat{b}$	$\hat{c}$
1	.4000	−1.1000	.1000	.5000	−1.5000	.1000
2	1.7000	.9000	.2000	1.6000	.5000	.2000
3	1.2000	2.2000	.1000	1.0000	2.0000	.1000
Mean	1.1000	.6667	.1333	1.0333	.3333	.1333
Sd	.5354	1.3573	.0471	.4497	1.4337	.0471
	Mean/Sigma	Mean/Mean				
A	1.0563	1.0645				
B	−.3709	−.3763				

Item	Scale I Converted to Scale J Using Mean/Sigma Results			Scale I Converted to Scale J Using Mean/Mean Results		
	$\hat{a}$	$\hat{b}$	$\hat{c}$	$\hat{a}$	$\hat{b}$	$\hat{c}$
1	.3787	−1.5328	.1000	.3758	−1.5473	.1000
2	1.6094	.5798	.2000	1.5970	.5817	.2000
3	1.1360	1.9530	.1000	1.1273	1.9656	.1000
Mean	1.0414	.3333	.1333	1.0333	.3333	.1333
Sd	.5069	1.4337	.0471	.5030	1.4449	.0471

Estimated probability of correct response given $\theta_i = 0$

Item	Original Scale J	Mean/Sigma	Mean/Mean
1	.8034	.7556	.7559
2	.3634	.3359	.3367
3	.1291	.1202	.1203
sum	1.2959	1.2118	1.2130

Similarly, for the mean/mean method,

$$Hdiff(\theta_i = 0) = (.8034 - .7559)^2 + (.3634 - .3367)^2 + (.1291 - .1203)^2$$
$$= .003047.$$

$Hdiff(\theta_i = 0)$ is smaller for the mean/mean method than for the mean/sigma method, indicating that the mean/mean method is somewhat "better" than the mean/sigma method at $\theta_i = 0$ based on $Hdiff(\theta_i)$.

To calculate $SLdiff(\theta_i = 0)$ using equation (6.13), the estimated probabilities are summed over items, resulting in the sums listed at the bottom of the table. These sums represent the value of the test characteristic curve at $\theta_i = 0$. For the mean/sigma method,

$$SLdiff(\theta_i = 0) = (1.2959 - 1.2118)^2 = .007073.$$

For the mean/mean method,

$$SLdiff(\theta_i = 0) = (1.2959 - 1.2130)^2 = .006872.$$

$SLdiff(\theta_i = 0)$ is smaller for the mean/mean method than for the mean/ sigma method, indicating that the mean/mean method is somewhat "better" than the mean/sigma method at $\theta_i = 0$. Thus, the mean/mean method is "better" at $\theta_i = 0$ for both criteria. In using these methods, differences would actually need to be calculated at many values of θ_i.

If the scaling were actually done using the characteristic curve methods, $Hcrit$ and $SLcrit$ would be calculated by summing $Hdiff(\theta_i)$ and $SLdiff(\theta_i)$ over different values of θ_i. Also, the iterative minimization algorithms described by Haebara (1980) and Stocking and Lord (1983) would be used to find the A and B that minimized $Hcrit$ and $SLcrit$. Typically, the mean/ mean or mean/sigma method estimates of A and B would be used as starting values in the minimization process.

Comparison of Criteria. Little empirical work has been done that compares the $Hcrit$- and $SLcrit$-based methods to each other, although Way and Tang (1991) found that methods based on the two criteria produced similar results. Theoretically, the $Hcrit$ methods might be argued to be superior to the $SLcrit$ methods for the following reason: $Hdiff(\theta_i)$ can be 0 only if the item characteristic curves are identical at θ_i, whereas $SLdiff(\theta_i)$ could be 0 even if the item characteristic curves differed. In this sense, $Hdiff(\theta_i)$ might be viewed as being more stringent than $SLdiff(\theta_i)$. On the other hand, it might be argued the $SLcrit$-based methods are preferable theoretically, because they focus on the difference between test characteristic curves.

Little empirical work has been done that compares different ways of defining the examinee group used to cumulate the differences between the item characteristic curves over examinees. If the methods were shown not to differ much from one another, then the use of an equally spaced set of abilities that covers the likely range of abilities might be easiest to implement. However, the different methods will produce somewhat different results. What would be the theoretically preferable way to define the group used to cumulate the differences? The procedure discussed by Haebara (1980), which sums over estimated abilities of examinees who were administered either the new form or the old form, seems preferable because it will produce a symmetric function. This procedure also uses data on examinees who were administered both forms. However, this procedure might be more difficult to implement. Empirical research needs to be conducted to compare these methods.

One potential limitation of the characteristic curve methods is that they do not explicitly account for the error in estimating item parameters. [See Divgi (1985), for a method that takes into account such error.] The failure to take into account error in estimating item parameters, explicitly, might not be that crucial when the sample size is large and the item characteristic curves are well estimated. However, there are situations in which problems might arise. For example, if considerably larger sample sizes are used to estimate parameters for one form than for another, then ignoring the error in parameter estimates might lead to problems in estimating A and B, and in estimating equating relationships. Empirical research is needed to address this issue.

Comparisons Among Scale Transformation Methods

Research has been conducted that compares the mean/sigma or mean/mean methods to the characteristic curve methods (Baker and Al-Karni, 1991; Hung *et al.*, 1991; Way and Tang, 1991). The general conclusion from this research is that the characteristic curve methods produce more accurate results. However, Baker and Al-Karni (1991) found that the Stocking and Lord and mean/mean methods produce very similar results for most comparisons; although when the IRT parameter estimation was problematic, the Stocking and Lord procedure was more accurate.

Given these findings, implementations of the characteristic curve methods might be safest. If software for these methods is unavailable, the following strategy might produce acceptable results. Construct a scatterplot of the IRT a-parameter estimates by plotting the parameter estimates for the common items for both groups. Construct similar scatterplots for the b- and c-parameter estimates. Examine the scatterplots and identify any items that appear to be outliers. The identification of outliers is necessarily a subjective process. Estimate the A- and the B-constants with the mean/sigma and mean/mean methods both with and without including the item or items with parameter estimates that appear to be outliers. If the mean/sigma and mean/mean procedures give very different results with the outliers included but similar results with the outliers removed, then consider removing these items. If the results from this procedure are not clear, then the use of the characteristic curve procedure might be the best alternative. Note that even when the characteristic curve procedures are used, it is best to use more than one method, and to examine scatterplots to consider eliminating items with very different parameter estimates. In practice, it might be best to implement each of the methods and evaluate the effects of the differences between the methods on equating relationships and on resulting scale scores. Procedures for choosing among equating results are considered in Chapter 8.

Equating and Scaling

When a test is scored using estimated IRT abilities, there is no further need to develop a relationship between scores on Form X and Form Y. Still, the estimated abilities can be converted to scale scores. The ability estimates can be converted so that the reported scores are positive integers, which are presumably easier for examinees to interpret than are scores that may be negative and noninteger, as is the case with estimated IRT abilities. This conversion might involve a linear conversion of estimated abilities, followed by truncating the converted scores so that they are in a specified range of positive numbers, and then rounding the scores to integers for reporting purposes.

However, using estimated IRT abilities results in several practical problems, which might be why they are not used that often. One issue is that, to use estimated abilities with the three-parameter logistic model, the whole 0/1 response string, rather than the number-correct score, is used to estimate θ. Thus, examinees with the same number-correct score often receive different estimated abilities, which can be difficult to explain to examinees. In addition, estimates of θ are difficult to compute (they typically cannot be computed by hand) and can be costly to obtain. Another concern is that the estimated θ-values with the three-parameter logistic model typically are subject to relatively greater amounts of measurement error for high and low ability examinees than for middle ability examinees. Lord (1980, p. 183) indicated that the measurement error variability for examinees of extreme ability could be 10 or even 100 times that of middle ability examinees, which can create problems in interpreting summary statistics such as means and standard deviations. For these practical reasons, tests are often scored number-correct, even when they are developed and equated using IRT. When tests are scored with number-correct scores, an additional step is required in the IRT equating process. The two methods that have been proposed are to equate true scores and to equate observed scores. These procedures are considered next.

Equating True Scores

After the item parameters are on the same scale, IRT true score equating can be used to relate number-correct scores on Form X and Form Y. In this process, the true score on one form associated with a given θ is considered to be equivalent to the true score on another form associated with that θ.

Test Characteristic Curves

In IRT, the number-correct true score on Form X that is equivalent to θ_i is defined as

$$\tau_X(\theta_i) = \sum_{j:X} p_{ij}(\theta_i; a_j, b_j, c_j), \qquad (6.16)$$

where the summation $j:X$ is over items on Form X. The number-correct true score on Form Y that is equivalent to θ_i is defined as

$$\tau_Y(\theta_i) = \sum_{j:Y} p_{ij}(\theta_i; a_j, b_j, c_j), \tag{6.17}$$

where the summation $j:Y$ is over items on Form Y. Equations (6.16) and (6.17) are referred to as *test characteristic curves* for Form X and Form Y. These test characteristic curves relate IRT ability to number-correct true score.

When using the three-parameter logistic model of equation (6.1), very low true scores are not attainable with the three-parameter logistic IRT model, because as θ approaches $-\infty$ the probability of correctly answering item j approaches c_j and not 0. Therefore, true scores on Forms X and Y are associated with a value of θ only over the following ranges (recall that K_X and K_Y are the numbers of items on Forms X and Form Y, respectively):

$$\sum_{j:X} c_j < \tau_X < K_X \qquad \text{and} \qquad \sum_{j:Y} c_j < \tau_Y < K_Y. \tag{6.18}$$

True Score Equating Process

In IRT true score equating, for a given θ_i, true scores $\tau_X(\theta_i)$ and $\tau_Y(\theta_i)$ are considered to be equivalent. The Form Y true score equivalent of a given true score on Form X is

$$irt_Y(\tau_X) = \tau_Y(\tau_X^{-1}), \qquad \sum_{j:X} c_j < \tau_X < K_X, \tag{6.19}$$

where τ_X^{-1} is defined as the θ_i corresponding to true score τ_X. Equation (6.19) implies that true score equating is a three-step process:

1. Specify a true score τ_X on Form X (typically an integer $\sum_{j:X} c_j < \tau_X < K_X$).
2. Find the θ_i that corresponds to that true score (τ_X^{-1}).
3. Find the true score on Form Y, τ_Y, that corresponds to that θ_i.

Form Y equivalents of Form X integer number-correct scores typically are found using these procedures.

Whereas Step 1 and Step 3 are straightforward, Step 2 requires the use of an iterative procedure. For example, suppose that the Form Y equivalent of a Form X score of 5 is to be found. Implementation of Step 2 requires finding the θ_i that results in the right-hand side of equation (6.16) equaling 5. Finding this value of θ_i requires the solution of a nonlinear equation using an iterative process. This process is described in the next section.

The Newton-Raphson Method

The Newton-Raphson method is a general method for finding the roots of nonlinear functions. To use this method, begin with a function that is set to 0. Refer to that function as $func(\theta)$, which is a function of the variable θ. Refer to the first derivative of the function with respect to θ as $func'(\theta)$. To apply the Newton-Raphson method, an initial value is chosen for θ, which is referred to as θ^-. A new value for θ is calculated as

$$\theta^+ = \theta^- - \frac{func(\theta)}{func'(\theta)}. \tag{6.20}$$

Typically, θ^+ will be closer to the root of the equation than θ^-. The new value then is redefined as θ^-, and the process is repeated until θ^+ and θ^- are equal at a specified level of precision or until the value of $func$ is close to 0 at a specified level of precision.

When using the Newton-Raphson method, the choice of the initial value is an important consideration, because a poor choice can lead to an erroneous solution. Press *et al.* (1989) describe modifications to the Newton-Raphson method that are more robust than the Newton-Raphson method to the choice of poor initial values.

Using the Newton-Raphson Method in IRT Equating. To apply this method to IRT true score equating, let τ_X be the true score whose equivalent is to be found. From equation (6.16) it follows that θ_i is to be found such that the expression

$$func(\theta_i) = \tau_X - \sum_{j:X} p_{ij}(\theta_i; a_j, b_j, c_j) \tag{6.21}$$

equals 0. The Newton-Raphson method can be employed to find this θ_i using the first derivative of $func(\theta_i)$ with respect to θ_i, which is

$$func'(\theta_i) = -\sum_{j:X} p'_{ij}(\theta_i; a_j, b_j, c_j) \tag{6.22}$$

where $p'_{ij}(\theta_i; a_j, b_j, c_j)$ is defined as the first derivative of $p_{ij}(\theta_i; a_j, b_j, c_j)$ with respect to θ_i. Lord (1980, p. 61) provided this first derivative:

$$p'_{ij}(\theta_i; a_j, b_j, c_j) = \frac{1.7a_j(1 - p_{ij})(p_{ij} - c_j)}{(1 - c_j)}, \tag{6.23}$$

where $p_{ij} = p_{ij}(\theta_i; a_j, b_j, c_j)$. The resulting expressions for $func(\theta_i)$ and $func'(\theta_i)$ are substituted into equation (6.20).

A Hypothetical Example. A hypothetical example using this procedure is presented in Table 6.3. In this example, a five-item Form X is to be equated to a five-item Form Y. Parameters (not estimates) are given, and

Table 6.3. Hypothetical Example for IRT True Score Equating.

Form X Item Parameters

Item Parameter	Item 1	Item 2	Item 3	Item 4	Item 5
a_j	.60	1.20	1.00	1.40	1.00
b_j	-1.70	-1.00	.80	1.30	1.40
c_j	.20	.20	.25	.25	.20

Solve for $\tau_X = 2$ Using Starting Value $\theta_i = -2$

Iteration		Item 1	Item 2	Item 3	Item 4	Item 5	sum	θ_i^+
1	p_{ij}	.5393	.2921	.2564	.2503	.2025	1.5405	$-.7941$
	p'_{ij}	.1993	.1662	.0107	.0007	.0042	.3811	
2	p_{ij}	.7727	.6828	.2968	.2551	.2187	2.2261	-1.1295
	p'_{ij}	.1660	.3905	.0746	.0121	.0311	.6743	
3	p_{ij}	.7132	.5475	.2772	.2523	.2107	2.0009	-1.1308
	p'_{ij}	.1877	.4010	.0446	.0055	.0180	.6566	
4	p_{ij}	.7130	.5469	.2771	.2523	.2107	2.0000	-1.1308
	p'_{ij}	.1877	.4008	.0445	.0055	.0179	.6564	

Therefore, $\tau_X = 2$ corresponds to $\theta_i = -1.1308$.

Form Y Item Parameters

Item Parameter	Item 1	Item 2	Item 3	Item 4	Item 5
a_j	.70	.80	1.30	.90	1.10
b_j	-1.50	-1.20	.00	1.40	1.50
c_j	.20	.25	.20	.25	.20

Form Y True Score Equivalent of $\theta_i = -1.1308$

	Item 1	Item 2	Item 3	Item 4	Item 5	τ_Y
p_{ij}	.6865	.6426	.2607	.2653	.2058	2.0609

Therefore, $\tau_X = 2$ corresponds to $\tau_Y = 2.0609$.

assume that the parameters for the two forms are on the same scale. Table 6.3 shows how to find a Form Y equivalent of a Form X score of 2. The item parameters for Form X are presented in the top portion of the table. To find the Form Y equivalent, the θ_i must be found that corresponds to a Form X score of 2. That is, the θ_i must be found such that, when it is substituted into the right-hand side of equation (6.16), it results in a 2 on the left-hand side.

The second portion of Table 6.3 illustrates how to find θ_i using the Newton-Raphson method. First, a starting value of $\theta_i^- = -2.0$ is chosen (this value is an initial guess). Using $\theta_i^- = -2$, the item characteristic curve value from equation (6.1) is calculated for each item. For example, the probability of an examinee with an ability of -2 correctly answering item 1 is .5393. The first derivative is also calculated. For example, for the first item, the first derivative of this item evaluated at an ability of -2 can be calculated using equation (6.23) as

$$p'_{ij} = \frac{1.7(.60)(1 - .5393)(.5393 - .20)}{(1 - .20)} = .1993,$$

which is also presented in the table.

Next, $func(\theta_i^-)$ from equation (6.21) is calculated using 2 for τ_X and the tabled value of 1.5405 as the sum of the item characteristic curves at an ability of -2. So, $func(\theta_i^-) = 2 - 1.5405$. Then, the negative of the sum of the first derivatives is $func'(\theta_i^-) = -.3811$ from equation (6.22). Finally, using equation (6.20), the updated ability is

$$\theta_i^+ = \theta_i^- - \frac{func(\theta_i^-)}{func'(\theta_i^-)} = -2 - \frac{2 - 1.5405}{-.3811} = -.7943.$$

The value of $-.7943$ differs in the fourth decimal place from the tabled value because of rounding error; the tabled value is more accurate. The value of $-.7941$ then is used as θ_i^- in the next iteration. The iterations continue until the values of θ stabilize. Note that after the fourth iteration, θ_i^+ equals θ_i^+ after the third iteration, to four decimal places. Also, the sum of the p_{ij} is 2.0000 when $\theta_i = -1.1308$. Thus, a true score of 2 on Form X corresponds to a θ_i of -1.1308.

The Form Y equivalent of a Form X score of 2 is found next. The Form Y item parameter estimates are needed and are shown in Table 6.3. (Note that the item parameters for Form X and Form Y must be on the same θ-scale.) To find the Form Y equivalent of a Form X score of 2, calculate the value of the item characteristic curve for each Form Y item at $\theta_i = -1.1308$ and sum these values over items. This process is illustrated at the bottom of the table. As shown, a score of 2 on Form X corresponds to a score of 2.0609 on Form Y.

Using the procedures outlined, the reader can verify that a true score of 3 on Form X corresponds to a θ_i of .3655 and a Form Y true score of 3.2586. Also, a true score of 4 on Form X corresponds to a θ_i of 1.3701 and a Form Y true score of 4.0836. Note that a Form X true score of 1 is below the sum of the c-parameters for that form, so the Form Y true score equivalent of a Form X true score of 1 cannot be calculated by the methods described so far.

Sometimes Form Y true score equivalents of all Form X integer scores that are between the sum of the c-parameters and all correct need to be found. The recommended procedure for finding these is to begin with the smallest Form X score that is greater than the sum of the c-parameters. Use a small value of θ as a starting value (e.g., $\theta_i^- = -3$), and then solve for the Form Y true score equivalent. The θ that results from this process can be used as the starting value for solving for the next highest true score. This process continues for all integer true scores on Form X that are below a score of all correct. Sometimes even this procedure can have convergence problems. In this case, the starting values might need to be modified or the modified Newton-Raphson method described by Press *et al.* (1989) could be tried.

Using True Score Equating with Observed Scores

The true score relationship is appropriate for equating true scores on Form X to true scores on Form Y. However, true scores of examinees are never known, because they are parameters. In practice, the true score equating relationship often is used to convert number-correct observed scores on Form X to number-correct observed scores on Form Y. However, no theoretical reason exists for treating scores in this way. Rather, doing so has been justified in item response theory by showing that the resulting true score conversions are similar to observed score conversions (Lord and Wingersky, 1984).

Recall from equation (6.18) that the lowest possible true score for the three-parameter IRT model is the sum of the c_j, not 0. Therefore, when using true score equating with observed scores, a procedure is needed for converting Form X scores outside the range of possible true scores on Form X. Lord (1980) and Kolen (1981) presented ad hoc procedures to handle this problem. The Kolen (1981) ad hoc procedure is as follows:

1. Set a score of 0 on Form X equal to a score of 0 on Form Y.
2. Set a score of the sum of the c_j-parameters on Form X equal to the sum of the c_j-parameters on Form Y.
3. Use linear interpolation to find equivalents between these points.
4. Set a score of K_X on Form X equal to a score of K_Y on Form Y.

To formalize this procedure, define τ_X^* as a score outside the range of possible true scores, but within the range of possible observed scores. Equivalents then are defined by the following equation:

$$irt_Y(\tau_X^*) = \frac{\sum_{j:Y} c_j}{\sum_{j:X} c_j} \tau_X^*, \qquad 0 \le \tau_X^* \le \sum_{j:X} c_j,$$

$$= K_Y, \qquad \tau_X^* = K_X \qquad (6.24)$$

The use of Kolen's (1981) ad hoc procedure can be illustrated using the hypothetical example presented in Table 6.3. For the item parameters presented, the sum of the c_j-parameters is 1.1 for Form X and 1.1 for Form Y. To apply the procedure to find Form Y equivalents of Form X scores at or below 1.1, take

$$irt_Y(\tau_X^*) = \frac{\sum_{j:Y} c_j}{\sum_{j:X} c_j} \tau_X^* = \frac{1.1}{1.1} \tau_X^* = \tau_X^*.$$

Thus, for example, a score of 1 on Form X is considered to be equivalent to a score of 1 on Form Y. Note that a score of 1 on Form X would have been considered to be equivalent to a score other than 1 on Form Y if the sum of the c_j-parameters was different for the two forms.

In practice, for IRT true score equating, estimates of the item parameters are used to produce an *estimated true score relationship*. Then the estimated true score conversion is applied to the observed scores.

Equating Observed Scores

Another procedure, *IRT observed score equating*, uses the IRT model to produce an estimated distribution of observed number-correct scores on each form, which then are equated using equipercentile methods. For Form X, the *compound binomial distribution* (see Lord and Wingersky, 1984) is used to generate the distribution of observed number-correct scores for examinees of a given ability. These observed score distributions then are cumulated over a population of examinees to produce a number-correct observed score distribution for Form X. Similar procedures are followed to produce a number-correct observed score distribution for Form Y. The resulting number-correct score distributions then are equated using conventional equipercentile methods. IRT observed score equating requires explicit specification of the distribution of ability in the population of examinees.

Consider a group of examinees all of ability θ_i who have been administered a three-item test with p_{ij} defined by equation (6.1). Assuming local independence, the probability that examinees of ability equal to θ_i will incorrectly answer all three items and earn a raw score of 0 is $f(x = 0|\theta_i) = (1 - p_{i1})(1 - p_{i2})(1 - p_{i3})$. To earn a score of 1, an examinee could answer item 1 correctly and items 2 and 3 incorrectly, or the examinee could answer item 2 correctly and items 1 and 3 incorrectly, or the examinee could answer item 3 correctly and items 1 and 2 incorrectly. That is, there are three ways to earn a score of 1 on a three-item test. The probability of earning a 1 is as follows:

$$f(x = 1|\theta_i) = p_{i1}(1 - p_{i2})(1 - p_{i3}) + (1 - p_{i1})p_{i2}(1 - p_{i3})$$
$$+(1 - p_{i1})(1 - p_{i2})p_{i3}.$$

The probabilities of correctly answering two and three items can be constructed similarly as follows:

$$f(x = 2|\theta_i) = p_{i1}p_{i2}(1 - p_{i3}) + p_{i1}(1 - p_{i2})p_{i3} + (1 - p_{i1})p_{i2}p_{i3},$$

and

$$f(x = 3|\theta_i) = p_{i1}p_{i2}p_{i3}.$$

Based on the hypothetical example in Table 6.1, for examinees with ability equal to that of Person 1 ($\theta_{I1} = -2.0$),

$$f(x = 0|\theta_1) = (1 - .26)(1 - .27)(1 - .18) = .4430,$$

$$f(x = 1|\theta_1) = (.26)(1 - .27)(1 - .18) + (1 - .26)(.27)(1 - .18)$$
$$+(1 - .26)(1 - .27)(.18)$$
$$= .4167,$$

$$f(x = 2|\theta_1) = (.26)(.27)(1 - .18) + (.26)(1 - .27)(.18) + (1 - .26)(.27)(.18)$$
$$= .1277,$$

$$f(x = 3|\theta_1) = (.26)(.27)(.18) = .0126.$$

Note that these values sum to 1, which is consistent with their being probabilities.

A recursion formula (Lord and Wingersky, 1984) can be used to generalize this procedure to more than three items. To implement the recursion formula, define $f_r(x|\theta_i)$ as the distribution of number-correct scores over the first r items for examinees of ability θ_i. Define $f_1(x = 0|\theta_i) = (1 - p_{i1})$ as the probability of earning a score of 0 on the first item and $f_1(x = 1|\theta_i) = p_{i1}$ as the probability of earning a score of 1 on the first item. For $r > 1$, the recursion formula is as follows:

$$
\begin{aligned}
f_r(x|\theta_i) &= f_{r-1}(x|\theta_i)(1 - p_{ir}), & x = 0 \\
&= f_{r-1}(x|\theta_i)(1 - p_{ir}) + f_{r-1}(x - 1|\theta_i)p_{ir}, & 0 < x < r, \\
&= f_{r-1}(x - 1|\theta_i)p_{ir}, & x = r \quad (6.25)
\end{aligned}
$$

An example of the use of this recursion formula is presented in Table 6.4. An abbreviated notation is used in this table to simplify the presentation. Specifically, θ_i is dropped and p_r means p_{ir}. To find the distribution for a particular value of r, equation (6.25) and Table 6.4 indicate that the distribution for $r - 1$ and the probability of correctly answering item r are needed. Although expressions are only presented up to $r = 4$, the table readily generalizes to higher values of r using the recursion formula. The probabilities listed for the example under $r = 3$ (e.g., .4430, .4167, .1277, and .0126) are identical to results presented earlier.

The procedures presented thus far give the observed score distribution for examinees of a given ability. To find the observed score distribution for examinees of various abilities, the observed score distribution for examinees at each ability is found and then these are accumulated. When the ability distribution is continuous, then

$$f(x) = \int_\theta f(x|\theta)\psi(\theta)d\theta, \quad (6.26)$$

where $\psi(\theta)$ is the distribution of θ.

Table 6.4. IRT Observed Score Distribution Recursion Formula Example.

r	x	$f_r(x)$ for $r \leq 4$			Example (Using Table 6.1 test for persons with $\theta_i = -2$)		
1	0	$f_1(0) = (1 - p_1)$			$= (1 - .26)$		$= .74$
	1	$f_1(1) = p_1$					$= .26$
2	0	$f_2(0) = f_1(0)(1 - p_2)$			$= .74(1 - .27)$		$= .5402$
	1	$f_2(1) = f_1(1)(1 - p_2)$	$+$	$f_1(0)p_2$	$= .26(1 - .27)$	$+ .74(.27)$	$= .3896$
	2	$f_2(2) =$		$f_1(1)p_2$	$=$	$.26(.27)$	$= .0702$
3	0	$f_3(0) = f_2(0)(1 - p_3)$			$= .5402(1 - .18)$		$= .4430$
	1	$f_3(1) = f_2(1)(1 - p_3)$	$+$	$f_2(0)p_3$	$= .3896(1 - .18)$	$+ .5402(.18)$	$= .4167$
	2	$f_3(2) = f_2(2)(1 - p_3)$	$+$	$f_2(1)p_3$	$= .0702(1 - .18)$	$+ .3896(.18)$	$= .1277$
	3	$f_3(3) =$		$f_2(2)p_3$	$=$	$.0702(.18)$	$= .0126$
4	0	$f_4(0) = f_3(0)(1 - p_4)$					
	1	$f_4(1) = f_3(1)(1 - p_4)$	$+$	$f_3(0)p_4$			
	2	$f_4(2) = f_3(2)(1 - p_4)$	$+$	$f_3(1)p_4$			
	3	$f_4(3) = f_3(3)(1 - p_4)$	$+$	$f_3(2)p_4$			
	4	$f_4(4) =$		$f_3(3)p_4$			

To implement this procedure in practice, some method is needed to perform the integration in equation (6.26). Some form of numerical integration is one possibility. When BILOG is used, the distribution of ability typically is characterized by a discrete distribution on a finite number of equally spaced points as a method of approximating the integral. Using this characterization,

$$f(x) = \sum_i f(x|\theta_i)\psi(\theta_i). \tag{6.27}$$

When the distribution of ability is characterized by a finite number of abilities for N examinees, then

$$f(x) = \frac{1}{N}\sum_i f(x|\theta_i). \tag{6.28}$$

This characterization can be used, for example, with a set of abilities that are estimated using LOGIST or BILOG.

To conduct observed score equating, observed score distributions are found for Form X and for Form Y. For example, assume that the characterization of the ability distribution associated with equation (6.27) is used. The following distributions could be specified using this equation:

1. $f_1(x) = \sum_i f(x|\theta_i)\psi_1(\theta_i)$ is the Form X distribution for Population 1.
2. $f_2(x) = \sum_i f(x|\theta_i)\psi_2(\theta_i)$ is the Form X distribution for Population 2.
3. $g_1(y) = \sum_i g(y|\theta_i)\psi_1(\theta_i)$ is the Form Y distribution for Population 1.
4. $g_2(y) = \sum_i g(y|\theta_i)\psi_2(\theta_i)$ is the Form Y distribution for Population 2.

These quantities then are weighted using synthetic weights described in Chapters 4 and 5 to obtain the distributions of X and Y in the synthetic population. Conventional equipercentile methods then are used to find score equivalents.

When BILOG is used, the number-correct observed score distributions can be estimated by using the estimated posterior distribution of ability in place of $\psi(\theta_i)$ in equation (6.27) along with estimates of $f(x|\theta_i)$ based on substituting estimates for parameters in equation (6.25). An alternative is to use the set of estimated abilities in place of the abilities in equation (6.28). However, the use of estimates of θ might create systematic distortions in the estimated distributions and lead to inaccurate equating (Han, 1993; Lord, 1982a). The estimation of observed score distributions using IRT models is an area for further research.

IRT True Score Versus IRT Observed Score Equating

Compared to IRT observed score equating, IRT true score equating has the advantages of (a) easier computation and (b) a conversion that does not depend on the distribution of ability. However, IRT true score equating has the disadvantage that it equates true scores, which are not available in practice. No justification exists for applying the true score relationship to observed scores. Also, with the three-parameter logistic model, equivalents are undefined at very low scores and at the top number-correct score.

IRT observed score equating has the advantage that it defines the equating relationship for observed scores. Also, assuming reasonable model fit, the distribution of Form X scores converted to the Form Y scale is approximately equal to the distribution of Form Y scores for the synthetic population of examinees. There is no theoretical reason to expect this property to hold for IRT true score equating. Also, using posterior distributions of θ from BILOG, the computational burden of IRT observed score equating is acceptable.

IRT observed score and IRT true score equating methods were found by Kolen (1981) and Han (1993) to produce somewhat different results using the random groups design with achievement tests. However, Lord and Wingersky (1984) concluded that the two methods produce very similar results in a study using the common-item nonequivalent groups design in the SAT. Further research comparing these methods is need.

Larger differences between IRT true and observed score equating might be expected to occur near a number-correct score of all correct and near number-correct scores below the sum of the c-parameter estimates, because these are the regions where IRT true score equating does not produce equivalents. In practice, both methods should be applied with special attention paid to equating results near these regions. Procedures for choosing among the results from equating methods are considered in Chapter 8.

Illustrative Example

The real data example from Chapters 4 and 5 is used to illustrate some aspects of IRT equating, using the common-item nonequivalent groups design. Two forms of a 36-item multiple-choice test, Form X and Form Y, are used in this example. Every third item on the test forms is a common item, and the common items are in the same position on each form. Thus, items $3, 6, 9, \ldots, 36$ on each form represent the 12 common items. Form X was administered to 1655 examinees and Form Y to 1638 examinees. As was indicated in Chapters 4 and 5, the examinees who were administered Form X had a number-correct score mean of 5.11 and a standard deviation of 2.38 on the common items. The examinees who were administered Form Y had a number-correct score mean of 5.87 and a standard deviation of 2.45 on the common items. Thus, on the common items, the group taking Form Y was higher achieving than the group taking Form X.

Item Parameter Estimation and Scaling

Item parameters were estimated using BILOG 3 separately for each form. (Default BILOG parameter settings were used, except for the FLOAT option.) The parameter estimates are given in Table 6.5. The proportion of examinees correctly answering each item (p-value) is also presented.

The Form X item parameter estimates need to be rescaled. The Macintosh computer program ST that is described in Appendix B was used to conduct the scaling. The common items are tabulated separately in the upper portion of Table 6.6. Because the items appeared in identical positions in the two forms, item 3 on Form X is the same as item 3 on Form Y, and so forth.

The parameter estimates for the common items are plotted in Figure 6.2 to look for outliers—items with estimates that do not appear to lie on a straight line. In this figure, one item appears to be an outlier for the a-parameter estimate. This item, which is item 27, has a-parameter estimates of 1.8811 on Form X and 1.0417 on Form Y. Because item 27 appears to function differently in the two forms, this item might need to be eliminated from the common-item set. (The c-parameter estimates for item 21 might also be judged to be an outlier, so that item 21 could be considered for elimination as well. This item was not considered for elimination in the present example because it does not seem to be as clearly an outlier as item 27.) Removal of items that appear to be outliers is clearly a judgmental process.

Table 6.5. Item Parameter Estimates for Common-Item Equating.

	Form X				Form Y			
Item	p-value	$\hat{a}$	$\hat{b}$	$\hat{c}$	p-value	$\hat{a}$	$\hat{b}$	$\hat{c}$
1	.8440	.5496	−1.7960	.1751	.8527	.8704	−1.4507	.1576
2	.6669	.7891	−.4796	.1165	.6161	.4628	−.4070	.1094
3	**.7025**	**.4551**	**−.7101**	**.2087**	**.7543**	**.4416**	**−1.3349**	**.1559**
4	.5405	1.4443	.4833	.2826	.7145	.5448	−.9017	.1381
5	.6723	.9740	−.1680	.2625	.8295	.6200	−1.4865	.2114
6	**.7412**	**.5839**	**−.8567**	**.2038**	**.7946**	**.5730**	**−1.3210**	**.1913**
7	.5895	.8604	.4546	.3224	.6351	1.1752	.0691	.2947
8	.6475	1.1445	−.1301	.2209	.6094	.4450	.2324	.2723
9	**.5816**	**.7544**	**.0212**	**.1600**	**.6852**	**.5987**	**−.7098**	**.1177**
10	.5296	.9170	1.0139	.3648	.6644	.8479	−.4253	.1445
11	.4825	.9592	.7218	.2399	.7439	1.0320	−.8184	.0936
12	**.5574**	**.6633**	**.0506**	**.1240**	**.6076**	**.6041**	**−.3539**	**.0818**
13	.5411	1.2324	.4167	.2535	.5685	.8297	−.0191	.1283
14	.4051	1.0492	.7882	.1569	.6094	.7252	−.3155	.0854
15	**.4770**	**1.0690**	**.9610**	**.2986**	**.5532**	**.9902**	**.5320**	**.3024**
16	.5139	.9193	.6099	.2521	.5092	.7749	.5394	.2179
17	.5175	.8935	.5128	.2273	.4786	.5942	.8987	.2299
18	**.4825**	**.9672**	**.1950**	**.0535**	**.5587**	**.8081**	**−.1156**	**.0648**
19	.4909	.6562	.3853	.1201	.6265	.9640	−.1948	.1633
20	.4081	1.0556	.9481	.2036	.4908	.7836	.3506	.1299
21	**.3404**	**.3479**	**2.2768**	**.1489**	**.3655**	**.4140**	**2.5538**	**.2410**
22	.4299	.8432	1.0601	.2332	.5905	.7618	−.1581	.1137
23	.3839	1.1142	.5826	.0644	.5092	1.1959	.5056	.2397
24	**.4063**	**1.4579**	**1.0241**	**.2453**	**.4774**	**1.3554**	**.5811**	**.2243**
25	.3706	.5137	1.3790	.1427	.4976	1.1869	.6229	.2577
26	.3077	.9194	1.0782	.0879	.5055	1.0296	.3898	.1856
27	**.2956**	**1.8811**	**1.4062**	**.1992**	**.3771**	**1.0417**	**.9392**	**.1651**
28	.2612	1.5045	1.5093	.1642	.3851	1.2055	1.1350	.2323
29	.2727	.9664	1.5443	.1431	.3894	.9697	.6976	.1070
30	**.1820**	**.7020**	**2.2401**	**.0853**	**.2231**	**.6336**	**1.8960**	**.0794**
31	.3059	1.2651	1.8759	.2443	.3166	1.0822	1.3864	.1855
32	.2146	.8567	1.7140	.0865	.3356	1.0195	.9197	.1027
33	**.1826**	**1.4080**	**1.5556**	**.0789**	**.2634**	**1.1347**	**1.0790**	**.0630**
34	.1814	.5808	3.4728	.1399	.1760	1.1948	1.8411	.0999
35	.1288	.9257	3.1202	.1090	.1424	1.1961	2.0297	.0832
36	**.1530**	**1.2993**	**2.1589**	**.1075**	**.1950**	**.9255**	**2.1337**	**.1259**

Note: Common-item numbers and parameter estimates are in **boldface** type.

Table 6.6. Common-Item Parameter Estimates and Scaling Constants.

	Form X				Form Y			
Item	p-value	$\hat{a}$	$\hat{b}$	$\hat{c}$	p-value	$\hat{a}$	$\hat{b}$	$\hat{c}$
3	.7025	.4551	−.7101	.2087	.7543	.4416	−1.3349	.1559
6	.7412	.5839	−.8567	.2038	.7946	.5730	−1.3210	.1913
9	.5816	.7544	.0212	.1600	.6852	.5987	−.7098	.1177
12	.5574	.6633	.0506	.1240	.6076	.6041	−.3539	.0818
15	.4770	1.0690	.9610	.2986	.5532	.9902	.5320	.3024
18	.4825	.9672	.1950	.0535	.5587	.8081	−.1156	.0648
21	.3404	.3479	2.2768	.1489	.3655	.4140	2.5538	.2410
24	.4063	1.4579	1.0241	.2453	.4774	1.3554	.5811	.2243
27	.2956	1.8811	1.4062	.1992	.3771	1.0417	.9392	.1651
30	.1820	.7020	2.2401	.0853	.2231	.6336	1.8960	.0794
33	.1826	1.4080	1.5556	.0789	.2634	1.1347	1.0790	.0630
36	.1530	1.2993	2.1589	.1075	.1950	.9255	2.1337	.1259
Mean	.4252	.9657	.8602	.1595	.4879	.7934	.4900	.1510
Sd	.1917	.4464	1.0658	.0707	.1960	.2837	1.2458	.0736
	Mean/	Mean/						
	Sigma	Mean						
$A =$	1.1689	1.2173						
$B =$	−.5156	−.5572						
Eliminating Item #27								
Mean	.4370	.8825	.8106	.1559	.4980	.7708	.4491	.1498
Sd	.1961	.3665	1.0999	.0728	.2018	.2858	1.2935	.0768
	Mean/	Mean/						
	Sigma	Mean						
$A =$	1.1761	1.1449						
$B =$	−.5042	−.4790						

The mean and standard deviation of the item parameter estimates for the common items are shown in Table 6.6. These means and standard deviations were used to estimate the A- and B-constants for transforming the θ-scale of Form X to the θ-scale of Form Y. For example, using equations (6.8a) and (6.9a) for the mean/sigma method,

$$A = \frac{1.2458}{1.0658} = 1.1689 \quad \text{and} \quad B = .4900 - (1.1689).8602 = -.5155.$$

The B-value differs from the tabled value in the fourth decimal place because of rounding error; the tabled values are more accurate.

Because item 27 appeared to be an outlier, the A- and B-constants were estimated again eliminating item 27. The means and standard deviations after eliminating this item are shown in Table 6.6 as are the new A- and

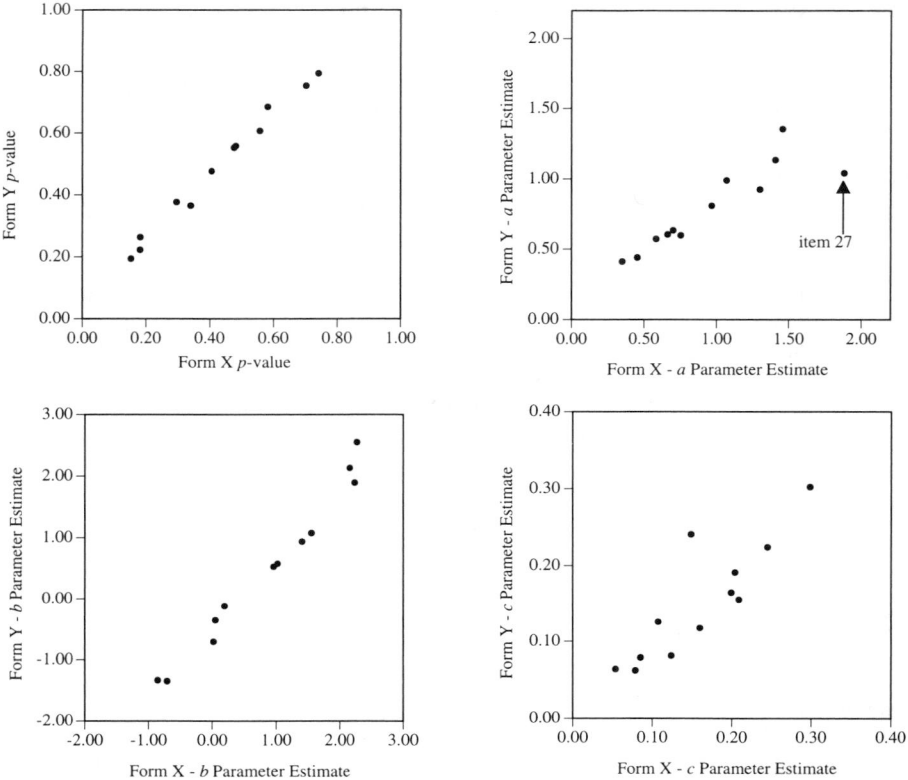

Figure 6.2. Plots of item parameter estimates on Form X versus Form Y.

B-constants. Eliminating item 27 results in the estimates of the *A*- and *B*-constants for the mean/sigma and mean/mean methods being closer to one another than when item 27 is included. In the present example, the scalings based on removing item 27 only are considered for ease of exposition. In practice, however, equating based on scalings with item 27 removed and included could be conducted and the results of the equating compared.

The rescaled Form X item parameter estimates for the common items are shown in Table 6.7. The mean/sigma method was used, excluding item 27. Because the Form Y item parameter estimates are not being transformed, they are identical to those in Table 6.6. To verify the tabled Form X *b*-parameter estimate for item 3, take $1.1761(-.7101) - .5042 = -1.3393$, which differs in the fourth decimal place because of rounding. To find the tabled Form X *a*-parameter estimate for this item, take $.4551/1.1761 = .3870$.

The means and standard deviations of the rescaled parameter estimates for the common items are shown at the bottom of Table 6.7. Because the mean/sigma method was used, the mean and standard deviation of the

Table 6.7. Common-Item Parameter Estimates Rescaled Using Mean/Sigma Method's *A* and *B* with All Common Items (Excluding Item 27).

	Form X				Form Y			
Item	*p*-value	$\hat{a}$	$\hat{b}$	$\hat{c}$	*p*-value	$\hat{a}$	$\hat{b}$	$\hat{c}$
3	.7025	.3870	−1.3394	.2087	.7543	.4416	−1.3349	.1559
6	.7412	.4965	−1.5118	.2038	.7946	.5730	−1.3210	.1913
9	.5816	.6414	−.4793	.1600	.6852	.5987	−.7098	.1177
12	.5574	.5640	−.4447	.1240	.6076	.6041	−.3539	.0818
15	.4770	.9089	.6260	.2986	.5532	.9902	.5320	.3024
18	.4825	.8224	−.2749	.0535	.5587	.8081	−.1156	.0648
21	.3404	.2958	2.1735	.1489	.3655	.4140	2.5538	.2410
24	.4063	1.2396	.7002	.2453	.4774	1.3554	.5811	.2243
30	.1820	.5969	2.1304	.0853	.2231	.6336	1.8960	.0794
33	.1826	1.1972	1.3253	.0789	.2634	1.1347	1.0790	.0630
36	.1530	1.1048	2.0349	.1075	.1950	.9255	2.1337	.1259
Mean	.4370	.7504	.4491	.1559	.4980	.7708	.4491	.1498
Sd	.1961	.3116	1.2935	.0728	.2018	.2858	1.2935	.0768

rescaled *b*-parameter estimates for Form X are equal to those for Form Y. Note, however, that the mean of the *a*-parameter estimates for Form X differs from the mean for Form Y. These two means would have been the same if the mean/mean method was used. How would the means and standard deviations of the parameter estimates compare if a characteristic curve method was used? All of these statistics would likely differ from Form X to Form Y. These results illustrate that the different methods of scaling using parameter estimates can produce different results, which in turn would affect the equating.

Test characteristic curves for the common items after the common-item parameter estimates were placed on the same θ-scale using the mean/sigma method are shown in Figure 6.3. The Form X curve is the test characteristic curve for the 11 common items (excluding item 27) estimated using BILOG on the examinees who took Form X. The Form Y curve is the test characteristic curve for these same items estimated using BILOG on the examinees who took Form Y. In general, the test characteristic curves appear to be similar. However, if the Stocking-Lord method had been used, then these test characteristic curves likely would have been even closer, because the Stocking-Lord procedure finds the *A*- and *B*-constants that minimize the difference between these characteristic curves. However, if the Stocking-Lord method had been used, then the means and standard deviations of both the *a*-parameter and the *b*-parameter estimates for the common items would have differed from Form X to Form Y.

Even after transformation to a common scale, however, the common items have different parameter estimates on Form X than they do on Form

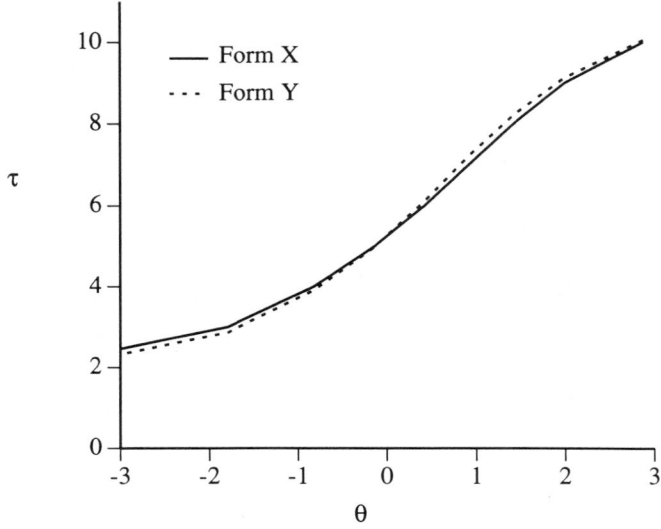

Figure 6.3. Estimated test characteristic curves for common items.

Y. These differences must be due to error in estimating the item parameters or failure of the IRT model to hold, because the items are identical on the two forms.

The rescaled Form X item parameter estimates for all of the items are shown in Table 6.8. The same transformation that is used for the common items on Form X is also used for the other items.

IRT True Score Equating

The rescaled item parameter estimates then are used to estimate the true score equating function; this process is illustrated in Table 6.9 and Figure 6.4. Figure 6.4 presents the test characteristic curves for Form X and Form Y, and Table 6.9 presents the conversion table. The equating was conducted using the *PIE* Macintosh computer program described in Appendix B. Suppose, for example, interest focuses on finding the Form Y equivalent of a Form X score of 25. First, find the θ that is associated with a true score of 25. In Figure 6.4, begin at a vertical axis value of 25 and go over to the Form X curve. Going down to the horizontal axis, the score of 25 is associated with a θ of approximately 1.1. With greater precision, from Table 6.9 this θ is 1.1022. This tabled value was found using the Newton-Raphson procedure that was described earlier. Next, find the Form Y true score that is associated with a θ of 1.1022. Graphically, this Form Y score is approximately 26.4. With greater precision, from Table 6.9 this true score is 26.3874. These procedures are repeated with each of the Form X

Table 6.8. Form X Item Parameter Estimates Rescaled with the Mean/Sigma Method's A and B Using All Common Items Except Item 27.

	Form X				Form Y			
Item	p-value	$\hat{a}$	$\hat{b}$	$\hat{c}$	p-value	$\hat{a}$	$\hat{b}$	$\hat{c}$
1	.8440	.4673	−2.6165	.1751	.8527	.8704	−1.4507	.1576
2	.6669	.6709	−1.0683	.1165	.6161	.4628	−.4070	.1094
3	**.7025**	**.3870**	**−1.3394**	**.2087**	**.7543**	**.4416**	**−1.3349**	**.1559**
4	.5405	1.2280	.0641	.2826	.7145	.5448	−.9017	.1381
5	.6723	.8282	−.7018	.2625	.8295	.6200	−1.4865	.2114
6	**.7412**	**.4965**	**−1.5118**	**.2038**	**.7946**	**.5730**	**−1.3210**	**.1913**
7	.5895	.7316	.0304	.3224	.6351	1.1752	.0691	.2947
8	.6475	.9731	−.6572	.2209	.6094	.4450	.2324	.2723
9	**.5816**	**.6414**	**−.4793**	**.1600**	**.6852**	**.5987**	**−.7098**	**.1177**
10	.5296	.7797	.6882	.3648	.6644	.8479	−.4253	.1445
11	.4825	.8156	.3446	.2399	.7439	1.0320	−.8184	.0936
12	**.5574**	**.5640**	**−.4447**	**.1240**	**.6076**	**.6041**	**−.3539**	**.0818**
13	.5411	1.0479	−.0141	.2535	.5685	.8297	−.0191	.1283
14	.4051	.8921	.4228	.1569	.6094	.7252	−.3155	.0854
15	**.4770**	**.9089**	**.6260**	**.2986**	**.5532**	**.9902**	**.5320**	**.3024**
16	.5139	.7817	.2130	.2521	.5092	.7749	.5394	.2179
17	.5175	.7598	.0989	.2273	.4786	.5942	.8987	.2299
18	**.4825**	**.8224**	**−.2749**	**.0535**	**.5587**	**.8081**	**−.1156**	**.0648**
19	.4909	.5580	−.0511	.1201	.6265	.9640	−.1948	.1633
20	.4081	.8976	.6109	.2036	.4908	.7836	.3506	.1299
21	**.3404**	**.2958**	**2.1735**	**.1489**	**.3655**	**.4140**	**2.5538**	**.2410**
22	.4299	.7169	.7425	.2332	.5905	.7618	−.1581	.1137
23	.3839	.9473	.1809	.0644	.5092	1.1959	.5056	.2397
24	**.4063**	**1.2396**	**.7002**	**.2453**	**.4774**	**1.3554**	**.5811**	**.2243**
25	.3706	.4368	1.1176	.1427	.4976	1.1869	.6229	.2577
26	.3077	.7817	.7639	.0879	.5055	1.0296	.3898	.1856
27	**.2956**	**1.5995**	**1.1495**	**.1992**	**.3771**	**1.0417**	**.9392**	**.1651**
28	.2612	1.2792	1.2708	.1642	.3851	1.2055	1.1350	.2323
29	.2727	.8217	1.3120	.1431	.3894	.9697	.6976	.1070
30	**.1820**	**.5969**	**2.1304**	**.0853**	**.2231**	**.6336**	**1.8960**	**.0794**
31	.3059	1.0757	1.7020	.2443	.3166	1.0822	1.3864	.1855
32	.2146	.7285	1.5115	.0865	.3356	1.0195	.9197	.1027
33	**.1826**	**1.1972**	**1.3253**	**.0789**	**.2634**	**1.1347**	**1.0790**	**.0630**
34	.1814	.4939	3.5801	.1399	.1760	1.1948	1.8411	.0999
35	.1288	.7871	3.1654	.1090	.1424	1.1961	2.0297	.0832
36	**.1530**	**1.1048**	**2.0349**	**.1075**	**.1950**	**.9255**	**2.1337**	**.1259**

Note: Common-item numbers and parameter estimates are in **boldface** type.

Table 6.9. Form Y Equivalents of Form X
Scores Using IRT Estimated True Score
Equating.

Form X Score	θ-Equivalent	Form Y Equivalent
0		.0000
1		.8880
2		1.7760
3		2.6641
4		3.5521
5		4.4401
6		5.3282
7	−4.3361	6.1340
8	−2.7701	7.1858
9	−2.0633	8.3950
10	−1.6072	9.6217
11	−1.2682	10.8256
12	−.9951	12.0002
13	−.7633	13.1495
14	−.5593	14.2803
15	−.3747	15.3995
16	−.2043	16.5135
17	−.0440	17.6271
18	.1088	18.7429
19	.2562	19.8612
20	.3998	20.9793
21	.5409	22.0926
22	.6805	23.1950
23	.8197	24.2806
24	.9598	25.3452
25	1.1022	26.3874
26	1.2490	27.4088
27	1.4031	28.4138
28	1.5681	29.4083
29	1.7491	30.3977
30	1.9533	31.3844
31	2.1916	32.3637
32	2.4824	33.3179
33	2.8604	34.2096
34	3.3992	34.9799
35	4.3214	35.5756
36		36.0000

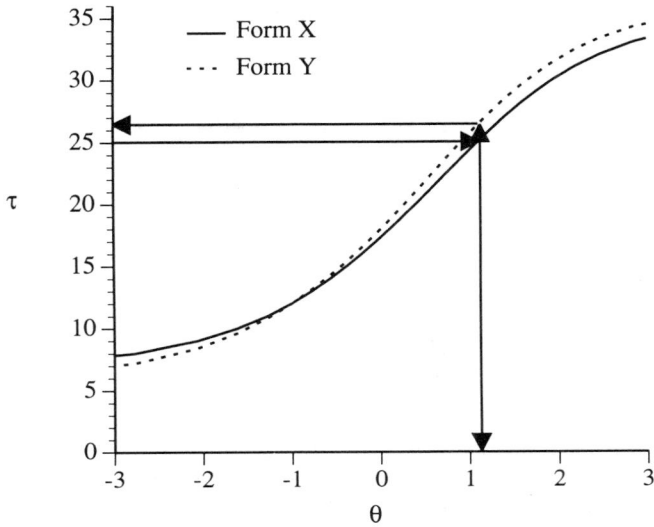

Figure 6.4. Estimated test characteristic curves for Form X and Form Y.

integer scores, and the resulting equivalents are plotted in Figure 6.5. The arrows in this figure illustrate that a Form X score of 25 corresponds to a Form Y score of 26.4 (26.3874 with greater precision). Based on this graph, Form Y is easier than Form X, except at the lower scores, because the curve for true score equating is higher than the line for identity equating at all but the low scores.

In Table 6.9 θ equivalents are not given for very low Form X scores or for a Form X score of 36. The sum of the c-parameter estimates on Form X equals 6.5271, so that true score equivalents for Form X integer scores at or below a score of 6 are undefined. Kolen's (1981) ad hoc method was used to find the Form Y equivalents for these scores.

IRT Observed Score Equating

Estimates of the distribution of θ are needed to conduct observed score equating in this example. The posterior distributions of θ that were estimated using BILOG are presented in Table 6.10. As was noted earlier, BILOG treats the posterior distribution as a discrete distribution on a finite number (10 in this example) of points. For Form X, the posterior distribution of θ needs to be converted to the ability scale of the group that took Form Y. Because the distribution is discrete, the scale conversion can be accomplished by using equation (6.2) to linearly transform the θ-values using the A- and B-constants that were estimated earlier using the mean/sigma methods. For example, to transform the first tabled θ-value using

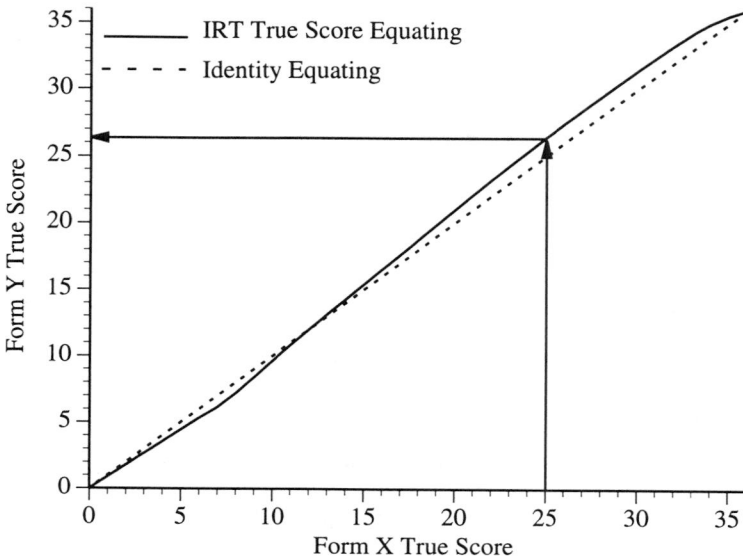

Figure 6.5. Estimated Form Y true score equivalents of Form X true scores using IRT true score equating.

Table 6.10. Distributions of θ Estimated Using BILOG.

Group Taking Form X		Group Taking Form X Converted to Form Y Scale		Group Taking Form Y	
θ_I	$\widehat{\psi}_1(\theta_I)$	θ_J	$\widehat{\psi}_1(\theta_J)$	θ_J	$\widehat{\psi}_2(\theta_J)$
−4.0000	.000101	−5.2086	.000101	−4.0000	.000117
−3.1110	.002760	−4.1630	.002760	−3.1110	.003242
−2.2220	.030210	−3.1175	.030210	−2.2220	.034490
−1.3330	.142000	−2.0720	.142000	−1.3330	.147100
−.4444	.314900	−1.0269	.314900	−.4444	.314800
.4444	.315800	.0184	.315800	.4444	.311000
1.3330	.154200	1.0635	.154200	1.3330	.152600
2.2220	.035960	2.1090	.035960	2.2220	.034060
3.1110	.003925	3.1546	.003925	3.1110	.002510
4.0000	.000186	4.2001	.000186	4.0000	.000112

the constants from the mean/sigma method, take $1.1761(-4.0000)$ $-.5042 = -5.2086$, which is the tabled value. The discrete densities (ψ) do not need to be transformed.

To continue the equating process, the number-correct observed score distributions need to be estimated for the synthetic group. To simplify the presentation, the synthetic group is chosen to be the group taking Form X, so that $w_1 = 1$. In this case, estimates of $f_1(x)$ and $f_1(y)$ are needed.

The distribution of Form X number-correct scores for Group 1 can be estimated directly from the data. However, equation (6.27) can be used to obtain a smoothed estimate of the distribution of $f_1(x)$ by using (a) the item parameter estimates for Form X converted to the Form Y scale shown in Table 6.8 and (b) the distribution of θ for the group taking Form X converted to the Form Y scale shown in Table 6.10. (In Table 6.10, the distribution of θ is approximated using 10 points, which is the number produced by BILOG if the user does not specify a different number. The use of 10 points reduces the computational time when running BILOG and makes it easier to display the distribution in the present example. However, the distribution of θ can be more accurately represented by 20 or even 40 points.)

The distribution of Form Y number-correct scores in Group 1 is not observed directly. To estimate this distribution use (a) the item parameter estimates for Form Y shown in Table 6.8 and (b) the distribution of θ for the group taking Form X converted to the Form Y scale shown in Table 6.10.

The distributions estimated using the IRT model are shown in Table 6.11 along with the equipercentile equivalents that are obtained using these distributions. The equivalents were calculated using the *PIE* computer program described in Appendix B. (These smoothed distributions are still somewhat irregular, which might be due to the use of only 10 quadrature points in BILOG. For example, modes are present at Form X scores of 11 and 25 and at Form Y scores of 11, 17, and 26.) Moments for these distributions are shown in Table 6.12, where the moments labeled "Actual" are those that came from the data without any IRT estimation. These moments were presented in Chapters 4 and 5. The moments in the next section of Table 6.12 are for the distributions estimated using the IRT model. For example, the mean of 15.8169 for Group 1 on Form X is the mean of the distribution for Group 1 on Form X shown in the second column of Table 6.11. The Group 1 Form X moments from the two sources are quite similar. The actual mean, without any IRT estimation, was 15.8205, whereas the mean for the estimate of the distribution using the IRT model was 15.8169. Similarly, the moments for Group 2 Form Y from the two sources are similar.

Because $w_1 = 1$, the moments for Group 1 are the only ones needed. In Group 1, for example, Form X is $16.1746 - 15.8169 = .3577$ points more difficult than Form Y.

Table 6.11. IRT Observed Score Results
Using $w_1 = 1$.

Score	$\hat{f}_1(x)$	$\hat{g}_1(y)$	$\hat{e}_Y(x)$
0	.0000	.0000	−.3429
1	.0001	.0002	.6178
2	.0005	.0011	1.5800
3	.0018	.0034	2.5457
4	.0050	.0081	3.5182
5	.0110	.0155	4.5021
6	.0201	.0248	5.5042
7	.0315	.0349	6.5309
8	.0437	.0446	7.5848
9	.0548	.0527	8.6604
10	.0626	.0585	9.7464
11	.0660	.0606	10.8345
12	.0651	.0589	11.9282
13	.0615	.0545	13.0431
14	.0579	.0501	14.1945
15	.0560	.0480	15.3672
16	.0555	.0488	16.5109
17	.0541	.0505	17.5953
18	.0498	.0502	18.6416
19	.0424	.0459	19.6766
20	.0338	.0379	20.7364
21	.0271	.0290	21.8756
22	.0240	.0221	23.1020
23	.0245	.0195	24.2897
24	.0261	.0209	25.3624
25	.0262	.0242	26.3651
26	.0233	.0264	27.3440
27	.0182	.0251	28.3226
28	.0132	.0205	29.3203
29	.0102	.0147	30.3521
30	.0092	.0106	31.3787
31	.0087	.0093	32.3473
32	.0072	.0092	33.2818
33	.0049	.0083	34.2001
34	.0027	.0060	35.0759
35	.0012	.0035	35.8527
36	.0003	.0014	36.3904

Table 6.12. Moments for Equating Form X and Form Y.

Group	Score	$\hat{\mu}$	$\hat{\sigma}$	$\widehat{sk}$	$\widehat{ku}$
Actual					
1	X	15.8205	6.5278	.5799	2.7217
2	Y	18.6728	6.8784	.2051	2.3028
Estimated Using IRT Observed Score Methods					
1	X	15.8177	6.5248	.5841	2.7235
1	Y	16.1753	7.1238	.5374	2.5750
2	X	18.0311	6.3583	.2843	2.4038
2	Y	18.6659	6.8788	.2270	2.3056
Group 1 Form X Converted to Form Y Scale Using IRT True,					
IRT Observed, and Frequency Estimation Methods					
1	$\hat{\tau}_Y(x)$	16.1784	7.2038	.4956	2.5194
1	$\hat{e}_Y(x)$ IRT	16.1794	7.1122	.5423	2.5761
1	$\hat{e}_Y(x)$ Freq. Est.	16.8329	6.6017	.4622	2.6229

The bottom portion of Table 6.12 shows the moments of converted scores for Group 1 examinees for IRT true score, IRT observed score, and frequency estimation (from Chapter 5) equating. For example, the mean of the Form X scores converted to the Form Y scale using IRT true score equating is 16.1808; using IRT observed score equating the mean is 16.1807. The mean for frequency estimation equating is 16.8329, which was given in Table 5.10. The moments of converted scores are very similar for the two IRT methods, although the moments differ noticeably from those for frequency estimation.

The conversions are plotted in Figure 6.6. In this plot, the relationship for both IRT methods differs noticeably from the frequency estimation relationship. This difference is likely a result of the very different statistical assumptions used in frequency estimation as compared to IRT. Also, the true and observed score methods relationships are similar over most of the score range. The largest differences occur around the sum of the c-parameter estimates and at the very high scores, which are near the regions of the score scale where true scores are undefined. This figure illustrates that if interest is in accurately estimating equivalents at very high scores or near the sum of the c-parameter estimates, such as when a passing score is at a point in one of these score scale regions, then distinctions between the IRT true and observed score methods need to be considered.

Rasch Equating

The fit of the Rasch model to these data might not be good because these multiple-choice items are possibly subject to the effects of guessing, and the

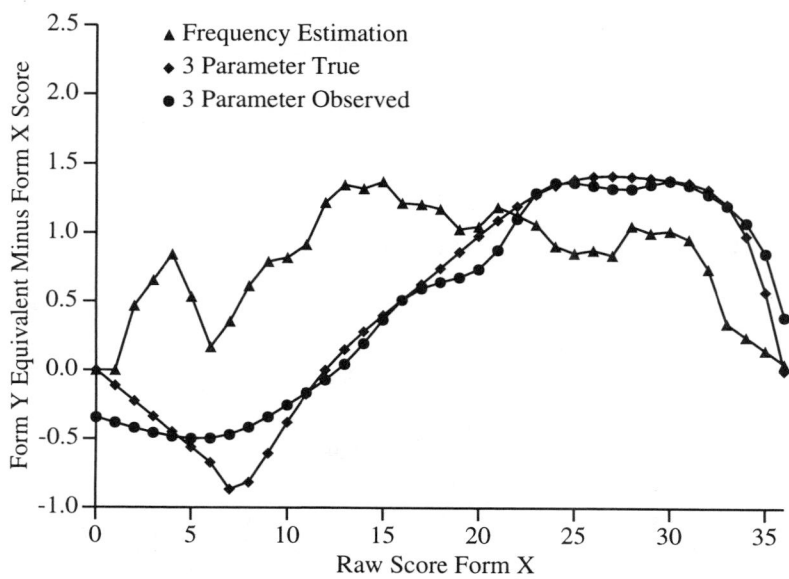

Figure 6.6. Estimated equating relationships for IRT true and IRT observed score equating.

items on these forms are not built to be equally discriminating. Still, these data can be used to examine equating with the Rasch model. As was described earlier in this chapter, the Rasch model can be viewed as a special case of the three-parameter logistic model, where $D = 1.0$, all $a_j = 1$, and all $c_j = 0$.

BILOG was used to estimate the item parameters and posterior distributions of θ using the Rasch model. After being placed on a common scale, the Rasch item difficulty parameter estimates are shown in Table 6.13. The item difficulty estimates for the common items (after scaling) are shown in Figure 6.7. There appear to be no outliers.

Rasch true score and observed score (with $w_1 = 1$) equating results are shown in Table 6.14, and moments are shown in Table 6.15. The equating relationships for the Rasch and three-parameter model are plotted in Figure 6.8.

Overall, the Rasch results appear to differ from the three-parameter model results shown earlier. The Rasch observed score and true score results differ slightly at the lower scores.

These results demonstrate that Rasch observed score equating and Rasch true score equating methods are distinct. Even though Rasch true score equating is typically used in practice, Rasch observed score equating also should be considered, especially when interest is in ensuring comparability of observed score distributions. Issues in choosing among results when conducting equating in practice are discussed in Chapter 8. Because

Table 6.13. Rasch Item Difficulty
Estimates.

Item	Form X	Form Y
1	−2.2593	−2.0388
2	−1.1559	−.5748
3	**−1.3429**	**−1.3275**
4	−.5455	−1.0935
5	−1.1838	−1.8460
6	**−1.5596**	**−1.5901**
7	−.7757	−.6703
8	−1.0582	−.5412
9	**−.7384**	**−.9317**
10	−.4947	−.8215
11	−.2756	−1.2651
12	**−.6246**	**−.5325**
13	−.5484	−.3414
14	.0903	−.5417
15	**−.2502**	**−.2675**
16	−.4217	−.0569
17	−.4386	.0893
18	**−.2757**	**−.2943**
19	−.3150	−.6273
20	.0757	.0306
21	**.4129**	**.6461**
22	−.0285	−.4484
23	.1936	−.0570
24	**.0844**	**.0948**
25	.2594	−.0015
26	.5861	−.0396
27	**.6525**	**.5864**
28	.8508	.5463
29	.7831	.5246
30	**1.3792**	**1.4673**
31	.5958	.9051
32	1.1458	.8025
33	**1.3750**	**1.2106**
34	1.3835	1.8085
35	1.8361	2.0944
36	**1.6137**	**1.6644**

Note: Common-item numbers and parameter
estimates are in **boldface** type.

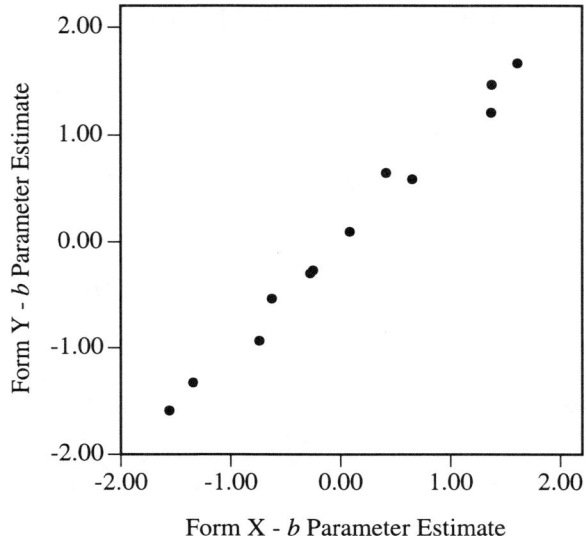

Figure 6.7. Plots of Rasch difficulty estimates on Form X versus Form Y.

the Rasch model has relatively modest sample size requirements, this model might be considered when the sample size is small.

Using IRT Calibrated Item Pools

A *calibrated item pool* is a group of items that have item parameter estimates which have all been placed on the same θ-scale. One potential benefit of using IRT is that calibrated item pools can be constructed, and the item parameter estimates can be used directly in equating. Equating designs that use calibrated item pools often allow for greater flexibility in constructing test forms than the other designs that have been described previously. In this section, the development of IRT calibrated item pools, and how they are used in equating, are described.

Common-Item Equating to a Calibrated Pool

Consider the following simplified example of how an IRT calibrated item pool might evolve. Form Y is constructed and then administered. A transformation is developed to convert scores on Form Y to scale scores, and the item parameters for Form Y also are estimated. So far, equating has not been considered, because there is only a single form.

Table 6.14. Rasch True and Observed
Score Equating Results.

x	$\hat{t}_Y(x)$	$\hat{e}_Y(x)$
0	.0000	.6995
1	1.0780	1.2612
2	2.1550	2.3202
3	3.2280	3.3782
4	4.2953	4.4318
5	5.3563	5.4739
6	6.4107	6.5024
7	7.4586	7.5207
8	8.5002	8.5419
9	9.5358	9.5670
10	10.5655	10.5914
11	11.5896	11.6098
12	12.6083	12.6206
13	13.6218	13.6257
14	14.6302	14.6275
15	15.6336	15.6280
16	16.6322	16.6274
17	17.6260	17.6239
18	18.6150	18.6153
19	19.5994	19.6010
20	20.5793	20.5809
21	21.5546	21.5560
22	22.5257	22.5272
23	23.4925	23.4956
24	24.4554	24.4628
25	25.4147	25.4269
26	26.3707	26.3891
27	27.3241	27.3486
28	28.2754	28.3047
29	29.2255	29.2587
30	30.1757	30.2137
31	31.1275	31.1711
32	32.0827	32.1302
33	33.0439	33.0914
34	34.0142	34.0572
35	34.9978	35.0302
36	36.0000	36.0113

Table 6.15. Moments for Equating Form X and Form Y Using Rasch Equating.

Group	Score	$\hat{\mu}$	$\hat{\sigma}$	$\widehat{sk}$	$\widehat{ku}$
Actual					
1	X	15.8205	6.5278	.5799	2.7217
2	Y	18.6728	6.8784	.2051	2.3028
Estimated Using Rasch Observed Score Methods					
1	X	15.8307	6.4805	.3658	2.5974
1	Y	16.3808	6.4388	.3107	2.5542
2	X	18.1342	6.9291	.1328	2.3458
2	Y	18.6553	6.8406	.0810	2.3438
Group 1 Form X Converted to Form Y Scale Using Rasch True,					
Rasch Observed, and Frequency Estimation Methods					
1	$\hat{\tau}_Y(x)$	16.3554	6.4685	.5212	2.6521
1	$\hat{e}_Y(x)$ Rasch	16.3830	6.4266	.3156	2.5559
1	$\hat{e}_Y(x)$ Freq. Est.	16.8329	6.6017	.4622	2.6229

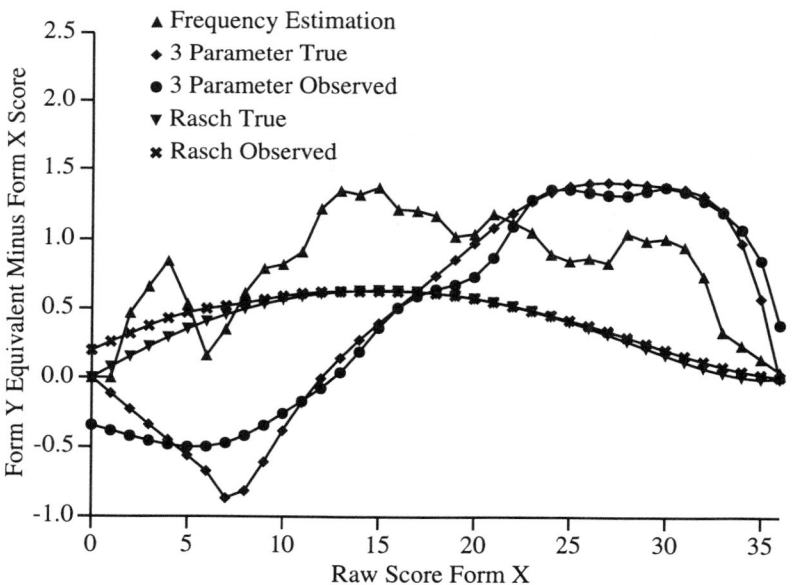

Figure 6.8. Estimated equating relationships for three-parameter and Rasch true and observed score equating.

Form X_1 is constructed next. Form X_1 contains some new items and some items in common with Form Y. Form X_1 is administered to a new group of examinees, and the item parameters are estimated for the new form. Form X_1 can be equated to Form Y using the common-item equating procedures described earlier in this chapter. Along with a conversion table for Form X_1 scores, this common-item equating procedure results in item parameter estimates for Form X_1 which are on the ability scale that was established with Form Y. Actually, there is now a calibrated pool of items, some of which were in Form Y only, some of which were in Form X_1 only, and some of which were in both forms. Refer to Table 6.8. The item parameter estimates in this table are all on the same θ-scale. The items in this table could be considered to be an IRT calibrated item pool.

The use of an IRT calibrated item pool makes possible the use of an equating design that is similar to the common-item nonequivalent groups design. However, in this new design, the common items are drawn from the pool rather than from a single old form. This new design is referred to here as *common-item equating to a calibrated pool*.

To describe this design, suppose that another new form, Form X_2, is constructed. This form consists of a set of common items from the IRT calibrated item pool and some new items. Assume that Form X_2 is administered to a group of examinees. Procedures described earlier can be used to transform the IRT scale that results from estimating Form X_2 item parameters to the scale that was established for the pool. To implement these procedures, the item parameter estimates from the calibrated pool for the common items are considered to be on Scale J, and the item parameter estimates from the calibration of Form X_2 are considered to be on Scale I.

After the new form item parameter estimates are transformed to the θ-scale for the calibrated pool, IRT estimated true score or observed score equating could be conducted. Estimated true score equating for Form X_2 could be implemented as follows. First, find the θ that corresponds to each Form X_2 integer number-correct score. Finding these θ values requires an iterative procedure as described earlier. Second, find the Form Y true score equivalent of each of the θ-values. Following this step results in a true score equating of Form X_2 to Form Y. Use the Form Y scale score transformation to convert the Form X_2 integer number-correct scores to scale scores. These procedures are very similar to what is done in common-item equating, with the major difference being that the common items are taken from a calibrated item pool rather than from a single previously equated form.

After the equating is completed, the new Form X_2 items have item parameter estimates on the θ-scale that was established for the pool. These new items can be added to the IRT calibrated item pool. In this way, the pool can be continually expanded. The common-item sets for new forms are constructed from a continually increasing IRT calibrated item pool. A

Common Item Equating to a Calibrated Pool

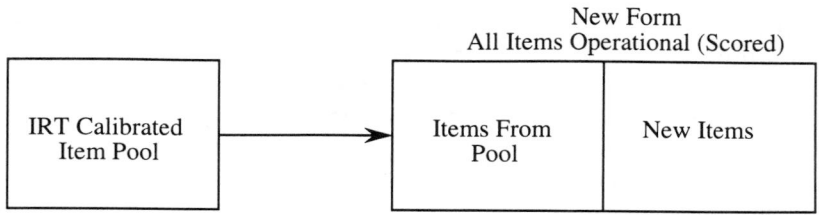

Item Preequating

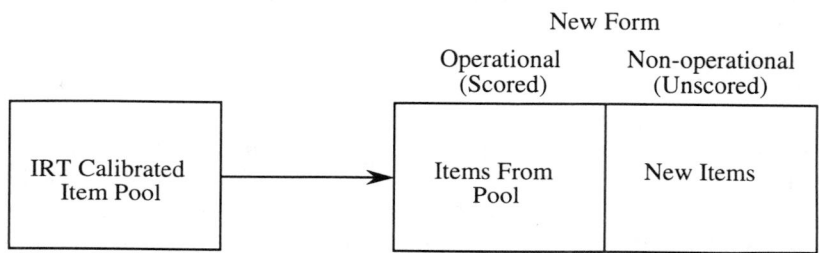

Figure 6.9. Equating designs that use an IRT calibrated item pool.

diagram representing common-item equating to a calibrated pool is presented in the top portion of Figure 6.9.

Many practical issues affect the implementation of IRT calibrated item pools in practice. For example, items might be removed from a pool because their content becomes dated or for test security purposes. Also, when items are used more than once, procedures need to be considered for updating the parameter estimates that are associated with each item in the pool. (For example, two sets of item parameter estimates exist for each common item in Table 6.8.) These are among the issues that are considered when using item pools in a testing program.

Common-item equating to a calibrated pool is more flexible than the common-item nonequivalent groups design, because it allows for the common-item set to be chosen from many previous test forms rather than from a single test form. The effects of violations of IRT assumptions need to be considered, however, when using this design. For example, the position of items can affect their performance. For this reason, the position of each common item on the new form should be as close as possible to its position on the form in which it appeared previously.

Also, real tests are typically not strictly unidimensional. To guard against multidimensionality causing problems with equating, as with traditional equating, the set of common items should be built to the same content specifications, proportionally, as the total test. In this way, the violations of assumptions might affect the common items in the same way that they affect the total scores. Also, a large enough number of common items should be chosen to represent fully the content of the total test.

IRT might be the only procedure that could be used when equating using common-item equating with a calibrated item pool. What if the IRT assumptions are severely violated? Then adequate equating might be impossible with this design. For this reason, if common-item equating to a calibrated item pool is being considered for use, the common-item nonequivalent groups design should be used for a few administrations. The results for the IRT method and traditional methods could be compared and the effects of multidimensionality could be assessed. Switching to common-item equating with a calibrated item pool should be done only if no problems are found with that procedure.

Item Preequating

The use of IRT calibrated item pools also makes an *item preequating design* possible. The goal of item preequating is to be able to produce raw-to-scale score conversion tables *before* a form is administered intact. If a conversion table is produced before the test form is administered, then scores can be reported to examinees without the need to wait for equating to be conducted. Item preequating is possible if the items that contribute to examinees' scores have been previously administered and calibrated.

Consider the following example of how an item preequating design might evolve. Form Y is developed. Form Y contains *operational items*, which are items that contribute to examinees' scores. Form Y also contains *nonoperational items*, which are items that do not contribute to examinees' scores. A conversion of Form Y number-correct scores on the operational items to scale scores is constructed. (The scale could be defined either before or after administration of Form Y.) Form Y is administered and item parameters of the operational and nonoperational items are estimated. At this point, the IRT calibrated item pool consists of the operational and the nonoperational Form Y items that have parameter estimates on the same IRT scale. So far, equating has not been considered, because there is only a single form.

The operational component of a new form, Form X_1, could be constructed from this calibrated pool of items. If so, the operational component of Form X_1 would consist of some combination of Form Y operational and Form Y nonoperational items. Because the operational items in Form X_1 already have estimated item parameters, a conversion table could be constructed for the operational component of Form X_1 before Form X_1 was ever administered intact. That is, the operational portion of Form X_1

could be "preequated." Form X_1 also would contain nonoperational items, which would be newly written items that were not yet part of the item pool. After Form X_1 was administered, the item parameters for all Form X_1 items (operational and nonoperational) could be estimated. The operational Form X_1 items then could be used as the set of common items for transforming the item parameter estimates for the nonoperational items to the θ-scale that was established with Form Y. These nonoperational Form X_1 items then would be added to the calibrated item pool. The operational portion of subsequent test forms would be constructed from the calibrated pool. The nonoperational portion of subsequent test forms would consist of new items, and would be used to expand the item pool continually.

A diagram representing the item preequating design is presented in the bottom portion of Figure 6.9. The item preequating design and common-item designs differ as to whether or not scores on the new items contribute to examinee scores. These designs also differ in whether or not conversion tables can be produced before the new form is administered.

A variety of issues need to be considered when using item preequating in practice. Suppose it is found that the answer key for an operational item needs to be modified (e.g., an item needs to be double-keyed) after the test is administered. Then the preequating would need to be modified.

In addition, to ensure that items will behave the same on each administration, items should appear in contexts and positions when they appear operationally that are similar to those used when they appear nonoperationally. Problems can occur when the nonoperational items are presented in a separate section. For example, Eignor (1985), Eignor and Stocking (1986), and Stocking and Eignor (1986) conducted a series of studies that suggested problems with item preequating if it were used with the SAT. Kolen and Harris (1990) found similar problems with item preequating if it were used with the ACT tests. Context effects and multidimensionality were suggested as reasons for these problems. In situations where the context and positions of items cannot be fixed from one testing to the next, formal studies need to be conducted to make sure that the resulting data will produce fairly robust parameter estimates and equated scores.

The use of item preequating can cause difficulties in estimating the item parameters for the nonoperational items. For example, assume that a test is not strictly unidimensional. In this case, IRT estimation procedures will estimate some composite of multidimensional abilities. The appropriate composite for a test will be the composite for forms that are all built to the test specifications. Estimates that are based only on considering the operational items would estimate this composite. Consider a situation in which the nonoperational items do not represent well the test content specifications. What would happen if the nonoperational item parameters were estimated in the same run of an IRT computer program as the operational items? The composite that is estimated might differ from the composite that would result if only the operational items were used. The use of a different composite might lead to bias in the estimation of item parameters. Although it might be possible to handle estimation problems in practice, the scenario just described suggests that estimation can be quite complicated when estimating parameters for nonoperational items. The problems just described can affect parameter estimates whenever nonoperational items are used in tests that are equated using IRT methods under any of the equating designs described in this book, such as whenever items are tried out (pretested) for inclusion in future forms.

On the surface, item preequating seems straightforward. However, its implementation can be quite complicated. Context effects and dimensionality issues need to be carefully considered, or misleading results will be likely.

Many variations on designs for equating using IRT calibrated item pools exist. For example, new forms might consist of items in common with the pool, new operational items, and nonoperational items. Such pools can be used to produce computer administered and computer adaptive tests (see Chapter 8 for a brief discussion). No attempt will be made here to enumerate all of these variations. However, context effects and dimensionality issues that arise with each variation need to be carefully considered when using item pools in operational testing programs.

Practical Issues and Caveat

We recommend the following when using IRT to conduct equating in practice:

1. Use both the Stocking and Lord and the Haebara methods for scale transformation as well as the mean/sigma and mean/mean methods.
2. When equating number-correct scores, use both IRT true score equating and IRT observed score equating.
3. Whenever possible, conduct traditional equipercentile or linear methods on the forms that are being equated as a check.

Often all of the methods applied provide similar equating results and conversion to scale scores (where appropriate), which is reassuring. However, when the results for the different methods diverge, then a choice must be made about which results to believe. The assumptions required and the effects of poor parameter estimates need to be considered in these cases. The issue of choosing among results in equating is discussed in more detail in Chapter 8.

Unidimensional IRT methods assume that the test forms are unidimensional and that the relationship between ability and the probability of correct response follows a specified form. These requirements are difficult to justify for many educational achievement tests, although the methodology might be robust to violations in many practical situations. The general approach taken in this chapter, and in this book as a whole, is to recommend that equating studies be designed to minimize the effects of violations of assumptions. In this regard, the following advice from Cook and Petersen (1987) is especially relevant:

> Regardless of whether IRT true-score or conventional equating procedures are being used, common items should be selected that are a miniature of the tests to be equated and these items should remain in the same relative position when administered to the new- and old-form groups. It would also seem prudent to evaluate the differential difficulty of the common items administered to the equating samples, particularly when equating samples come from different administration dates. (p. 242)

Exercises

6.1. For the test in Table 6.1, find the probability of correctly answering each of the three items for examinees with ability $\theta_{Ii} = .5$.

6.2. For the test in Table 6.1, find the distribution of observed scores for examinees with ability $\theta_{Ii} = .5$.

6.3. Prove the following:
   a. $A = (b_{Jj} - b_{Jj^*})/(b_{Ij} - b_{Ij^*})$ from equation (6.6). [Hint: The proof can be done by setting up a pair of simultaneous equations for b_{Jj^*} and b_{Jj} using equation (6.4) and solving for A.]
   b. $A = a_{Ij}/a_{Jj}$ from equation (6.6). [Hint: Use equation (6.3).]
   c. $A = \sigma(b_J)/\sigma(b_I)$ in equation (6.8a). [Hint: Use equation (6.4).]
   d. $A = \mu(a_I)/\mu(a_J)$ in equation (6.8b). [Hint: Use equation (6.3).]

6.4. For the test in Table 6.1, what is the value of the test characteristic curve at $\theta_{Ii} = -2.00, .5$, and 1.00? How about at $\theta_{Ji} = -1.50$ and 0.00?

6.5. For the hypothetical example in Table 6.3, conduct observed score equating for a population of examinees with equal numbers of examinees at three score levels: $\theta = -1, 0, 1$. [Hints: Use equation (6.25) to find $f(x|\theta)$ and $g(y|\theta)$ for $\theta = -1$, 0, and 1. Then apply equation (6.26). Finally, do conventional equipercentile equating. *Warning*: This problem requires considerable computation.]

6.6. For the example in Table 6.4, provide the probabilities of earning scores 0, 1, 2, 3, and 4 for $r = 4$ assuming that the probability of correctly answering the fourth item for an examinee of ability $\theta_i = -2$ equals .4.

6.7. For the example in Table 6.2, calculate *Hdiff* and *SLdiff* for $\theta = 1$ on Scale J using the mean/sigma and mean/mean methods.

6.8. Why is IRT equating to a particular old form important if all items are in an IRT calibrated item pool?

6.9. The following are some of the steps involved in equating (assume that number-correct scoring is used and that scale scores are reported to examinees): (a) select the design for data collection and how to implement it; (b) construct, administer, and score the test; (c) estimate equating relationships; (d) construct a conversion table of raw-to-scale scores; (e) apply the conversions to examinees; and (f) report scores to examinees. At each of these steps, what would be the differences in equating a new form using the IRT methods described in Chapter 6 versus the traditional methods described in Chapters 2–5?

CHAPTER 7

Standard Errors of Equating

Two general sources of error in estimating equating relationships are present whenever equating is conducted using data from an equating study: *random error* and *systematic error*. Random equating error is present when the scores of examinees who are considered to be samples from a population or populations of examinees are used to estimate equating relationships. When only random equating error is involved in estimating equating relationships, the estimated equating relationship differs from the equating relationship in the population because data were collected from a sample, rather than from the whole population. If the whole population were available, then no random equating error would be present. Thus, the amount of random error in estimating equating relationships becomes negligible as the sample size increases.

The focus of the present chapter is on estimating random error, rather than systematic error. The following examples of systematic error are intended to illustrate the concept of systematic error, and to distinguish systematic from random error. One way that systematic error can occur in estimating equating relationships is when the estimation method introduces bias in estimating the equating relationship. As was indicated in Chapter 3, smoothing techniques can introduce systematic error—a useful smoothing method results in a reduction in random error that exceeds the amount of systematic error which is introduced. Another way that systematic error in estimating equating relationships can occur is when the statistical assumptions that are made in an equating method are violated. For example, systematic error would be introduced if the Tucker method described in Chapter 4 was used in a situation in which the regression of X on V differed from Population 1 to Population 2. Similarly, systematic error would be introduced if IRT true score equating, as described in

Chapter 6, was used to equate multidimensional tests. A third way that systematic error could occur is if the design used to collect the data for equating is improperly implemented. For example, suppose that in the random groups design, the test center personnel assigned Form X to examinees near the front of the room and Form Y to examinees near the back of the room. This distribution pattern likely would lead to systematic differences between examinees who were administered the forms, unless the examinees were seated randomly. As another example, suppose that in the common-item nonequivalent groups design the common items appeared near the beginning of the test in Form X and near the end of the test in Form Y. In this case, the common items might behave very differently on the two forms, because of the different placement. A fourth way that systematic error could occur is if the group(s) of examinees used to conduct equating differ substantially from the group who takes the equated form. It is important to note that the use of large sample sizes would not reduce the magnitude of these systematic error components. Thus, a major distinguishing factor between random and systematic error is that as the sample size increases, random error diminishes, whereas systematic error does not diminish.

Standard errors of equating index random error in estimating equating relationships only—they are not directly influenced by systematic error. Standard errors of equating approach 0 as the sample size increases, whereas systematic errors of equating are not directly influenced by the sample size of examinees. Only random error in estimating equating relationships is considered in the present chapter; systematic error is a prominent consideration in Chapter 8. In the present chapter, standard errors of equating are defined, and both bootstrap and analytic standard errors are considered. We describe procedures for estimating standard errors of equating for many of the methods described in Chapters 2 through 6, including standard errors for raw and scale scores. We show how the standard errors can be used to estimate sample size requirements and to compare the precision of different equating methods and designs.

Definition of Standard Error of Equating

The standard error of equating is a useful index of the amount of equating error. The standard error of equating is conceived of as the standard deviation of equated scores over hypothetical replications of an equating procedure in samples from a population or populations of examinees. In one hypothetical replication, specified numbers of examinees would be randomly sampled from the population(s). Then the Form Y equivalents of Form X scores would be estimated at various score levels using a particular equating method. The standard error of equating at each score level is the standard deviation, over replications, of the Form Y equivalents at each score level on Form X. Standard errors typically differ across score levels.

To define standard errors of equating, each of the following need to be specified:

- the design for data collection (e.g., common-item nonequivalent groups);
- the definition of equivalents (e.g., equipercentile);
- the method used to estimate the equivalents (e.g., unsmoothed equipercentile);
- the population(s) of examinees;
- the sample sizes (e.g., 2000 for the old form and 3000 for the new form);
- the score level or score levels of interest (e.g., each integer score from 0 to K_X).

Given a particular specification, define $\widehat{eq}_Y(x_i)$ as an estimate of the Form Y equivalent of a Form X score in the sample and define $\mathbf{E}[\widehat{eq}_Y(x_i)]$ as the expected equivalent, where $\mathbf{E}$ is the expectation over random samples from the population(s). For a given sample estimate, equating error at a particular score level on Form X is defined as the difference between the sample Form Y equivalent and the expected equivalent. That is, equating error at score x_i for a given equating is

$$\widehat{eq}_Y(x_i) - \mathbf{E}[\widehat{eq}_Y(x_i)]. \tag{7.1}$$

Suppose that the equating is replicated a large number of times, such that for each replication the equating is based on random samples of examinees from the population(s) of examinees who take Form X and Form Y, respectively. The equating error variance at score point x_i is

$$var[\widehat{eq}_Y(x_i)] = \mathbf{E}\{[\widehat{eq}_Y(x_i) - \mathbf{E}[\widehat{eq}_Y(x_i)]\}^2, \tag{7.2}$$

where the variance is taken over replications. The standard error of equating is defined as the square root of the error variance,

$$se[\widehat{eq}_Y(x_i)] = \sqrt{var[\widehat{eq}_Y(x_i)]} = \sqrt{\mathbf{E}\{[\widehat{eq}_Y(x_i) - \mathbf{E}[\widehat{eq}_Y(x_i)]\}^2}. \tag{7.3}$$

The error indexed in equations (7.1)–(7.3) is random error that is due to the sampling of examinees to estimate the population quantity, $eq_Y(x_i) = \mathbf{E}[\widehat{eq}_Y(x_i)]$.

Standard errors can be considered for specific data collection designs. In a random groups design, a single population of examinees is considered. A random sample of size N_X is drawn from the population and administered Form X, another random sample of size N_Y is drawn from the population and administered Form Y, and equating is conducted using these data. Conceptually, the hypothetical sampling and equating process is repeated a large number of times, and the variability at each score point is tabulated to obtain standard errors for this design. Recall from Chapter 3 that a conceptual scheme for considering standard errors of equipercentile equating using the random groups design was presented in Figure 3.1.

How would this hypothetical sampling/equating process proceed for the common-item nonequivalent groups design? In this design, on each replication N_X examinees from Population 1 who took Form X and N_Y examinees from Population 2 who took Form Y would be sampled. On each replication, the equivalents would be found using an equating method appropriate for this design, such as the frequency estimation method. The standard error at a particular Form X score would be the standard deviation of the Form Y equivalents over replications.

In the present chapter, the population of examinees is assumed to be infinite (or at least very large) in size. Often it makes sense to conceive of the population as being infinite in size, such as when the population is conceived of as all potential past, current, and future examinees. The examinees in a current sample could be considered as a sample from this population. Although not the approach taken here, it might be argued that the group of examinees *is* the whole population. In this case, there can be no random error in estimating equating relationships because no sampling of examinees is involved.

In practice, data are available from a single sample or pair of samples of examinees. Two general types of procedures have been developed for estimating the standard errors from such data collection designs. The first type is computationally intensive resampling procedures. In these procedures, many samples are drawn from the data at hand and the equating functions estimated on each sampling. Standard errors are calculated using the data from these many resamplings. The resampling method that is considered in this chapter is the bootstrap. The second type of procedures is analytic in that they result in an equation that can be used to estimate the standard errors using sample statistics. The development of the equations in these analytic methods can be very time-consuming, and the resulting equations can be very complicated. The analytic method that is described in this chapter is referred to as the delta method. Both types of methods are useful, depending on the information desired and the uses to be made of the standard errors.

The Bootstrap

The *bootstrap method* (Efron, 1982; Efron and Tibshirani, 1993) is a computer-intensive method for estimating standard errors of a wide variety of statistics. As is described subsequently in more detail, the bootstrap involves taking multiple random samples with replacement from the sample data at hand. A computer is used to draw random samples using a pseudorandom number generator when applying the bootstrap in practice. Refer to Press *et al.* (1989) for a discussion of pseudorandom number generation. To introduce the bootstrap method, a simple example is used in which the standard error of a sample mean is estimated. Then applications to equating are described.

Standard Errors Using the Bootstrap

The steps in estimating standard errors of a statistic using the bootstrap from a single sample are as follows:

1. Begin with a sample of size N.
2. Draw a random sample, *with replacement*, of size N from this sample data. Refer to this sample as a *bootstrap sample*.
3. Calculate the statistic of interest for the bootstrap sample.
4. Repeat Steps 2 and 3 R times.
5. Calculate the standard deviation of the statistic of interest over the R bootstrap samples. This standard deviation is the estimated bootstrap standard error of the statistic.

Of special importance is that the random sample in Step 2 is drawn *with replacement*.

Consider a simple hypothetical example for illustrative purposes. Suppose that an investigator is interested in estimating the standard error of a mean using the bootstrap method. Assume that a sample of size $N = 4$ is drawn from the population and the sample values are 1, 3, 5, and 6. To estimate the standard error of the mean using the bootstrap, bootstrap samples would be drawn with replacement from these four sample values and the mean calculated for each bootstrap sample. Suppose that the following four random bootstrap samples were drawn with replacement from the sample values 1, 3, 5, and 6:

Sample 1: 6 3 6 1	*Mean* = 4.00	
Sample 2: 1 6 1 3	*Mean* = 2.75	
Sample 3: 5 6 1 5	*Mean* = 4.25	
Sample 4: 5 1 6 1	*Mean* = 3.25	

The same sample value may be chosen more than once because bootstrap sampling is done *with replacement*. For example, the score of 6 was chosen twice in bootstrap Sample 1, even though there was only one 6 in the data. The bootstrap estimate of the standard error of the mean is the standard deviation of the means over the four bootstrap samples. To calculate the standard deviation, note that the mean of the four means is $(4.00 + 2.75 + 4.25 + 3.25)/4 = 3.5625$. Using $R - 1 = 3$ as the divisor, the standard deviation of the four means is

$$\sqrt{\frac{(4.00 - 3.5625)^2 + (2.75 - 3.5625)^2 + (4.25 - 3.5625)^2 + (3.25 - 3.5625)^2}{3}}$$

$$= .6884.$$

Thus, using these four bootstrap samples, the estimated standard error of the mean is .6884. In practice, many more than four samples would be chosen. Efron and Tibshirani (1993) recommended using between 25 and 200 bootstrap samples for estimating standard errors. In practice, however, as many 1000 bootstrap replications are common.

In this situation, standard statistical theory would have been easier to implement than the bootstrap. Noting that the sample standard deviation (using $N - 1$ in the denominator) of the original sample values $(1, 3, 5, 6)$ is 2.2174, the estimated standard error of the mean using standard procedures is $2.2174/\sqrt{4} = 1.1087$. The bootstrap estimate would likely be similar to this value if a large number of bootstrap replications were used for estimating the standard error for the population.

In equating, analytic procedures are not always available for estimating standard errors, or the analytic procedures that are available might make assumptions that are thought to be questionable. The bootstrap can be used in such cases. Although computationally intensive, the bootstrap can be readily implemented using a computer, often with much less effort than it would take to derive analytic standard errors.

Standard Errors of Equating

Now consider using the bootstrap to equate two forms using the random groups design. To implement this method, begin with sample data. For equipercentile equating with the random groups design, the samples would consist of N_X examinees with scores on Form X and N_Y examinees with scores on Form Y. To estimate the $se[\hat{e}_Y(x_i)]$:

1. Draw a random bootstrap sample with replacement of size N_X from the sample of N_X examinees.
2. Draw a random bootstrap sample with replacement of size N_Y from the sample of N_Y examinees.
3. Estimate the equipercentile equivalent at x_i using the data from the random bootstrap samples drawn in Steps 1 and 2, and refer to this estimate as $\hat{e}_{Yr}(x_i)$.
4. Repeat Steps 1 through 3 R times, obtaining bootstrap estimates $\hat{e}_{Y1}(x_i)$, $\hat{e}_{Y2}(x_i), \ldots, \hat{e}_{YR}(x_i)$.
5. The standard error is estimated by

$$\widehat{se}_{boot}[\hat{e}_Y(x_i)] = \sqrt{\frac{\sum_r [\hat{e}_{Yr}(x_i) - \hat{e}_{Y.}(x_i)]^2}{R - 1}}, \tag{7.4}$$

where

$$\hat{e}_{Y\cdot}(x_i) = \frac{\sum_r \hat{e}_{Yr}(x)}{R}. \tag{7.5}$$

These procedures can be applied at any x_i. Typically, the same R bootstrap samples are used to estimate standard errors for all integer values of x_i between 0 and K_X, because the interest is in estimating standard errors for the whole range of scores.

The equipercentile equating of the ACT Mathematics test forms that was described in Chapter 2 is used to illustrate the computation of bootstrap standard errors. In this example, Form X and Form Y of the 40-item test were equated using equipercentile methods. The sample sizes were 4329 for Form X and 4152 for Form Y. Unsmoothed equipercentile results were presented in Table 2.7.

To compute bootstrap standard errors in this example, 4329 Form X scores and 4152 Form Y scores were sampled with replacement from their respective distributions. Form Y equipercentile equivalents at each Form X integer score were found. $R = 500$ bootstrap replications were used, and the estimated standard errors were calculated at each score point using equation (7.4). The Macintosh computer program *Equating Error* listed in Appendix B was used to conduct these and the subsequent bootstrap analyses described in this chapter.

The resulting bootstrap standard errors are graphed in Figure 7.1. For comparison purposes, the estimated analytic standard errors that were presented in Table 3.2 also are graphed. [These analytic standard errors were calculated using equation (7.12), which is presented later in the present chapter.] In this figure, the standard errors tend to be smallest around Form X scores in the range of 8 to 12. These scores tend to be the most frequently occurring Form X scores, as can be seen in Figure 2.8. Also, the analytic and bootstrap standard errors are very similar. Empirical studies have found that the two methods produce very similar results in both linear and equipercentile equating of number-correct scores when a large number of bootstrap replications are used (e.g., Kolen, 1985; Jarjoura and Kolen, 1985). Finally, the graph of the standard errors is irregular in appearance, which is presumably due to the relatively small numbers of examinees earning each score.

The bootstrap can be readily applied in the common-item nonequivalent groups design. In this design, a sample of N_X examinees would be drawn from the examinees who were administered Form X, and a sample of N_Y examinees would be drawn from among the examinees who were administered Form Y. An appropriate method, such as the Tucker linear method or the frequency estimation equipercentile method, then would be used to find the equivalents. The sampling process would be repeated a large number of times, and the standard error again would be the standard deviation of the estimates over samples.

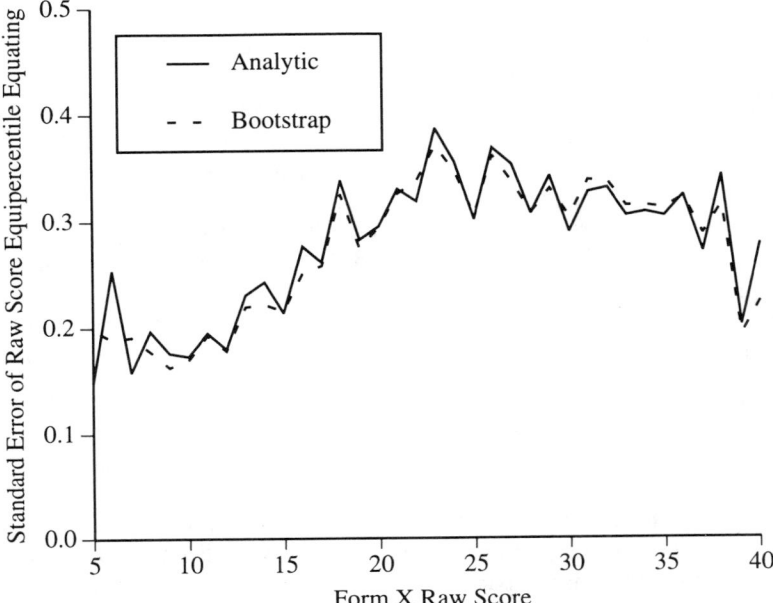

Figure 7.1. Bootstrap and analytic standard errors of equipercentile equating for raw scores.

Parametric Bootstrap

One problem that can be encountered in estimating standard errors in equipercentile equating is that estimates of standard errors might not be very accurate, especially at score points with very small frequencies, as was illustrated by the irregular graphs in Figure 7.1. Efron and Tibshirani (1993) suggested using the *parametric bootstrap* in these situations. In the parametric bootstrap, a parametric model is fit to the data. The standard errors are estimated by treating the fitted parametric model as if it appropriately described the population and simulating standard errors by sampling from the fitted model. Because populations are assumed to be infinite in size, sampling with or without replacement is considered to be the same. As an example, the following steps could be used to apply the parametric bootstrap to estimate the standard errors of equipercentile equating using the random groups design:

1. Fit the Form X empirical distribution using the log-linear method. Choose C using the techniques described in Chapter 3.
2. Fit the Form Y empirical distribution using the log-linear method. Choose C using the techniques described in Chapter 3.
3. Using the fitted distribution from Step 1 as the population distribution

for Form X, randomly select N_X scores from this population distribution. The distribution of these scores is the parametric bootstrap sample distribution of scores on Form X.

4. Using the fitted distribution from Step 2 as the population distribution for Form Y, randomly select N_Y scores from this population distribution. The distribution of these scores is the parametric bootstrap sample distribution of scores on Form Y.

5. Conduct equipercentile equating using the sampled parametric bootstrap distributions from Steps 3 and 4, and tabulate the equipercentile equivalent at score x_i.

6. Repeat Steps 3 through 5 a large number of times. The estimated standard error is the standard deviation of the equivalents at x_i over samples.

In the parametric bootstrap, samples are taken from fitted distributions. In the bootstrap, samples are taken from the empirical distribution. The parametric bootstrap leads to more stable estimates of standard errors than the bootstrap. However, the parametric bootstrap can produce biased estimates of the standard errors if the fitted parametric model is not an accurate estimate of the population distribution. Very little research has been conducted on the parametric bootstrap in equating. More research would be required for the parametric bootstrap to be recommended as a general procedure, although it seems promising.

Results from the use of the parametric bootstrap are shown in Figure 7.2. The bootstrap standard errors are the same as those shown in Figure 7.1. To calculate the parametric bootstrap standard errors in Figure 7.2, a log-linear model with $C = 6$ was fit to the Form X and Form Y distributions. Each parametric bootstrap replication involved drawing a random sample from the fitted distributions and conducting unsmoothed equipercentile equating. As can be seen in Figure 7.2, the parametric bootstrap results in a more regular graph of the standard errors than the bootstrap. In addition, the parametric bootstrap results are more regular than the analytic results shown in Figure 7.1

Standard Errors of Smoothed Equipercentile Equating

Smoothed equivalents can be used in place of the unsmoothed equivalents in the preceding procedures to estimate standard errors of smoothed equipercentile equating. A comparison of standard errors of smoothed and unsmoothed equipercentile equating is presented in Figure 7.3. The parametric bootstrap was used in these comparisons. (The regular bootstrap could have been used here also.) The standard errors of unsmoothed equipercentile equating shown in Figure 7.3 are identical to those shown in Figure 7.2 for the parametric bootstrap. To calculate the standard errors for smoothed equating, the distributions on each parametric bootstrap

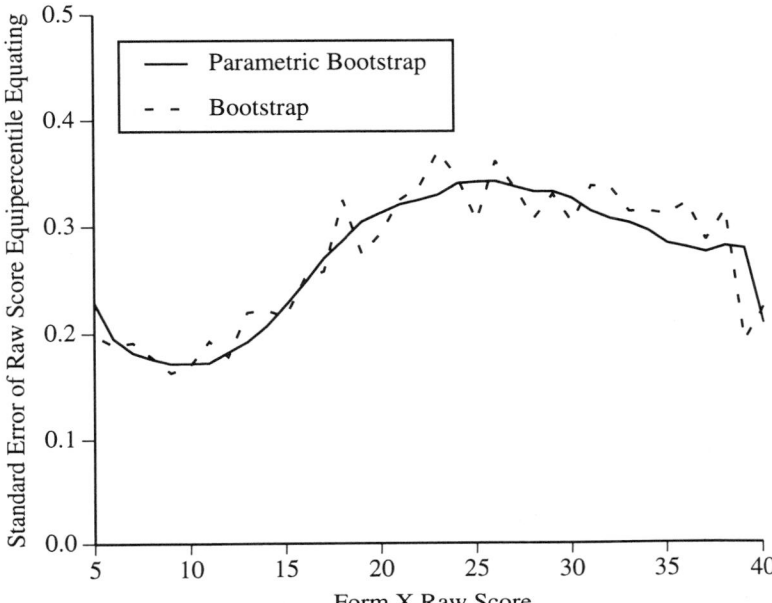

Figure 7.2. Bootstrap and parametric bootstrap standard errors of equipercentile equating for raw scores.

replication were smoothed using the log-linear model with $C = 6$. The smoothed distributions on each replication then were equated using equipercentile methods. Over most of the score range the standard errors for smoothed equipercentile equating were less than those for unsmoothed, indicating that smoothing reduces the standard error of equating. Note, however, that the standard errors only take into account random error; systematic error is not indexed. Thus, as was stressed in Chapter 3, a smoothing method that results in lower standard errors still could produce more total error than unsmoothed equipercentile equating.

Standard Errors of Scale Scores

So far, the bootstrap has been presented using equated raw scores. The bootstrap can be readily applied to scale scores, as well, by transforming the raw score equivalents to scale score equivalents on each replication. The standard error is the standard deviation of the scale score equivalents over replications. Standard errors of both unrounded and rounded scale score equivalents can be estimated using the bootstrap procedure.

Scale score standard errors of equipercentile equating are shown in Figure 7.4. First consider the standard errors for unrounded scale scores.

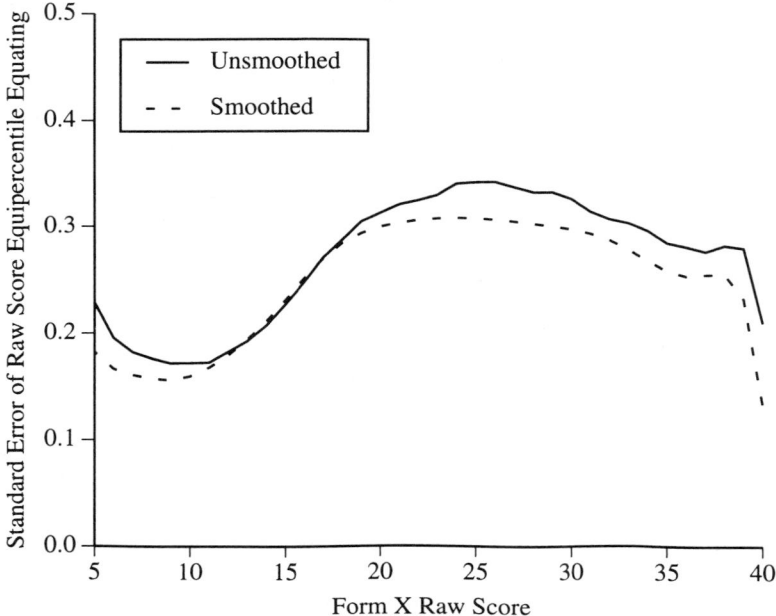

Figure 7.3. Parametric bootstrap standard errors of equipercentile equating for raw scores.

The standard errors tend to be relatively large in the range of raw scores of 36 to 39, which results because the raw-to-scale score transformation is steeper than at other ranges. (The raw-to-scale score transformation for equipercentile equating is shown in Table 2.8.)

Next consider the standard errors for rounded scale scores. These standard errors tend to be greater than those for the unrounded scores, because the rounding process introduces error. When the decimal portion of the unrounded scale scores is close to 1/2, there tends to be a larger difference between the unrounded and rounded standard errors. For example, from Table 2.8, the unrounded scale score at a Form X score of 22 is 20.5533, and the standard error for rounded scale scores for a Form X score is much larger than the standard error for unrounded scale scores. When the decimal portion of the unrounded scale score is close to 0, the standard errors for the rounded and unrounded scale scores tend to be similar.

Standard Errors of Equating Chains

Often equating involves a chain of equating so that scores can be reported on the scale of an earlier form or in terms of scale scores. For example, in

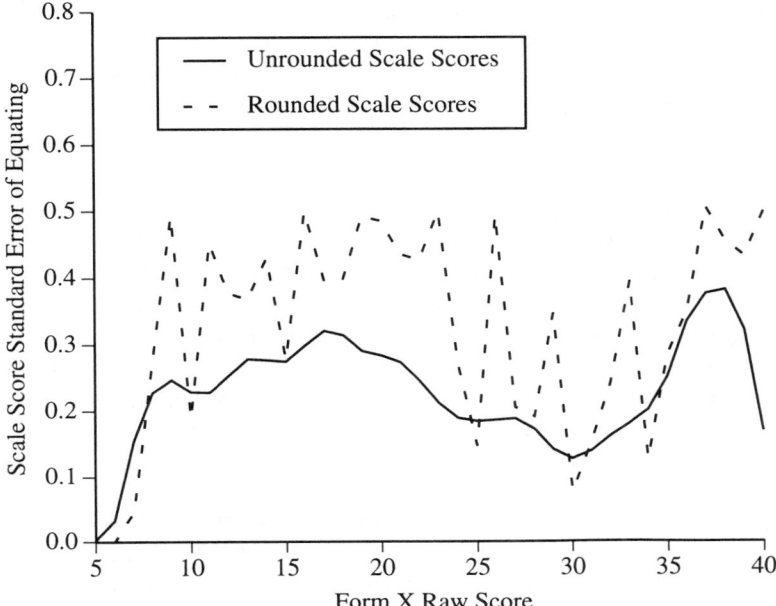

Figure 7.4. Parametric bootstrap standard errors of equipercentile equating for scale scores.

ACT Assessment (American College Testing Program, 1989) equating, new forms are equated to the score scale using a chain of equating which goes to the score scale which was developed for use in 1989. This chain could include numerous test forms. (Also refer to the discussion of the scaling and equating process described with Table 1.1.) Error in a chain of equating can be estimated using the bootstrap.

Consider an example where Form X_2 is to be equated to Form Y through Form X_1. The chaining process involves equating Form X_2 to Form X_1, which can be symbolized as $eq_{X1}(X_2)$, and equating Form X_1 to Form Y, which can be symbolized as $eq_Y(X_1)$. The chain can be symbolized as $eq_{Y(chain:X1)}(X_2) = eq_Y[eq_{X1}(X_2)]$. The notation "chain:X1" in the subscript is used to indicate that the equating function is for a chain that involves Form X_1. The equating chain expression implies that to convert a Form X_2 score to Form Y, the Form X_2 score first is converted to the Form X_1 scale using $eq_{X1}(X_2)$. Then take this converted score and convert it to the Form Y scale using $eq_Y(X_1)$. In practice, estimates of the equating relationships would be available. In the example, each of the two equatings that need to be estimated has error which needs to be incorporated into an estimate of the standard error of the equating chain.

To further develop the example, assume the following: (a) the equating relationships are to be estimated using the random groups design; (b) Form X_1 and Form Y are spiraled in Administration A, and the resulting data are used to equate these forms; and (c) Form X_2 and Form X_1 are spiraled in Administration B, and the resulting data are used to equate these forms. Given this situation, the following steps could be used to estimate bootstrap standard errors of the equating chain:

1. Take a bootstrap sample of the examinees from Administration A who were administered Form X_1. Take a bootstrap sample of examinees from Administration A who were administered Form Y. Equate Form X_1 to Form Y using these bootstrap samples. Refer to the estimated equating relationship from bootstrap samples r as $\hat{eq}_{Yr}(X_1)$.
2. Take a bootstrap sample of the examinees from Administration B who were administered Form X_2. Take a bootstrap sample of examinees from Administration B who were administered Form X_1. Equate Form X_2 to Form X_1 using these bootstrap samples. Refer to the estimated equating relationship from bootstrap samples r as $\hat{eq}_{X1r}(X_2)$.
3. Find the conversion of Form X_2 scores to Form Y scores through the equating chain using the equating relationships developed in Steps 2 and 3. Refer to the estimated equating chain from bootstrap samples r as $\hat{eq}_{Yr(\text{chain}:X1)}(X_2)$.
4. Repeat Steps 1–3 a large number of times. The standard deviation of the converted scores from Step 3 at a particular Form X_2 score is the bootstrap standard error of the equating chain at that score.

This procedure could be generalized to longer chains, although the process can become extremely computationally intensive as the length of the chain increases. This process also could be adapted to the single group and common-item nonequivalent groups designs, and to other equating methods, such as linear methods.

Mean Standard Error of Equating

Sometimes it is useful to have an aggregate value for the standard error of equating, such as when an index of the overall effect of smoothing is desired. One way to get an aggregate value is to find the square root of the average equating error variance over examinees from the population that was administered Form X. In this way, the average standard error of equipercentile equating is defined as:

$$\sqrt{\sum_i f(x_i) se^2[\hat{e}_Y(x_i)]}.$$

In this equation, the error variance at each score point is weighted by its density, $f(x_i)$, and then summed over score points. Weighting by the den-

Table 7.1. Average Standard Errors of Equipercentile Equating.

Score	Bootstrap	Parametric Bootstrap	Analytic
Raw Score			
Unsmoothed	.2713	.2674	.2767
Smoothed	.2536	.2519	
Unrounded Scale Score			
Unsmoothed	.2549	.2501	
Smoothed	.2373	.2385	
Rounded Scale Score			
Unsmoothed	.3636	.3632	
Smoothed	.3526	.3494	

sity is done so that the error variance for each examinee in the population is weighted equally.

For the equipercentile equating example, the average standard errors estimated by substituting estimates for the parameters are shown in Table 7.1. Average analytic standard errors were calculated only for raw scores without smoothing. The averages for the bootstrap and parametric bootstrap are very similar. For raw scores, the average standard error is somewhat lower for smoothed equating than for unsmoothed equating. The same is true for scale scores and rounded scale scores. The average standard error for rounded scale scores is considerably larger than the average for unrounded scale scores in this example.

Caveat

A major concern in using the bootstrap is that it is computationally very intensive. If, for example, 500 bootstrap replications are to be conducted, then samples need to be drawn and equating needs to be conducted 500 times. Stable standard error estimates might require using 1000 or more replications. However, with modern computer equipment, this many replications often can be accomplished without great expense, at least for the mean, linear, and equipercentile methods considered in Chapters 2–5.

Bootstrap standard errors of equating also can be used with item response theory methods. However, to use the bootstrap with IRT methods, random samples would need to be drawn *and* item parameters estimated many times. Although possible, the computation of bootstrap standard errors of IRT equating, especially using the three-parameter logistic model, might be difficult due to the extensive computational requirements.

The Delta Method

Equations for estimating standard errors can be useful when computational time needs to be minimized or when estimating the desired sample sizes for an equating study. The *delta method* is a commonly used statistical method for deriving standard error expressions. The delta method is used to derive the approximate standard error of a statistic that is a function of statistics for which expressions for the standard errors already exist. As a simple example, the standard error of the sample mean squared could be estimated using the delta method, because an expression for the standard error of a sample mean is known. For the mean, linear, and equipercentile equating methods that were considered in Chapters 2 through 5, the estimated equating relationships are functions of sample moments and cumulative probabilities that have standard errors which can be estimated directly. Thus, the delta method can be used to estimate standard errors of scores equated using mean, linear, and equipercentile equating methods.

The delta method (Kendall and Stuart, 1977) is based on a Taylor series expansion. Define for the population $eq_Y(x_i; \Theta_1, \Theta_2, \ldots, \Theta_t)$ as an equating function of test score x_i and parameters $\Theta_1, \Theta_2, \ldots, \Theta_t$. In linear equating, $\Theta_1, \Theta_2, \ldots, \Theta_t$ are moments. In equipercentile equating, $\Theta_1, \Theta_2, \ldots, \Theta_t$ are cumulative probabilities. By the delta method, an approximate expression for the sampling variance is

$$var[\widehat{eq}_Y(x_i)] \cong \sum_j eq'^2_{Yj} \, var(\hat{\Theta}_j) + \sum_{j \neq k} \sum eq'_{Yj} eq'_{Yk} \, cov(\hat{\Theta}_j, \hat{\Theta}_k). \quad (7.6)$$

In this equation, $\hat{\Theta}_j$ is a sample estimate of Θ_j and eq'_{Yj} is the partial derivative of eq_{Yj} with respect to Θ_j and evaluated at $x_i, \Theta_1, \Theta_2, \ldots, \Theta_t$. This equation requires that expressions for the sampling variances (*var*) and sampling covariances (*cov*) of the $\hat{\Theta}_j$ be available. The standard error is the square root of *var* in equation (7.6).

The following steps are used to apply the delta method:

1. Specify the error variances and covariances for each $\hat{\Theta}_j$.
2. Find the partial derivative of the equating equation with respect to each Θ_j.
3. Substitute the variances and partial derivatives into equation (7.6).

The resulting standard errors are expressed in terms of parameters. Estimates of the parameters are used in place of the parameters to obtain the estimated standard errors.

Mean Equating Using Single Group and Random Groups Designs

For illustrative purposes, consider a simple example using mean equating in the single group design with no counterbalancing (use of counterbalancing would make this example more complicated). In this design, for the population,

$$m_Y(x_i) = x_i - \mu(X) + \mu(Y),$$

which is estimated by

$$\hat{m}_Y(x_i) = x_i - \hat{\mu}(X) + \hat{\mu}(Y).$$

To apply the delta method, note that from standard statistical theory,

$$var[\hat{\mu}(X)] = \sigma^2(X)/N, var[\hat{\mu}(Y)] = \sigma^2(Y)/N, \qquad \text{and}$$

$$cov[\hat{\mu}(X), \hat{\mu}(Y)] = \sigma(X, Y)/N.$$

Also, the required partial derivatives are as follows:

$$\partial\hat{m}/\partial\hat{\mu}(X) = -1, \qquad \partial\hat{m}/\partial\hat{\mu}(Y) = 1.$$

Define Θ_1 as $\mu(X)$ and Θ_2 as $\mu(Y)$. Substituting the sampling variances and covariances and partial derivatives into equation (7.6) results in

$$var[\hat{m}_Y(x_i)] \cong (-1)^2\sigma^2(X)/N + (1)^2\sigma^2(Y)/N + 2(-1)(1)\sigma(X, Y)/N$$

$$= [\sigma^2(X) + \sigma^2(Y) - 2\sigma(X, Y)]/N, \tag{7.7}$$

for the single group design without counterbalancing.

What if a random groups design were used for mean equating with $N_X = N_Y = N$? In this case, the covariance between X and Y is 0 because independent samples of examinees are administered the two forms. Thus, for random groups,

$$var[\hat{m}_Y(x_i)] \cong [\sigma^2(X) + \sigma^2(Y)]/N. \tag{7.8}$$

As can be seen by comparing equations (7.7) and (7.8), if scores on Form X and Form Y have a positive covariance for the single group design, then the error variance for the single group design will be smaller than the error variance for the random groups design.

Linear Equating Using the Random Groups Design

In implementing the delta method for linear equating with the random groups design, $\mu(X)$, $\sigma(X)$, $\mu(Y)$, and $\sigma(Y)$ need to be estimated. Because Form X and Form Y are given to independent random samples, estimates of the moments for Form X are independent of estimates of the moments for Form Y.

Braun and Holland (1982, p. 33) presented the necessary partial derivatives and standard errors and covariances between the moments to apply the delta method. They showed that:

$$var[\hat{l}_Y(x_i)] \cong \sigma^2(Y)\left\{ \frac{1}{N_X} + \frac{1}{N_Y} + \left[\frac{sk(X)}{N_X} + \frac{sk(Y)}{N_Y} \right]\left[\frac{x_i - \mu(X)}{\sigma(X)} \right] \right.$$

$$\left. + \left[\frac{ku(X) - 1}{4N_X} + \frac{ku(Y) - 1}{4N_Y} \right]\left[\frac{x_i - \mu(X)}{\sigma(X)} \right]^2 \right\} \tag{7.9}$$

This equation indicates that the standard error of equating depends on the skewness and kurtosis of the population distribution.

Inspection of this equation leads to some observations about the standard errors for the random groups design. First, as the sample sizes increase, the error variance decreases. In this equation, this observation is made by noting that the sample sizes are always in the denominators of the expressions. Second, error variance tends to be smallest near the mean. This observation is based on noting that the term

$$\left[\frac{x_i - \mu(X)}{\sigma(X)} \right]^2$$

becomes larger as x_i moves farther from the mean, and this term is multiplied by a term that is almost always positive (because kurtosis is positive as defined here). Third, error variance tends to be larger in the direction that a distribution is skewed. This observation follows because, if both distributions are positively skewed, then the term

$$\left[\frac{sk(X)}{N_X} + \frac{sk(Y)}{N_Y} \right]\left[\frac{x_i - \mu(X)}{\sigma(X)} \right]$$

is positive for all x_i that are above the mean and negative for all x_i that are below the mean. The reverse is true for negatively skewed distributions.

As can be seen, the error variance expression in equation (7.9) is fairly complicated, even in the simple situation in which linear equating is used with the random groups design. Also, this expression requires computing skewness and kurtosis terms. Equation (7.9) can be simplified. If X and Y are assumed to be normally distributed, then skewness is 0 and kurtosis is 3. In this case, equation (7.9) simplifies to

$$var[\hat{l}_Y(x_i)] \cong \frac{\sigma^2(Y)}{2}\left[\frac{1}{N_X} + \frac{1}{N_Y} \right]\left\{ 2 + \left[\frac{x_i - \mu(X)}{\sigma(X)} \right]^2 \right\}. \tag{7.10}$$

This expression is presented in Petersen et al. (1989) and is similar to the one presented by Angoff (1971).

A further simplification is possible if sample sizes for the two forms are assumed to be equal. If $N_{tot} = N_X + N_Y = 2N_X = 2N_Y$, then equation (7.10) further simplifies to

$$var[\hat{l}_Y(x_i)] \cong \frac{2\sigma^2(Y)}{N_{tot}}\left\{ 2 + \left[\frac{x_i - \mu(X)}{\sigma(X)} \right]^2 \right\}. \tag{7.11}$$

From equation (7.11) it can readily be seen that error variance becomes larger as x_i departs farther from the mean.

As Braun and Holland (1982) pointed out, if equation (7.10) or (7.11) for error variance is used with nonnormal distributions, then the estimates of the standard errors will be biased to some extent. However, the expressions that assume normality are easier to calculate and might be useful as approximations in some situations. For example, when planning sample size requirements for equating studies, data are unavailable on the forms that are to be equated. In this case, the approximations might provide sufficiently accurate estimates of equating error. Procedures for estimating sample size requirements are described later in this chapter.

Equipercentile Equating Using the Random Groups Design

Lord (1982) used the delta method to develop expressions for the standard error of equipercentile equating under the random groups design. Using the notation developed in Chapter 2, this error variance can be expressed as

$$var[\hat{e}_Y(x_i)] \cong \frac{1}{[G(y_U^*) - G(y_U^* - 1)]^2} \left\{ \frac{[P(x_i)/100][1 - P(x_i)/100](N_X + N_Y)}{N_X N_Y} \right.$$
$$\left. - \frac{[G(y_U^*) - P(x_i)/100][P(x_i)/100 - G(y_U^* - 1)]}{N_Y[G(y_U^*) - G(y_U^* - 1)]} \right\}. \tag{7.12}$$

To estimate the error variances, sample values can be substituted in place of the parameters in equation (7.12). The error variance depends on the proportion of examinees at scores on Form Y, as symbolized by $G(y_U^*) - G(y_U^* - 1)$. If this quantity were 0, then the error variance would be undefined because of a 0 term in the denominator. As an alternative to using sample values, the Form X and Form Y distributions could be smoothed using the log-linear method and the smoothed distribution values used in place of the parameters in equation (7.12). The use of smoothed distribution values in equation (7.12) would be similar to using the parametric bootstrap that was described earlier.

Lord (1982b) also presented an approximation to equation (7.12). Petersen et al. (1989) used Lord's approximation and made a normality assumption to provide the following approximation to the standard error of equipercentile equating under the random groups design:

$$var[\hat{e}_Y(x_i)] \cong \sigma^2(Y) \frac{[P(x_i)/100][1 - P(x_i)/100]}{\phi^2} \left(\frac{1}{N_X} + \frac{1}{N_Y} \right), \tag{7.13}$$

where ϕ is the ordinate of the standard normal density at the unit-normal score, z, below which $P(x_i)/100$ of the cases fall. If the sample sizes are equal, such that $N_{tot} = N_X + N_Y = 2N_X = 2N_Y$, then equation (7.13) simplifies to

$$var[\hat{e}_Y(x_i)] \cong \frac{4\sigma^2(Y)}{N_{tot}} \frac{[P(x_i)/100][1 - P(x_i)/100]}{\phi^2}. \tag{7.14}$$

Standard Errors for Other Designs

The derivations of standard errors of equating using the delta method can be very complicated, and the expression of the results can be particularly cumbersome. For example, Kolen (1985) derived the standard errors of Tucker equating. The presentation of the required partial derivatives took one full page in the article and the presentation of the sampling errors for the moments took another full page. The presentation of standard errors of frequency estimation equating by Jarjoura and Kolen (1985) took even more space to present. For this reason, a comprehensive presentation of standard errors is not provided in this book.

Table 7.2 contains references to articles that provide standard errors of equating for many of the methods and designs discussed in this book. These articles should be consulted for the standard error equations. See Lord (1975) and Zeng (1993) for descriptions of the use of numerical derivatives in using the delta method. Also, Angoff (1971), Lord (1950), and Petersen et al. (1989) provide standard errors using normality assumptions.

Table 7.2. References[a] to Delta Method Standard Errors.

Design and Method	Reference
Single Group	
Linear	Zeng and Cope (1995)
Equipercentile	Lord (1982b)
Random Groups	
Linear	Braun and Holland (1982)
Equipercentile	Lord (1982b)
Smoothed Equipercentile	Holland et al. (1989)
Common-Item Nonequivalent Groups	
Linear—Tucker	Kolen (1985)
Linear—Levine Observed Score	Hanson et al. (1993)
Linear—Levine True Score	Hanson et al. (1993)
Frequency Estimation	Jarjoura and Kolen (1985)
Equipercentile	
Smoothed Equipercentile	Holland et al. (1989)
IRT True Score	Lord (1982c)

[a] Lord (1950) and Angoff (1971) provided standard errors for linear methods based on a normality assumption. Petersen et al. (1989) also provided standard error expressions.

Note that standard errors for IRT methods are provided in Table 7.2 only for IRT true score equating in the common-item nonequivalent groups design, because these are the only IRT standard errors that have been derived. Even the standard errors for this method that are cited were derived without considering error in transforming the item parameters to be on a common ability scale, under the assumption that examinee ability parameters are known, and by making other simplifying assumptions. For these reasons, Lord (1982c) indicated that these standard errors are underestimates. Analytic standard errors of IRT true score equating have not been developed for the single group or random groups design or for IRT observed score equating under any of the designs. The derivation of such standard errors is an important area for further research. Standard errors also have not been derived for the Braun-Holland linear method.

For comparative purposes, estimated standard errors of equating for the real data example presented in Chapters 4 and 5 are presented in Table 7.3. In this example, Form X and Form Y were equated using the common-item nonequivalent groups design. These standard errors were calculated using the *CIPE* Macintosh computer program listed in Appendix B. The synthetic population weight $w_1 = 1$ is used in this example. Estimated standard errors for the Tucker method, the Levine observed score method, and the frequency estimation equipercentile method are presented. Average standard errors also were calculated. As can be seen from this example, the estimated standard errors are smaller near the middle of the distribution than at the extremes. Also, of the three methods, the Tucker method produced the smallest estimated standard errors. The Levine observed score method produced smaller estimated standard errors than the frequency estimation equipercentile method at most score points. Recall that standard errors account for random error only. Just because the Tucker method has smaller standard errors than the Levine method in this case does not necessarily mean that the Tucker method is preferable. More systematic error might be present with the Tucker method than with the Levine method in this case, although it is impossible to know for sure. In practice, a choice of method involves assessing the reasonableness of the statistical assumptions described in Chapter 4 for the equating at hand, as well as other practical issues that are described in Chapter 8.

Approximations

Approximations to standard error expressions that are less complicated than the expressions in the Table 7.2 references are useful in some situations. In this section, two approximations are considered which are useful for comparing designs and equating methods.

One approximation for the single group design was presented by Angoff (1971). This approximation ignores counterbalancing and assumes that X

Table 7.3. Standard Errors of Equating for the Common-Item
Nonequivalent Groups Example.

		Standard Error		
			Levine Observed Score	Frequency Estimation Equipercentile
x	$\hat{F}_1(x)$	Tucker		
0	.0000	.2643	.3615	
1	.0000	.2518	.3437	
2	.0006	.2395	.3261	
3	.0036	.2273	.3087	
4	.0091	.2154	.2915	.2880
5	.0169	.2038	.2746	.2665
6	.0387	.1925	.2580	.2592
7	.0695	.1816	.2419	.2603
8	.1160	.1712	.2262	.2499
9	.1680	.1613	.2111	.2351
10	.2236	.1521	.1967	.2172
11	.2918	.1437	.1832	.2199
12	.3692	.1363	.1709	.2188
13	.4236	.1300	.1598	.2123
14	.4918	.1250	.1505	.2041
15	.5402	.1214	.1432	.1995
16	.5952	.1193	.1381	.2072
17	.6477	.1190	.1357	.2160
18	.6918	.1203	.1359	.2336
19	.7221	.1232	.1388	.2308
20	.7662	.1276	.1443	.2349
21	.7988	.1334	.1520	.2506
22	.8314	.1404	.1617	.2487
23	.8562	.1484	.1730	.2614
24	.8773	.1572	.1855	.2321
25	.9027	.1668	.1992	.2022
26	.9215	.1770	.2137	.1639
27	.9402	.1877	.2289	.2299
28	.9541	.1988	.2447	.3578
29	.9674	.2103	.2610	.3377
30	.9776	.2221	.2776	.3207
31	.9825	.2341	.2946	.2777
32	.9909	.2464	.3118	.3864
33	.9952	.2589	.3292	.4707
34	.9988	.2715	.3468	
35	.9994	.2842	.3646	
36	1.0000	.2971	.3826	
Average		.1480	.1819	.2302

and Y have a bivariate normal distribution. Note also that N refers to the number of examinees who take both forms:

$$var[\hat{l}_Y(x_i)] \cong \frac{\sigma^2(Y)[1 - \rho(X, Y)]}{N} \left\{ 2 + [1 + \rho(X, Y)]\left[\frac{x_i - \mu(X)}{\sigma(X)}\right]^2 \right\}. \quad (7.15)$$

In this equation, $\rho(X, Y)$ is the correlation between scores on X and Y.

Another approximation was presented by Angoff (1971) for the common-item random groups design mentioned in Chapter 5, in which randomly equivalent groups of examinees are administered two forms that contain common items. This equation assumes that the populations taking X and Y are equivalent, that X and V are bivariate normally distributed in the population, that Y and V are bivariate normally distributed in the population, that the correlation between X and V is equal to the correlation between Y and V, and that the sample sizes for examinees taking the old and new forms are equal. This equation is

$$var[\hat{l}_Y(x_i)] \cong \frac{\sigma^2(Y)[1 - \rho^2(X, V)]}{N_{tot}} \left\{ 2 + [1 + \rho^2(X, V)]\left[\frac{x_i - \mu(X)}{\sigma(X)}\right]^2 \right\}. \quad (7.16)$$

In this equation, $\rho(X, V)$ is the correlation between common items and total score, and N_{tot} is the total number of examinees taking the forms (i.e., twice the number of examinees taking any one form).

The error variance in equation (7.16) can be rewritten as follows:

$$var[\hat{l}_Y(x_i)] \cong \frac{\sigma^2(Y)}{N_{tot}} \left\{ 2[1 - \rho^2(X, V)] + [1 - \rho^4(X, V)]\left[\frac{x_i - \mu(X)}{\sigma(X)}\right]^2 \right\}. \quad (7.17)$$

From equation (7.17), it can readily be seen that, as positive values of $\rho(X, V)$ increase, the error variance decreases. That is, the greater the correlation between the total score and the common-item score, the smaller the error variance. Equations (7.16) and (7.17) provide an approximation to the Kolen (1985) result for the Tucker method that might be useful when estimating sample size requirements for linear equating in the common-item nonequivalent groups design. However, the standard errors presented by Hanson *et al.* (1993) should be used whenever possible, and especially when documenting the amount of error in an equating.

Standard Errors for Scale Scores

Standard errors of equating for scale scores can be approximated based on the delta method standard errors for raw scores. A variation of the delta method can be used to estimate the scale score standard errors. To develop this variation, consider a situation in which a parameter Θ is being estimated, where the estimate is symbolized by $\hat{\Theta}$. Also assume that the error variance in estimating the parameter is known, which is symbolized by $var(\hat{\Theta})$. Finally, assume that the estimate is to be transformed using the function f. In this situation, Kendall and Stuart (1977) showed that, approximately,

$$var[f(\hat{\Theta})] \cong f'^2(\Theta)var(\hat{\Theta}),$$

where f' is the first derivative of f. That is, the error variance of the function of a random variable can be approximated by the product of the square of the derivative of the function at the parameter value and the error variance of the random variable.

This formulation can be applied to equating by substituting $eq_Y(x_i)$ for the parameter Θ, $\widehat{eq}_Y(x_i)$ for $\hat{\Theta}$, and the Form Y raw-to-scale score transformation s for the function f. To apply this equation directly, the first derivative of the Form Y raw-to-scale score transformation is needed at $eq_Y(x_i)$.

If the Form Y raw-to-scale score transformation is linear, then the derivative of the raw-to-scale score transformation is equal to the slope of the Form Y raw-to-scale score linear transformation, which is a constant at all $eq_Y(x_i)$. In this case, the scale score error variance can be approximated by taking the product of the raw score error variance and the squared slope of the Form Y raw-to-scale score transformation. If the Form Y raw-to-scale score transformation is nonlinear but continuous, then the scale score error variance can be approximated by taking the product of the squared first derivative of the estimated Form Y raw-to-scale score transformation at $eq_Y(x_i)$ and the estimated raw score error variance.

The Form Y raw-to-scale score transformation is often nonlinear and not continuous. In this case, the derivative of the Form Y raw-to-scale score conversion near $eq_Y(x_i)$ needs to be approximated. To approximate this derivative, the Form Y raw-to-scale score conversion can be viewed as a set of points connected by straight lines. The slope of the line near $\widehat{eq}_Y(x_i)$ can be used as an approximation of the derivative. For example, in the numerical example presented in Chapter 2 (see Table 2.7), under equipercentile equating a Form X raw score of 24 was estimated to be equivalent to a Form Y raw score of 23.9157. The slope of the Form Y raw-to-scale score conversion at a Form Y raw score of 23.9157 can be found by taking the difference between the Form Y scale score equivalents at Form Y raw scores of 24 and 23. From Table 2.8, these equivalents are 22.3220 and 21.7000. The difference between these equivalents is .6220, which can be taken as the slope of the raw-to-scale score conversion at a Form Y raw score of 23.9157. From Table 3.2, the estimated raw score standard error of unsmoothed equipercentile equating at a Form X score of 24 is .3555. Thus, the scale score error variance for unsmoothed equipercentile equating is approximately $.6220^2(.3555^2) = .0489$, and the scale score standard error is approximately $.6220(.3555) = .2211$. This process can be used to approximate scale score standard errors of equating at other score points as well. Because these standard errors are designed only for unrounded scale scores, the bootstrap or a similar procedure should be used to estimate standard errors for rounded scale scores.

Standard Errors of Equating Chains

Delta method standard errors can be used to estimate standard errors of equating chains. When the equatings are independent, as is typically the case with the random groups design, a delta method variant suggested by Braun and Holland (1982, p. 36) can be used. Suppose that in the equating chain, Form X_2 is equated to Form Y by equating Form X_2 to Form X_1 and Form X_1 to Form Y. The error variance of converted scores for an equating chain can be approximated as follows:

$$var[\widehat{eq}_{Y(\text{chain:}X1)}(x_2)] \cong var[\widehat{eq}_Y(x_1^*)] + eq'^2_{X1}(x_2) \cdot var[\widehat{eq}_{X1}(x_2)],$$

where $x_1^* = eq_{X1}(x_2)$ and $eq'^2_{X1}(x_2)$ is the squared first derivative of the function for equating Form X_2 to Form X_1. The standard error is the square root of this expression. If the equating function is not continuous, then an approximation to the derivative (e.g., the slope of the conversion at x_2) could be used in its place. Braun and Holland (1982) pointed out that when forms which are constructed to be parallel are equated, this derivative is generally close to 1. In this case, the derivative can be set equal to 1 and the error variance of the chain can be approximated by summing the error variances of the two component equatings.

The procedure just described is appropriate only when the equatings are independent, such as in a chain of equatings conducted using the random groups design. When using the common-item nonequivalent groups design, Zeng et al. (1994) suggested that equatings are dependent. As an example of this dependency in a chain, examinees who were administered Form X_1 would be involved in equating Form X_2 to Form X_1 and Form X_1 to Form Y. In this case, the dependency needs to be incorporated into the estimation. See Lord (1975) and Zeng et al. (1994) for details on how the effects of the dependency can be incorporated into the process of estimating standard errors.

Using Delta Method Standard Errors

The standard error expressions are useful for comparing the precision of equating designs and equating methods, and for estimating sample sizes. Because comparisons can become exceedingly complicated, in this section

Table 7.4. Selected Equating Error Variance Equations Assuming
Normality and Equal Sample Sizes Per Test Form.

Random Groups Linear

$$var[\hat{l}_Y(x_i)] \cong \frac{2\sigma^2(Y)}{N_{tot}} \left\{ 2 + \left[\frac{x_i - \mu(X)}{\sigma(X)} \right]^2 \right\} \tag{7.11}$$

Random Groups Equipercentile

$$var[\hat{e}_Y(x_i)] \cong \frac{4\sigma^2(Y)}{N_{tot}} \frac{[P(x_i)/100][1 - P(x_i)/100]}{\phi^2} \tag{7.14}$$

Single Group Linear

$$var[\hat{l}_Y(x_i)] \cong \frac{\sigma^2(Y)[1 - \rho(X, Y)]}{N} \left\{ 2 + [1 + \rho(X, Y)] \left[\frac{x_i - \mu(X)}{\sigma(X)} \right]^2 \right\} \tag{7.15}$$

only an idealized situation is examined in which normal distributions are assumed. Also, only the random groups and single group designs are studied, although the approach described can be generalized to other designs. Equipercentile equating is examined only for the random groups design. Lord (1950) and Crouse (1991) provided comparisons in addition to the ones presented here. For ease of reference, Table 7.4 lists the equations that are used in this section.

Random Groups Linear Versus Random Groups Equipercentile. One question that might be asked is how precise is equipercentile equating relative to linear equating when using the random groups design? This question can be addressed readily if the sample size is equal and a normality assumption is made. Under these assumptions, the linear error variances are given in equation (7.11), the equipercentile error variances are given in equation (7.14), and

$$z = \frac{x_i - \mu(X)}{\sigma(X)}$$

is a unit-normal score. To compare the error variances note that both equations have $2\sigma^2(Y)/N_{tot}$ as multipliers, so these terms can be ignored when comparing the *relative* magnitudes of the error variances by taking the ratio of one error variance to the other.

A comparison of the relative magnitudes is made in Table 7.5 at selected z-scores. The z-scores are used so that the table can be used with any test by converting the number-correct scores to z-scores. In this table, $P^{**} = P/100$. The rightmost column of the table presents the ratio of the error variances at selected z-scores. For scores near the mean, the values

Table 7.5. Comparison of Relative Magnitudes of Random Groups Linear and Equipercentile Error Variances.

z	P^{**}	$1 - P^{**}$	ϕ	$\dfrac{2P^{**}(1 - P^{**})}{\phi^2}$	$2 + z^2$	$\dfrac{2P^{**}(1 - P^{**})}{\phi^2(2 + z^2)}$
.0	.5000	.5000	.3989	3.14	2.00	1.57
.5	.6915	.3085	.3521	3.44	2.25	1.52
1.0	.8413	.1587	.2420	4.56	3.00	1.52
1.5	.9332	.0668	.1295	7.43	4.25	1.75
2.0	.9772	.0228	.0540	15.28	6.00	2.54
2.5	.9938	.0062	.0175	40.23	8.25	4.88
3.0	.9987	.0013	.0044	134.12	11.00	12.19

around 1.5 indicate that the error variance for equipercentile equating is approximately 1.5 times that of linear equating. The ratio becomes much larger farther away from the mean; for example, for a z-score of 2.5, the ratio is nearly 5.

The ratios in the table can be used to make statements about the relative sample sizes required in linear and equipercentile equating to achieve the same equating precision. For example, to achieve the equating precision near the mean that is achieved with a sample size of 1000 with linear equating, a sample size of 1570 (1000×1.57) would be needed with equipercentile equating. As another example, to achieve the equating precision at a z-score of 2.5 that is achieved with a sample size of 1000 with linear equating, a sample size of 4880 (1000×4.88) would be needed with equipercentile equating.

Do smaller standard errors for the linear method mean that the linear method is better than the equipercentile method? Not necessarily. Recall that standard errors account only for random error in equating. If the relationship is nonlinear, then equipercentile equating might provide a more accurate estimate of the population equivalent, even when it has a much larger standard error than the linear method, because of systematic error that could be introduced by using the linear method.

Random Groups Linear Versus Single Group Linear.   Table 7.6 presents the ratio of random groups to single group equating error variance for the linear method. Normal distributions are assumed. The values in this table were calculated by taking the ratio of equation (7.11) to equation (7.15) for selected values of z and $\rho(X, Y)$, where ρ is used to symbolize $\rho(X, Y)$ in the single group design. In taking the ratio, the total number of examinees for the single group design (N) cancels out the total number of examinees for the random groups design (N_{tot}).

These ratios indicate the relative precision of linear equating in the two

Table 7.6. Ratio of Linear Method Random Groups Equating Error
Variance to Single Group Equating Error Variance.

	$\rho = .0$	$\rho = .2$	$\rho = .5$	$\rho = .7$	$\rho = .9$
$z = .0$	2.00	2.50	4.00	6.67	20.00
$z = .5$	2.00	2.45	3.79	6.19	18.18
$z = 1.0$	2.00	2.34	3.43	5.41	15.38
$z = 1.5$	2.00	2.26	3.16	4.86	13.55
$z = 2.0$	2.00	2.21	3.00	4.55	12.50
$z = 2.5$	2.00	2.17	2.90	4.36	11.89
$z = 3.0$	2.00	2.15	2.84	4.24	11.52

designs. These ratios also indicate the relative number of examinees needed
to achieve a given level of precision. For example, in the unlikely event
that the correlation between X and Y is 0, the tabled ratio of 2.00 indicates
that twice as many examinees are needed in random groups design to get
the same precision that is achieved with the single group design. Thus, for
example, if $\rho(X, Y) = 0$, then 2000 examinees would be required in the
random groups design to achieve the same level of precision that could be
attained with 1000 examinees using the single group design.

In the single group design, however, each examinee takes Form X and
Form Y. In the random groups design, different examinees take Form X
and Form Y. Thus, in the preceding example, the 1000 examinees in the
single group design would take 2000 test forms (1000 Form X and 1000
Form Y). That is, if $\rho(X, Y) = 0$, then the same number of forms would
need to be administered under the two designs to achieve a given level of
precision. This example illustrates that, if interest is in estimating the relative number of test forms that need to be taken, rather than the relative
number of examinees that need to be tested, the values in Table 7.6 should
be divided by 2.

The quantity $\rho(X, Y)$ in the single group design is an alternate forms reliability coefficient. Of the tabled values, $\rho(X, Y) = .7$ or .9 are the most
realistic, because alternate forms to be equated can be expected to be positively correlated when administered to the same examinees. For $\rho = .70$
and $z = 0$, depending on the level of z, between 4.24 and 6.67 times as many
examinees would be needed for the random groups design to achieve the
same level of precision as for the single group design. For example, for $\rho =
.70$, 6670 examinees would be needed with the random groups design to ach-
ieve the same level of precision as would be achieved with 1000 examinees
in the single group design. For $\rho = .90$ and $z = 0$, 20,000 examinees would
be needed with the random groups design to achieve the same level of precision as would be achieved with 1000 examinees in the single group design. Therefore, for highly reliable tests, the sample size requirements for
the single group design can be considerably less than those for the random

groups design. Of course, it is possible that either of these sample sizes would lead to considerably more precision than would be necessary in an equating. (Estimating sample size requirements is considered in the next section.)

Counterbalancing issues and context effects, such as practice and fatigue, can introduce systematic error with the single group design. These issues are effectively ignored in Table 7.6. Using counterbalancing can lead to greater sample size requirements. Also, recall from Chapter 2 that when differential order effects are present in the single group with counterbalancing design, the data from the test taken second might need to be disregarded. In this case, the data that can actually be used to equate Form X and Form Y are from the form taken first, and the random groups standard errors would need to be used.

Estimating Sample Size for Random Groups Linear Equating. In addition to comparing equating error associated with different designs and methods, standard errors of equating also can be useful in specifying the sample size required to achieve a given level of equating precision for a particular equating design and method. In order to use standard errors in the process of estimating sample size requirements, the desired level of precision needs to be stated. Ideally, equating error should be small and not make a significant contribution to error in reported test scores. In practical situations, the significance of this contribution needs to be operationalized.

Consider the following example. Suppose that linear equating with the random groups design is to be used. Also suppose that, for a particular equating, a standard error of equating that is less than .1 standard deviation unit is judged not to make a significant contribution to error in reported scores. In this situation, what sample size would be required?

Equation (7.11) presents the error variance for this situation. Let u refer to the maximum proportion of standard deviation units that is judged to be appropriate for the standard error of equating. The value of N_{tot} is found that gives a specified value for $u\sigma(y)$ for the standard error of equating. In the example just presented, $u = .1$ standard deviation unit. Based on this specification, from equation (7.11),

$$u^2\sigma^2(Y) \cong \frac{2\sigma^2(Y)}{N_{tot}}\left\{2 + \left[\frac{x_i - \mu(X)}{\sigma(X)}\right]^2\right\}.$$

Solving for N_{tot},

$$N_{tot} \cong \frac{2}{u^2}\left\{2 + \left[\frac{x_i - \mu(X)}{\sigma(X)}\right]^2\right\}, \tag{7.18}$$

which represents the total sample size required for the standard error of equating to be equal to u standard deviation units on the old form. For example, if $u = .1$, then the sample size needed for a Form X unit-normal (z) score of 0 is

$$N_{tot} \cong \frac{2}{.1^2}(2 + 0) = 400.$$

Thus, a total of 400 examinees (200 per form) would be required at a unit-normal score of 0. How about at a z-score of 2? Using formula (7.18), $N_{tot} = 1200$ (600 per form).

What can be concluded? Over the range of Form X z-scores between -2 and $+2$, the standard error of equating will be less than .1 Form Y standard deviation unit if the total sample size is at least 1200. This specification requires a normality assumption, so it should be viewed as an approximation. In addition, the range of scores is stated in z-score units, which could be transformed to reported score units when describing how the necessary sample size was estimated.

Estimating Sample Size for Random Groups Equipercentile Equating. A similar question could be asked about equipercentile equating with the random groups design. Using the same logic that was used with linear equating, an expression for N_{tot} can be derived from equation (7.14) as

$$N_{tot} \cong \frac{4[P(x_i)/100][1 - P(x_i)/100]}{u^2\phi^2}. \tag{7.19}$$

Recall that this equation assumes that the scores on Form X are normally distributed. Consequently, $z = 0$ when $P(x_i) = 50$, and $z = 2$ when $P(x_i) = 97.72$ (see Table 7.5).

For example, if u is set at .1 for a Form X z-score of 0, this equation results in $N_{tot} = 628.45$. For a Form X z-score of 2, this equation results in $N_{tot} = 3056.26$. So, over the range of Form X z-scores between -2 and $+2$, the standard error of equating will be less than .1 Form Y standard deviation unit if the total sample size is at least 3057 (by rounding up) using equipercentile equating. No smoothing is assumed in deriving this result.

Refer to Table 7.5. The ratio of sample sizes for equipercentile and linear equating equals (within rounding error) the ratios given in Table 7.6. For example, for $z = 2$, the ratio of sample sizes is $3056.26/1200 = 2.55$, which is the value shown in Table 7.6, apart from rounding error.

Estimating Sample Size for Single Group Linear Equating. Sample size requirements also can be estimated for linear equating using the single group design. Using equation (7.15) and a process similar to that used to derive equation (7.18),

$$N \cong \frac{[1 - \rho(X, Y)]}{u^2}\left\{2 + [1 + \rho(X, Y)]\left[\frac{x_i - \mu(X)}{\sigma(X)}\right]^2\right\}. \tag{7.20}$$

To use equation (7.20) it is necessary to specify $\rho(X, Y)$.

To continue the example considered earlier, what sample size is required with linear equating for the single group design so that the standard error of equating is less than .1 Form Y standard deviation unit over the range of z-scores from -2 to $+2$? Assume that $\rho(X, Y) = .70$. In this case, application of equation (7.20) indicates that a sample size of $N = 60$ is required at $z = 0$ and a sample size of $N = 264$ is required at $z = 2$. At $z = 0$, the ratio of sample sizes for the linear random groups design to the linear single group design is 6.67 (400/60), which is the ratio shown for $z = 0$ and $\rho = .70$ in Table 7.6. Similarly, the ratio for $z = 2.0$ is 4.55 (1200/264), which also is shown in Table 7.6.

Specifying Precision in sem Units.   Sometimes, equating error is specified in terms of the standard error of measurement (*sem*) rather than the standard deviation, especially when the focus of test use is on individual examinees' scores. For example, an investigator might ask what sample size would be needed for the random groups design if the standard error of equating is to be less than .1 of the standard error of measurement? Using $\rho(X, Y)$ as alternate forms reliability, the standard error of measurement is

$$sem = \sigma(Y)\sqrt{1 - \rho(X, Y)}.$$

To use equations (7.18)–(7.20) to estimate sample size, it is necessary to relate error specified in terms of *sem* units to standard deviation units. Let u_{sem} represent *sem* units. Then, multiplying both sides of the preceding equation by u_{sem} results in

$$u_{sem}sem = u_{sem}\sigma(Y)\sqrt{1 - \rho(X, Y)}.$$

Because u was defined earlier as a multiplier for $\sigma(Y)$,

$$u = u_{sem}\sqrt{1 - \rho(X, Y)}.$$

In the example, assume that $\rho(X, Y) = .7$, as was done earlier. If the standard error of equating is to be less than .1 of the standard error of measurement, then

$$u = u_{sem}\sqrt{1 - \rho(X, Y)} \cong .1\sqrt{1 - .7} = .055.$$

In the example, finding the sample size for which the standard error of equating is .1 standard error of measurement unit is the same as finding the sample size for which the standard error of equating is .055 standard deviation unit. What would be the required sample size for the random groups design at $z = 2$? Applying equation (7.18),

$$N_{tot} \cong \frac{2}{.055^2}(2 + 2^2) \cong 3966.94.$$

For the single group design, applying equation (7.20) gives

$$N \cong \frac{1 - .7}{.055^2}[2 + (1 + .7)2^2] \cong 872.73.$$

The ratio of these two sample sizes is approximately the value of 4.55 shown in Table 7.6 for $z = 2.0$ and $\rho = .7$.

Using Standard Errors in Practice

Standard errors of equating are used as indices of random error in equating. As was discussed earlier in this chapter, the delta method standard errors of equating can be used to compare the amount of equating error variability in different designs and methods, and to estimate sample size requirements. In this process, the degree of precision needs to be stated, which is necessarily situation-specific. In some situations it is necessary to have considerable precision. For example, with the ACT Assessment (AAP, 1989), important decisions are made over most of the score range; this test is used to track educational trends, and small changes in the national mean from one year to the next make front-page news; and large samples can be made available for equating, so that high equating precision can be obtained. For tests where the decisions are viewed to be less critical, more equating error (as well as measurement error) might be judged to be acceptable. Or, it might be impossible to obtain large samples for equating, and more equating error might need to be present. For many certification and licensure tests, interest is primarily in deciding whether examinees exceed a passing score. Often with these tests, a passing score is set on one test form, and the primary purpose of equating is to ensure that an equivalent passing score is used on other test forms. In this case, scores near the passing score are of primary interest, and the focus would be on equating error near the passing score when comparing designs and estimating sample size requirements. For example, in finding the sample size, the standard error of equating at the passing score that would be desirable to achieve might be no more than .1 standard deviation unit.

In using equating error variability to compare different designs and methods, and to estimate sample size requirements, the delta method standard errors with the most restrictive assumptions (e.g., normality) were used in this chapter to provide reasonable approximations. The simplicity of these approximations facilitates these comparisons and sample size estimation. Also, more specific information about distributions, such as precise estimates of skewness and kurtosis, often is not available, providing further justification for using the approximations. However, these approximations should be used cautiously because they can be inaccurate, especially when the distributions are not normal or when the other simplifications used in these derivations are unrealistic.

Equating is a statistical procedure, and, as such, the amount of random error that is present in estimating equating relationships should be documented. Like measurement error, which is often indexed by the standard error of measurement, random equating error is potentially a significant source of error in scores that are reported to examinees. Therefore, it is important to have reasonable estimates of random equating error, and to be able to tell whether random equating error adds substantially to the amount of error in test scores. Bootstrap standard errors are useful for documentation purposes, and, as was indicated earlier in this chapter, bootstrap standard errors can be calculated for rounded scale scores. If available, delta method standard errors provide an analytic expression for the standard errors, although delta method standard errors have not been developed for rounded scale scores. When using delta method standard errors for documentation purposes, standard errors derived under the least restrictive assumptions (e.g., without a normality assumption) should, in general, be used unless the sample size is very small. With small samples, the standard errors derived under the least restrictive assumptions might be inaccurately estimated. For example, estimates of skewness and kurtosis are needed to apply the standard errors of linear equating derived under the least restrictive assumptions. Large samples are needed to estimate skewness and kurtosis precisely. In one simulation, Kolen (1985) found that the delta method standard errors with the normality assumption were preferable for estimating the standard errors of Tucker equating with a sample size of 100 examinees per form. In simulations with larger sample sizes conducted by Kolen (1985), the delta method standard errors without the normality assumption were more accurate. Standard errors also can be useful in the process of choosing among methods of equating. For example, in Chapter 3, standard errors were used to help choose between different degrees of smoothing in equipercentile equating.

As was indicated earlier, analytic standard errors of IRT equating have been derived only for true score equating. Even these standard errors are underestimates. The lack of analytic standard error expressions for IRT limits the usefulness of IRT equating, because IRT equating error cannot be documented accurately, and it is difficult to develop sample size requirements for IRT equating. Bootstrap methods could be used to document equating error, but, as was indicated earlier, calculating bootstrap standard errors for IRT equating is extremely computationally intensive. Clearly, much work needs to be done in developing standard errors for IRT equating.

Exercises

7.1. Assume that the four bootstrap samples of size $N_X = 4$ shown near the beginning of the chapter (see the section titled "Standard Errors Using the Bootstrap") and listed below were for Form X of a test. Also assume that, for Form Y, $N_Y = 3$ with values 1, 4, and 5 and that the following four bootstrap samples were drawn (use $N - 1$ to calculate sample variances):

Form X	Form Y
Sample 1: 6 3 6 1	Sample 1: 1 4 4
Sample 2: 1 6 1 3	Sample 2: 4 5 5
Sample 3: 5 6 1 5	Sample 3: 1 5 5
Sample 4: 5 1 6 1	Sample 4: 1 1 4

Also assume that Form X and Form Y were administered using the random groups design.

    a. What is the bootstrap estimated standard error of linear equating at Form X raw scores of 3 and 5?

    b. Assume that the following raw-to-scale score conversion equation for Form Y has already been developed: $s(y) = .4y + 10$. What is the bootstrap estimated standard error of linear equating of unrounded scale scores for Form X raw scores of 3 and 5?

    c. For the situation described in (b), what is the bootstrap estimated standard error of linear equating of scale scores, rounded to integers, for Form X raw scores of 3 and 5?

    d. What is the delta method (assume normality) estimated standard error of linear equating of raw scores for Form X raw scores of 3 and 5?

7.2. Verify that the standard error of equipercentile equating at a Form X raw score of 25 is approximately .30 for the data shown in Table 2.5. Use equation (7.12). How does this value compare to the value calculated using (7.13)? What possible factors would cause these values to differ?

7.3. A standard setting study was conducted on Form Y of a test, and the passing score was set at a score on Form Y that was approximately 1 standard deviation below the mean in a group of examinees who took the test earlier. Assume that the group of examinees to be used in an equating study is similar to the group of examinees that was administered Form Y earlier. What sample size would be needed in random groups linear equating to achieve a standard error of equating less than .2 standard deviation unit near the passing score? Use equation (7.18). What sample size would be needed to achieve comparable precision near the passing score using random groups equipercentile equating? Use equation (7.19). Suppose that the population equating relationship was truly linear. Which method would be preferable? Why?

7.4. Suppose that Form X scores and Form Y scores each had a population mean equal to 0 and standard deviation equal to 1. Also assume that, for the population, the Form Y equipercentile equivalent of a score of 1 on Form X was 1.2 and that the linear equivalent was 1.3. For estimating the equipercentile equivalent of a Form X score of 1, would it be better to use linear or equipercentile equating in this situation if the sample size was 100 examinees per form? How about if the sample size was 1000 examinees per form? What are the implications of your answers? [Use equations (7.11) and (7.14) as a means to simplify this situation. Hint: It is necessary to incorporate the notion of equating bias and provide an expression for mean squared equating error as discussed in Chapter 3 to answer this question. In this exercise, assume that equipercentile has no bias and that linear has bias of $.1 = 1.3 - 1.2$.]

7.5. For Form X and Form Y of a 50-item test, assume that $\mu(X) = 25$, $\mu(Y) = 27$, $\sigma(X) = 5$, and $\sigma(Y) = 4$.

a. Assume that a random groups design was used with $N_X = N_Y = 500$. Find the standard error of linear equating for $x = 23$ and 35. (Use normal distribution assumptions.)

b. Assume that a single group design was used with $N = 500$ and that $\rho(X, Y) = .75$. Find the standard error of linear equating for $x = 23$ and 35. (Use normal distribution assumptions.)

c. Assume that a random groups design was used with $N_X = N_Y = 500$. Find the standard error of equipercentile equating for $x = 23$ and 35. (Use normal distribution assumptions.)

d. Assuming that the reliability of the test is .75, what sample size would be needed for the standard error of random groups linear equating to be less than .3 standard errors of measurement on the Form Y scale for $x = 23$ and 35? (Use normal distribution assumptions.)

7.6. How would you estimate the standard error of the identity equating? What are the implications of your answer for using this method in practice?

CHAPTER 8

Practical Issues in Equating and Scaling to Achieve Comparability

Many of the practical issues that are involved in conducting equating are described in this chapter. We describe major issues and provide references that consider these issues in more depth. The early portions of this chapter focus on equating dichotomously scored paper and pencil tests. In later portions, the focus broadens to include practical issues in scaling to achieve comparability, including computerized testing and performance and other types of alternative assessments.

Various articles have been written that review practical issues in equating (Brennan and Kolen, 1987b; Cook and Petersen, 1987; Harris, 1993; Harris and Crouse, 1993; Skaggs, 1990; and Skaggs and Lissitz, 1986) in greater depth than those provided in this chapter. As was described in Chapter 1, situations can occur in which equating is not appropriate and procedures for scaling to achieve comparability might need to be used. Although procedures for scaling to achieve comparability often appear to be similar to procedures for equating, the interpretations of results from these processes are quite different. Mislevy (1992) and Linn (1993) recently presented a framework for considering such procedures, which are the subject of a section near the end of this chapter.

The practical issues described in this chapter follow from the discussion of equating and scaling to achieve comparability in Chapter 1. Chapter 1 indicated that equating should be considered when alternate forms of tests exist, scores on the alternate forms are to be compared, and the alternate forms are built to the same detailed specifications so that they are similar to one another in content and statistical characteristics. It was stressed that, under appropriate conditions, equating can be used to improve the accuracy of test scores used in making important individual level, institutional level, or public policy level decisions. When decisions might be made along the entire range of scores, equating at all score points is important. If only pass-fail decisions are to be made, then the accuracy of equating might be of concern mainly near the passing score.

Also, as was indicated in Chapter 1, a major consideration in designing and conducting equating is to minimize equating error. Although the purpose of equating is to decrease error, under some circumstances implementing an equating method can increase equating error, in which case it might be best not to equate. As was described in Chapter 7, random error is error due to sampling of examinees from a population of examinees. The use of large sample sizes, smoothing in equipercentile equating, and a judicious choice of an equating design can help control random error.

Systematic error results from violations of the conditions of equating or the statistical assumptions required; it is more difficult to control than random error. Some examples of situations where systematic error might be a problem include (1) the use of a regression method (refer to Chapter 2 for a discussion of the lack of symmetry of regression methods) to conduct equating, (2) the use of linear equating to estimate an equipercentile relationship when the linear relationship does not hold, (3) the use of the Levine observed score method when true scores on the common items are not perfectly correlated with true scores on the total test, and (4) *item context effects*, in which, for example, a common item appears as the first item in Form X and as the last item in Form Y, with consequent changes in the performance of that common item. Systematic error is difficult to quantify. In practice, whether or not equating reduces systematic error can be difficult to determine, and often no clear-cut criterion for evaluating the extent of the error exists. Systematic error can best be controlled through careful test development, adequate implementation of an equating design, and use of appropriate statistical techniques.

When conducting equating, judgments must be made that go beyond the statistical and design issues described in Chapters 2 through 7. Equating requires judgments about issues in test development, designing the data collection, implementing the design, analyzing the resulting data, and evaluating the results. As is discussed later in this chapter, sometimes practical constraints do not allow sound equating to be conducted, in which case it might be better not to equate. When equating is judged to be useful, many decisions need to be made. Prior to collecting data or applying statistical equating methods, choices need to be made, such as which data collection design to use, which form(s) to use as old form(s), and how many common items to use. Other choices about how to analyze the data need to be made as well, such as which operational definition(s) of equating to use and which statistical estimation method(s) to apply. Other decisions are made after the data are collected, such as which examinees to include in the equating process, which common items to retain, and which equating result to use. Clear-cut criteria and rules for making these decisions do not exist: The specific context of equating in the particular testing program dictates how these issues are handled. Equating involves compromises among various competing practical constraints. In this sense, an ideal equating likely has never occurred in practice.

Even when an equating study is well designed and statistical assumptions are met, an otherwise acceptable equating can be destroyed because of inadequate quality control procedures. Serious problems can result, for example, if an item is incorrectly keyed, if a common item differs from one form to another, or if a mistake is made in communicating the correct conversion table. In our experience, quality control procedures deserve considerable emphasis, because problems with quality control have serious consequences. If quality control procedures fail, then the data gathered in an equating study can lead to erroneous conclusions about the comparability of test forms. In major testing programs, quality control procedures often require considerably more effort than that expended in actually conducting the statistical equating.

The practical issues in equating described in this chapter are organized by topics in roughly the order that they might need to be considered: test development, equating designs, statistical procedures, evaluating results, and quality control and standardization procedures. Then, issues in scaling to achieve comparability, including comparability for computer-based tests and performance and other alternative assessments, are discussed.

Equating and the Test Development Process

According to Mislevy (1992),

> Test construction and equating are inseparable. When they are applied in concert, equated scores from parallel test forms provide virtually exchangeable evidence about students' behavior on the same general domain of tasks, under the same specified standardized conditions. When equating works, it is because of the way the *tests are constructed* ... (p. 37)

Thus, systematic test development procedures are vital to producing adequate equating. [See Millman and Greene (1989) for a general discussion of test development procedures.]

Test Specifications

Equated scores on alternate forms can be used interchangeably only if the alternate forms are built to carefully designed *content and statistical specifications*. Developing tests in this way can result in forms that are very similar in what they measure, with the only major difference being the particular items that appear on the alternate forms. No matter how careful the test construction process is, however, the forms that result will differ somewhat in difficulty. Equating is intended to adjust for these statistical differences.

When test construction procedures are functioning well in large-scale testing programs, considerable effort is made to ensure that alternate forms are similar. The content and statistical test specifications are detailed and forms are constructed to meet these specifications. Equating can be successful only if the test specifications are well defined and stable.

The content specifications are developed by considering the purposes of testing, and they provide an operational definition of the content that the test is intended to measure. The content specifications typically include the content areas to be measured and the item types to be used, with the numbers of items per content area and item types specified precisely. The content specifications are crucial for developing alternate forms that can be equated. A test form must be sufficiently long to be able to achieve the purposes of the test, and it must provide a large enough sample of the domain for the alternate forms to be similar. For example, a 10-item test that covers a content domain consisting of 20 areas could not be expected to sample the domain adequately. If each form is an inadequate sample, then the forms can differ considerably in what they measure, and scores on alternate forms might not be interchangeable, even after equating is attempted. One useful rule of thumb is that test length should be at least 30–40 items when equating educational tests with tables of specifications that reflect multiple areas of content, although the length of a test required depends on the purposes of testing, the heterogeneity of the content measured, and the nature of the test specifications.

Although not as crucial as content specifications, statistical specifications are also important. Statistical specifications often are based on classical statistics such as the target mean, standard deviation, and distribution of item difficulties and discriminations for a particular group of examinees. Correlations of items with other tests in a test battery also might be checked to maintain the same degree of association among tests in the new forms of the battery. Statistical specifications based on IRT often are used, such as target test characteristic curves and target information curves. For previously used items, the statistics are based on previous administrations. Statistics for the new items often are estimated using item *pretesting* procedures. If items cannot be pretested, then tests might need to be constructed without the benefit of item statistics, which can make it difficult to control the statistical characteristics of the test. Another benefit of pretesting is that previously undetected item flaws might be discovered before the item is used operationally. Often item statistics are adjusted to estimate the item characteristics for a particular group of examinees under operational testing conditions.

Characteristics of Common-Item Sets

When using the common-item nonequivalent groups design, common-item sets should be built to the same specifications, proportionally, as the total test if they are to reflect group differences adequately. In constructing common-item sections, the sections should be long enough to adequately represent test content.

The number of common items to use should be considered on both content and statistical grounds. Statistically, larger numbers of common items lead to less random equating error (Budescu, 1985; Wingersky et al., 1987). Petersen et al. (1983) indicated that too few items could lead to equating problems. Very small numbers of items were suggested as adequate in some of the studies reviewed by Harris (1993), although in most of the studies that supported the use of very few common items the recommendations were based on simulating data from a unidimensional model. Because educational tests tend to be heterogeneous, larger numbers of common items are likely required for equating to be adequate in practice. Experience suggests the rule of thumb that a common item set should be at least 20% of the length of a total test containing 40 or more items, unless the test is very long, in which case 30 common items might suffice. (Angoff, 1971, suggested a very similar rule of thumb.) In considering the numbers of common items to use in a particular testing program, the heterogeneity of the test content also should be considered.

As was suggested in Chapter 6, common-item statistics can be compared across examinee groups used in the equating to help decide whether or not the items are functioning differently in the two groups. IRT statistics and classical statistics can be used. For example, items might be identified with classical item difficulties that differ by .1, in absolute value, for the old and new groups. These items could be inspected, and explanations for the differences could be evaluated. An item might be dropped from the common-item section if it were found to have problems, such as: an item was printed differently in the two forms, an item became easier due to many repeating examinees having been administered the item previously, an item whose key had changed because of changes in the field of study, or an item for which a preceding item provided information that helped in answering the item. Harris (1993) suggested that differential item functioning statistics also might be used to screen items.

Dropping items from the common-item set might result in the set of common items not reflecting the test specifications. In this case, additional items might be dropped from the common-item set (but still retained as part of the total test) to achieve proportional content balance. For this reason, the common-item set should be of sufficient length to be able to tolerate removal of some items and still remain content and statistically representative. As an alternative to dropping items to achieve proportional content balance, Harris (1991a) suggested considering the use of statistical procedures to statistically weight item scores on the common items to help achieve such balance. In their review, Cook and Petersen (1987) reported that inadequate content representation of the common-item set creates especially serious problems when the examinee groups that take the alternate forms differ considerably in achievement level. (Refer also to Harris, 1991a, and Klein and Jarjoura, 1985). Serious problems can result if the context in which the common items appear differs from the old to the new form, as was the case with the NAEP example described in Chapter 1.

One way to help avoid having the common items function differently in the two groups is to administer common items in approximately the same position in the old and new forms (Cook and Petersen, 1987). Also, the response alternatives should appear in the same order in the old and new forms (Cizek, 1994). If a common item is associated with stimulus materials that were used with a set of items in the old form, then the entire set of items associated with those stimulus materials should be included on the new form to avoid context effects. If necessary to achieve content balance, some of these items could be treated as noncommon items for the purposes of equating. Other context effects and quality control issues (e.g., items changed from one administration to another) also should be controlled. Even after all the more obvious effects are controlled, common items might still perform differently across administrations. For example, Cook and Petersen (1987) reviewed research on a biology achievement test in which differential preparation of the groups taking the old and new forms led to differential functioning of some common items that caused serious problems with equating. In short, common items should be screened for differences in functioning across the groups taking the old and new forms.

Changes in Test Specifications

Test specifications often evolve over time. In a strict sense, any change in specifications leads to forms that might not be interchangeable. With minor changes, however, testing programs often continue to attempt to equate, often with only minimal problems. Sometimes the changes in the specifications are major. For example, in an achievement test, curriculum consultants might suggest that changes in instructional programs have altered the emphases in a subject matter area, thus requiring a change in the test. In professional certification examinations, the content specifications often change because of changes in the field of study. For example, a particular content area might become obsolete and be replaced by a new area. It is even possible for the answer key for an item to change, say, because of a change in law or a change in standard procedures.

When the test specifications are modified significantly, scores obtained before the test was modified cannot be considered interchangeable with scores obtained after the test was modified, even if an "equating" process is attempted. Indeed, in these situations it is better to refer to this process as *scaling to achieve comparability*. Instead of scaling to achieve comparability, the changes in content are often judged to be severe enough that the tests are rescaled. For example, when the SAT was revised, various technical issues associated with implications of changes in the test and the score scale were studied intensively (Lawrence *et al.*, 1994; Dorans, 1994a,b). When the ACT Assessment was rescaled (Brennan, 1989) concordance tables were developed that related scores on the new test to scores on the old test. In both of these cases, the ranges of scale scores stayed the same for political reasons, although choosing a distinct new score scale might have avoided confusion between the old and new scores. In practice, changes in specifications come in varying degrees, and the chosen approach should be tailored to the situation.

Data Collection: Design and Implementation

To conduct equating, a choice must be made about which equating design to use (see Chapters 1 and 6). Choices also need to be made about which previously administered form(s) are to be the old form(s) and what sample size to use. Adequate equating depends on having well-constructed tests, as was described earlier, and well-developed statistical and quality control procedures, as is described later in this chapter.

Choosing Among Equating Designs

The random groups, single group, single group with counterbalancing, and common-item nonequivalent groups designs were discussed in Chapter 1 and in subsequent chapters. In addition, designs that involve equating to an IRT calibrated item pool were described in Chapter 6.

The choice of an equating design involves a number of practical considerations that include test administration complications, test development complications, and statistical assumptions required to achieve the desired degree of equating precision. The relationship of these considerations to each of these designs is summarized in Table 8.1. As can be seen from this summary, the choice of a design requires making a series of trade-offs.

The random groups design typically results in the fewest test development complications, because there is no need to develop common-item sets that are representative of the content of the total test. (However, alternate forms still should be built to the same content and statistical specifications, and the forms must be developed in time to be equated in a special study.) Also, because group differences are handled by randomly assigning forms to examinees, and because there is no problem with order effects, this design results in the fewest problems with statistical assumptions.

Many equating situations exist, however, for which the random groups design cannot be used. If not enough examinees are available for using the random groups design, then the single group design might be preferable,

Table 8.1. Comparison of Equating Designs.

Design	Test Administration Complications	Test Development Complications	Statistical Assumptions Required
Random Groups	Moderate—more than one form needs to be spiraled	None	Minimal—that random assignment to forms is effective
Single Group with Counter-balancing	Major—each examinee must take two forms and order must be counter-balanced	None	Moderate—that order effects cancel out and random assignment is effective
Common-Item Nonequivalent Groups	None—forms can be administered in typical manner	Representative common-item sets need to be constructed	Stringent—that common items measure the same construct in both groups, the examinee groups are similar, and required statistical assumptions hold
Common Item to an IRT Calibrated Item Pool	None—forms can be administered in typical manner	Representative common-item sets need to be constructed	Stringent—that common items measure the same construct in both groups, the examinee groups are similar, and the IRT model assumptions hold

provided that a study can be undertaken in which two forms can be administered to each examinee and order can be counterbalanced effectively.

One situation that is often encountered in which the random groups and single group designs cannot be used is when only a single test form can be administered on a test date. Many of the reasons for using a single form revolve around test security. For example, administering a single form exposes fewer items than administering more than one form. Also, administering a form that is comprised mainly of new items minimizes the chances that examinees previously would have been exposed to the test items and minimizes the chances of a security breach in which items become known to examinees.

When only a single form can be administered on a test date and equating is to be conducted, the choice of equating design is restricted to a design that uses common items. When using these designs, representative common-item sets must be developed. Constructing representative common-item sets and incorporating them into the forms requires considerable effort during the test development process.

Test disclosure legislation also can complicate the choice of design (Marco, 1981). Such legislation often requires that all items which contribute directly to an examinee's score be released to the examinee soon after the test. When the items are released in this way, they cannot be used in future test forms because they are considered to be nonsecure. The typical legislation provides test developers with a way to conduct equating by not requiring that an unscored portion of a test be provided to examinees. Equating could be conducted, therefore, using the common-item non-equivalent groups design with external sets of common items, as is done with the SAT (Donlon, 1984). As was pointed out in Chapter 1, external common-item sets do not contribute directly to an examinee's raw score. Thus, these sections do not need to be released to examinees, even though the scored portion would be released to examinees.

Preequating methods also can be considered in test disclosure situations. In item preequating (see Chapter 6), an IRT calibrated item pool is developed. A new form is constructed from the items in this pool. Because all of the items have already been calibrated using an IRT model, the item parameter estimates for the new form are available and can be used to develop the conversion table before administering the new form intact. In using item preequating, new items are introduced by including some new items on the operational form, but not including these new items in the computation of examinees' scores. Research reviewed in Chapter 6 suggests that various context effects need to be controlled with item preequating. To minimize context effects, items should appear in a position and context when they are operational that is similar to the position and context in which they appeared when they were preequated.

Section preequating is another type of preequating methodology. In section preequating the operational portion of the test consists of sections of items that have been previously administered, with necessary item or section parameters estimated using data from the previous administrations. Using these results, the conversion table for the operational portion is estimated before it is ever administered as an intact form. Other sections administered to examinees are unscored, and are used to build up the pool of sections with estimated item or section parameters for use in subsequent forms. Linear methods, as well as IRT methods, can be used in section preequating. Linear methods can accommodate sections that measure different abilities. Petersen et al. (1989) provided a summary of section preequating procedures. Brennan (1992) illustrated that context effects involving the positioning of sections of items need to be controlled in section preequating designs. Harris (1993) presented a discussion, with many references, of practical issues in preequating.

Some situations require that tests be equated before being administered intact in a standard operational setting, such as when scores need to be reported to examinees immediately after they are administered a test. In this case, a conversion table needs to be available before the test administration. Preequating can be used in these situations.

Another way conversion tables could be made available prior to administering the test operationally is to use nonoperational administrations to conduct equating, so conversion tables will be available later for operational administrations. The equating results then are used when the form is used operationally. For example, a random groups design is used initially to equate new forms to an old form of the *Armed Services Vocational Aptitude Battery* (ASVAB) (Thomasson *et al.*, 1994) based on examinees who are already in the military. In a second random groups equating study, these new forms, along with a form that was equated previously, are administered operationally to examinees who want to be accepted into the military. Scores on the new forms for examinees in the second equating study are based on the initial equating. The conversion tables from the second equating are used subsequently, because the examinees in the second equating, as compared to examinees in the first equating, are likely to be more motivated and more similar to the examinees who are to be tested subsequently.

Another variation is used for equating the ACT Assessment (ACT, 1989). On most national test dates, the items on the ACT tests are released to examinees, in part, to meet test legislation requirements. However, on certain test dates the items are not released. On one of these test dates, one or more previously administered unreleased forms along with the new forms to be equated are administered using a random groups design. These forms are equated following this administration, and scores are reported to examinees who were administered the new forms. The conversion tables developed in the equating administration also are used when the new forms are administered later on.

Although not a comprehensive set of possibilities, the SAT, ASVAB, and ACT equating designs illustrate the use of the random groups design and the common-item nonequivalent groups design in situations that might suggest the need for an item or section preequating design.

Developing Equating Linkage Plans

When conducting equating, a choice is made about which old form or forms are to be used for equating a new form or forms. The choice of the old form or forms has a significant effect on how random and systematic equating error affects score comparisons across forms.

Random Groups Design. Consider the following example of a simple equating linkage plan. For the ACT Assessment (ACT, 1989), new forms are equated each year using a random groups design in which the new

Table 8.2. A Random Groups Equating Linkage Plan
that Uses a Different Old Form at Each Administration.

Process	Administration	Forms		
Construct Score Scale	1	A		
Equate Using Spiraling	2	$\boxed{\text{A}}$	B	C
Equate Using Spiraling	3	$\boxed{\text{C}}$	D	E
Equate Using Spiraling	4	$\boxed{\text{E}}$	F	G
Equate Using Spiraling	5	$\boxed{\text{G}}$	H	I

forms are spiraled along with one form that was equated in a previous year. This process allows the new form raw scores to be converted to scale scores by first equating raw scores on the new forms to raw scores on the old form. The raw-to-scale score conversion that was developed previously for the old form then is used to estimate raw-to-scale score conversions for the new forms.

A hypothetical example that displays a linkage plan which is similar to the ACT plan is shown in Table 8.2, where the old form is listed in a box. In Administration 1, the raw-to-scale score transformation for Form A establishes the score scale. In Administration 2, new Forms B and C are administered with Form A in a spiral administration. The data collected from this administration are used to develop the conversion that transforms Form B and Form C raw scores to scale scores through Form A. In Administration 3, Form C serves as the old form and Forms D and E as the new forms. The general plan is to spiral new forms along with an old form that was equated previously.

ASVAB (Thomasson *et al.*, 1994) also is equated using the random groups design. However, in the ASVAB program, the form that was used to conduct the original scaling is always the old form that is spiraled with the new forms. A hypothetical example that displays a linkage plan similar to the ASVAB plan is shown in Table 8.3.

Note that the major difference between the plans shown in Tables 8.2 and 8.3 is the old form that is used in the spiraling process. In Table 8.2, the old form is a form that was equated in the previous year. In Table 8.3, the old form is a form that was used initially in the scaling process. Both plans can be used to produce raw-to-scale score conversions. Is one plan preferable to the other? The answer depends on various practical issues having to do with the context of the equating.

One of these issues has to do with error in equating. As was suggested earlier, each time an equating is conducted, equating error is introduced. Error might accumulate over equatings. In Table 8.2, how many equatings separate Form I from Form A?

Table 8.3. A Random Groups Equating Linkage Plan
that Uses the Same Old Form at Each Administration.

Process	Administration	Forms		
Construct Score Scale	1	A		
Equate Using Spiraling	2	A	B	C
Equate Using Spiraling	3	A	D	E
Equate Using Spiraling	4	A	F	G
Equate Using Spiraling	5	A	H	I

Equating 1: Form I is equated to Form G.
Equating 2: Form G is equated to Form E.
Equating 3: Form E is equated to Form C.
Equating 4: Form C is equated to Form A.

Thus, four equatings separate Form I and Form A. Equating error from four different equatings would influence the comparison of scores between examinees who took Form A and those who took Form I.

How many equatings separate Form I from Form A in the example in Table 8.3? Just one. That is, error sources from only one equating influence the comparison of scores between examinees who took Form A and those who took Form I in the Table 8.3 plan. At least from this perspective, the plan in Table 8.3 is preferable.

However, there are at least two potential problems with the plan in Table 8.3. First, this plan requires Form A to be administered repeatedly. If the items became known to some examinees because of security breaches (e.g., test booklets stolen or students memorizing items and supplying them to coaching schools) or because many repeating examinees had seen the items in an earlier administration, then the equating could be severely compromised. Second, the content of Form A might become dated. For example, reading passages might become less relevant, causing examinee groups to respond differently to the passages over time. Also, an accumulation of minor changes in the way test specifications are applied over time might make Form A somewhat different from later forms. For these reasons, a plan like the one displayed in Table 8.3 must be used cautiously. Whether to use a plan like the one in Table 8.2 or one like that in Table 8.3 depends on weighing the problems associated with each of the plans and deciding which problems are more serious for the testing program at hand.

Compromise plans also could be constructed. For example, in the plan in Table 8.3, Form A could be used as the old form in Administrations 2 and 3. Then Form E could be used as the old form in Administrations 4 and 5. Compared to the plan in Table 8.3, this compromise plan would reduce the usage of Form A. Compared to the plan in Table 8.2, this compromise plan would lead to fewer equating error sources in comparing scores on Form A to scores on Form I.

In practice, constructing equating plans can be much more complicated than what has been considered in these hypothetical examples. A particular form might need to be ruled out as an old form because of security concerns or because many of the examinees to be included in the equating were administered the old form on a previous occasion. Also, an old form might be found to have bad items (e.g., items that are ambiguous, multiply keyed, or negatively discriminating), which could rule out its use in equating. These sorts of practical concerns often make it impossible to develop equating plans that are actually used very far into the future.

Double Linking with Random Groups.   One procedure that is often used to help solve the problems associated with developing linkage plans is to use two old forms to equate new forms. This process is referred to as *double linking*. As an example of double linking, the scheme in Table 8.2 could be modified by also administering Form B in Administration 4 and Form D in Administration 5. The resulting plan is shown in Table 8.4. In applying double linking, the new forms are equated separately to each of the old forms. The resulting equating relationships could then be averaged. For example, in Administration 5, one equating relationship could be developed to equate Form H to scale scores using Form D as the old form. A second equating relationship could be developed for equating Form H to scale scores using Form G as the old form. These two relationships likely would differ because of equating error. The two conversions could be averaged to produce a single conversion. (Braun and Holland, 1982, suggested using bisection instead of averaging when linear equating is used. Averaging and bisection produce very similar results, and averaging is simpler.)

The process of double linking has much to recommend it. It provides a built-in stability check on the equating process, which is in keeping with AERA, APA, NCME (1985) Standard 4.9. Two conversions that differ more than expected by chance might suggest problems with statistical as-

Table 8.4. A Random Groups Equating Linkage Plan that Uses Double Linking.

Process	Administration	Forms			
Construct Score Scale	1	A			
Equate Using Spiraling	2	[A]	B	C	
Equate Using Spiraling	3	[C]	D	E	
Equate Using Spiraling	4	[B]	[E]	F	G
Equate Using Spiraling	5	[D]	[G]	H	I

sumptions, quality control (e.g., scores incorrectly computed), administration (e.g., spiraling was not properly performed), or security (e.g., a security breach led to many examinees having access to one of the old forms). If such problems are suspected, then one of the links could be eliminated without destroying the ability to equate in the testing program. (Note, however, that if a security breach led to many examinees having had access to one of the old forms, then the scores of the examinees who took that old form might not be valid.) In addition, the use of double linking can provide for greater equating stability than the use of a single link, especially when the two old forms are chosen from different administrations, as was done in Administrations 4 and 5 in Table 8.4.

The average of two links also can be shown to contain less random equating error than the use of a single link. Consider the following situation. In one equating, Form C is equated directly to Form A; and in a second equating, Form C is equated first to Form B and then to Form A. For simplicity, assume that the error variance in equating is the same for any single equating. The equating of Form C to Form A contains the same amount of equating error variance as the equating of Form B to Form A. Refer to the error variance at a particular score point on Form C as var. Also assume that all equatings are independent.

In this case, the equating error in equating Form C to Form A equals var. Equating error variance in equating Form C to Form A through Form B equals $2var$. The average of the equivalents of the two equatings equals the sum of the equivalents divided by 2. In this case, equating error variance for the average can be shown to equal

$$\frac{1}{2^2} var + \frac{1}{2^2}(2)var = \frac{3}{4} var.$$

In this example, equating error variance for the average of the two links, $3/4 var$, is less than the equating error variance for either link taken by itself. This relationship illustrates that the use of double linking can reduce random equating error.

In practice, the double links might not be equally weighted. If one link is considered to have more error than another link, the first link might be weighted less than $1/2$. If substantial problems are present with one of the links, that link can be weighted 0.

Double linking does introduce complications into equating. More than one old form must be included in the study, which assumes the availability of another form and requires exposing more forms in the study, which might lead to security concerns. Using additional forms also requires that the overall sample size be larger, which in some cases might not be possible. For example, if the sample size needs to be 2000 examinees per form and 4000 examinees are available to do equating, then only one old form could be used when equating one new form. Even though there are complications in using two old forms, we recommend using double linking when feasible.

Common-Item Nonequivalent Groups Design.   Additional complications are present when developing equating plans with the common-item non-equivalent groups design. Group differences across administrations some-times are substantial. As was suggested earlier, the similarity between ex-aminee groups that are administered the old and new forms significantly affects the quality of equating: The more similar the groups, in general, the more adequate the equating.

The following situation illustrates some of these complications. A test is administered in the spring and in the fall every year, with a different form administered on each occasion. The group of examinees that tests in the spring tends to be different in its overall level of achievement than the group that tests in the fall. This difference in group level achievement sug-gests greater equating stability when a new form is equated to a form from the same time of year than to a form from a different time of year. A single section of common items is used to equate a new form.

Single Link Plan 1 in Figure 8.1 presents one possible single link pattern for this situation over a 5-year period. In this example, assume that the score scale was established on Form A. The arrows indicate which old form has items in common with the new form. For example, Form J is equated to the score scale using items that are in common with Form H. In this plan, Spring forms are always equated to Spring forms and Fall forms are always equated to Fall forms, with the exception of Form B. Note that in setting up equating patterns, all forms must link back to a single old form through an equating chain, so scores on all forms can be converted to scale scores. For this reason, Form B must be equated to Form A in the Figure 8.1 example.

This equating plan can be used to equate all forms to the score scale, because all forms eventually link back to Form A. This pattern makes as extensive use as possible of linking to forms that were previously adminis-tered in the same time of year, thus maximizing the similarity of groups used in the equatings. From the perspective of using similar groups, this plan is nearly ideal.

However, this plan has significant problems. Suppose that examinees tested in the Fall of Year 5 were to be compared to examinees tested in the Spring of Year 5. How many links would affect this comparison? Another way to ask this question is, how many arrows does it take to go from Form J to Form I in the linkage plan? By going from J to H, H to F, F to D, D to B, B to A, A to C, C to E, E to G, and G to I, there are nine of these arrows. Thus, nine links affect the comparison of scores on Form I and Form J. If this pattern were extended, the number of links for comparisons between forms administered in a given year increases by two each year. This linkage plan illustrates the development of what is sometimes referred to as an *equating strain*. Equating strains can lead to a situation in which examinees earn higher scale scores on one form than on another form. In developing equating linkage plans, equating strains should be avoided.

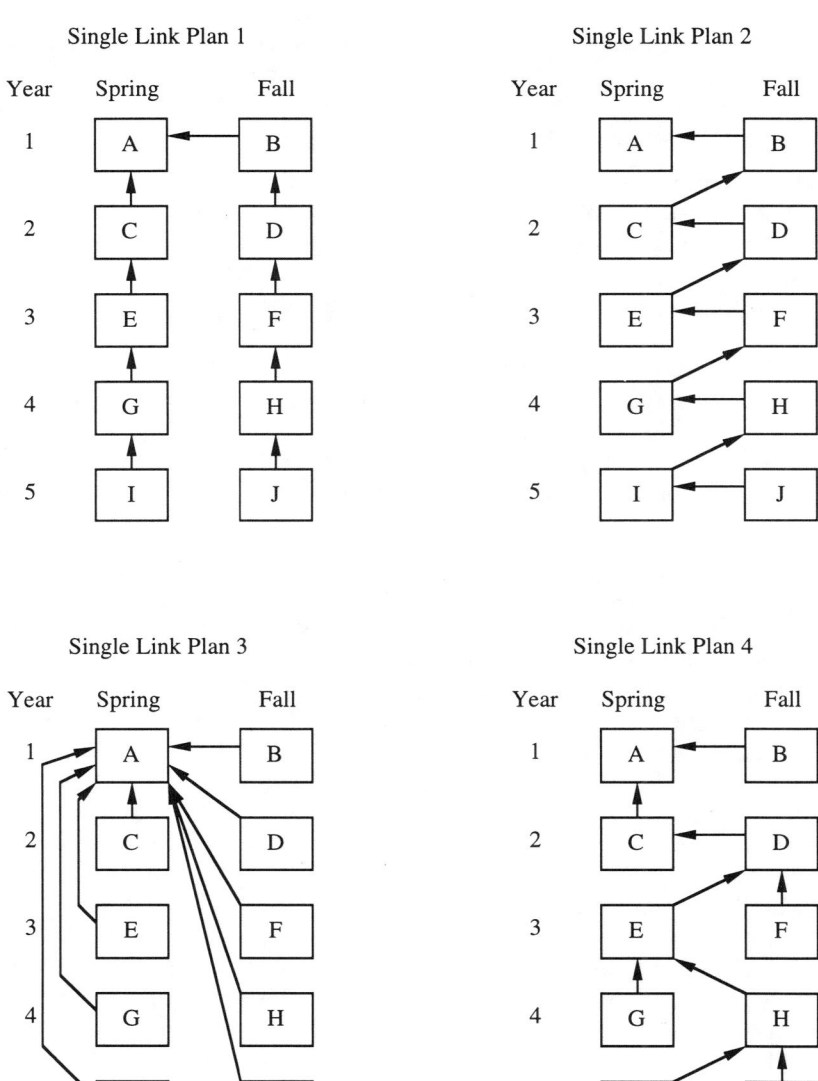

Figure 8.1. Four hypothetical single link plans.

The random groups and common-item nonequivalent groups examples considered so far illustrate the following four rules that can be used to construct equating linkage plans for the common-item nonequivalent groups design with internal common items:

Rule 1.   Avoid equating strains by minimizing the number of links that affect the comparison of scores on adjacent forms. (Single Link Plan 1 in Figure 8.1 violates this rule.)

Rule 2.   Use links to the same time of the year as often as possible. (Single Link Plan 1 in Figure 8.1 is an example of a plan that follows this rule.)

Rule 3.   Minimize the number of links back to the initial form. (The plan in Table 8.3, for the random groups design, is an example of a plan that follows this rule.)

Rule 4.   Avoid linking back to the same form too often. (The plan in Table 8.2, for the random groups design, is an example of a plan that follows this rule.)

Obviously, all of these rules cannot be followed simultaneously when constructing a plan that uses single links. Choosing a plan involves a series of compromises that must be made in the context of the testing program under consideration. For example, Rule 3 might be considered important when following trends in scores over time, but not otherwise.

Some additional examples can be used to explore these four rules more fully. Refer to Single Link Plan 2 in Figure 8.1. Rule 1 is followed as closely as possible, because forms are equated directly to the adjacent form. Rule 2 is violated as much as possible, because forms are always equated to a form from the other month. Rule 3 also is violated, in that the number of links back to Form A is as large as possible. Rule 4 is followed.

The Single Link Plan 3 in Figure 8.1 follows Rules 1 and 3. Rules 2 and 4 are not followed.

In the Single Link Plan 4 in Figure 8.1, Rule 1 is followed reasonably closely, in that there are no more than two links (arrows) separating adjacent forms. Rule 2 is followed for nearly 1/2 of the forms. Rule 3 is followed more closely for this plan than for Single Link Plan 2 in Figure 8.1, but less closely than for Single Link Plan 3 in Figure 8.1. Rule 4 is followed reasonably closely, although nearly 1/2 of the forms are equated back to twice. Although Single Link Plan 4 in Figure 8.1 is less than ideal, this plan might be a reasonable compromise.

The linkage plans in Figure 8.1 are presented for illustrative purposes only. Often, practical constraints make plans like these unworkable. For example, if many examinees repeat the test, a form that was administered within the last year or two might not be a good choice to use as a link form. The examinees who repeat the test could be unfairly advantaged by being administered the same items a second time. In other situations, scores might need to be comparable over a long period, in which case it probably would be desirable for at least one of the link forms to be a form that was originally administered in the more distant past. Sometimes problems exist with a potential old form which suggest that the form not be used as a link form. For example, the sample size for a potential link might have been very small when that form was equated, a potential link form might have had security problems, or a potential link form might have been found to have not been well constructed. Many testing programs have more than two test dates per year, which also complicates the design of equating plans. For an example, refer to the SAT linkage plan that is presented in Donlon (1984, pp. 16–17). Of necessity, linkage plans should be tailored to the particular testing program. However, the principles discussed here can be useful in designing and evaluating these plans.

Double Linking with the Common-Item Nonequivalent Groups Design. Double linking is useful in the common-item nonequivalent groups design because, as with the random groups design, it provides a built-in check on the equating process leading to greater equating stability, and it can be used to avoid equating strains. In addition, with two links, a second link still is available to be used for equating even if the strong statistical assumptions required under the common-item nonequivalent design are violated for one of the links. Also, if a significant number of common items on one link are found to have problems, or if security problems are discovered with one of the old forms, then a second link still exists that can be used to conduct the equating.

Double linking requires greater effort in test development and in equating than does equating using a single link. When using the common-item nonequivalent groups design, double linking requires that two sets of common items which are content representative be used in the development of new forms, which sometimes can be difficult. Using two links also creates a greater exposure of old forms in the random groups design and of common items with the common-item nonequivalent groups design. Double linking is most desirable in situations where form-to-form comparability is important over a long time, and might be less important in situations where periodic changes in test content require that the test be rescaled every few years. It is strongly recommended that double linking be used when feasible.

To capitalize on the benefits associated with double linking, the use of triple links has been suggested (McKinley and Schaeffer, 1989). However, triple linking can be difficult, practically, because it requires building three sets of common items that are content representative, and it can create even more exposure of forms and items than does double linking.

One beneficial way to use double linking in IRT equating to an item pool is for one link to be to a single old form and the other link to be to the overall pool. In this way, one of the links is an equating using the common-item nonequivalent groups design. This double linking process allows for use of the traditional methods as a check on the IRT methods.

Examinee Groups Used in Equating

Equating relationships typically are group dependent, so the group or groups of examinees used in equating affect the estimated equating relationship. For this reason, more adequate equating is expected when the examinees used in the equating study are as similar as possible to the entire group that is tested (Harris, 1993). Problems might occur when equating is conducted in a special study in which the groups are very different from the examinees who are to be tested later. In addition to differences in group characteristics, differences in examinee motivation between special studies and operational testing can affect equating. The ASVAB example presented in Chapter 1, in which the examinees were more motivated on the old form than the new form, is an extreme example of how motivation differences can cause significant problems.

The effect of the group used for equating depends on the data collection design. When carefully constructed alternate forms are equated using large samples in the random groups design, the equating relationship seems not to be too dependent on the group of examinees used to conduct the equating (Angoff and Cowell, 1986; Harris and Kolen, 1986). In the common-item nonequivalent groups design, however, large differences between the old and the new groups can cause significant problems in estimating equating relationships, both for traditional and IRT equating methods (for reviews of relevant research see Cook and Petersen, 1987; Harris, 1993; Skaggs, 1990; and Skaggs and Lissitz, 1986). Large group differences can lead to failure of the statistical assumptions for any equating method to hold. The research in the Dorans (1990a) special issue of *Applied Measurement in Education* assessed the use of matching procedures to make otherwise disparate groups more similar, but found that the procedures studied were not satisfactory.

The various statistical methods handle group differences differently. The Tucker, Braun-Holland, and frequency estimation methods require assumptions about the same regression holding across the different populations. These assumptions cannot be expected to hold when groups differ substantially. The IRT and Levine methods require that the common items and total scores measure the same construct in the two groups, in the sense that true scores are functionally related. This requirement places considerable emphasis on test development procedures, so that the same construct is measured in precisely the same way across alternate forms and common-item sets. If this requirement is met precisely, then the Levine and IRT methods might function more adequately than the other methods when there are large group differences. However, when the group differences become too large, no method likely will function well (see Cook and Petersen, 1987).

In our experience with the common-item nonequivalent groups design, mean differences between the two groups of approximately .1 or less standard deviation unit on the common items seem to cause few problems for any of the equating methods. Mean group differences of around .3 or more standard deviation unit can result in substantial differences among methods, and differences larger than .5 standard deviation unit can be especially troublesome. In addition, ratios of group standard deviations on the common items of less than .8 or greater than 1.2 tend to be associated with substantial differences among methods. Differences in group standard deviations have the potential to lead to differences among methods that are at least as great as those caused by differences in means. These rules of thumb are necessarily situation specific.

Repeating Examinees.   A consideration when conducting equating is whether or not to eliminate examinees who have taken the test previously. One argument for removing examinees who are repeating the test is that they might have seen the old form or common items, which could bias the equating. However, repeating examinees might not be identifiable in the time allowed for conducting equating. Also, excluding repeating examinees reduces sample size, which might lead to inadequate equating precision. Excluding repeaters might also lead to the group being included in the equating not being representative of the group tested, especially if many examinees repeat the test. Research on the effects of repeating examinees on equating (Andrulis *et al.*, 1978; Cope, 1986) produced mixed results. Decisions about whether or not to include repeating examinees in equating in a particular testing program depend on assessing how likely it is that examinees would have seen previously administered items or forms and whether or not it is possible to identify repeating examinees.

Editing Rules.   Another consideration is whether to delete examinees whose scores are very low or who omitted many items. For example, examinees who omit all the items on a test or earn a number-correct score of 0 often are excluded from equating. These are likely to be examinees who did not attempt the test and might have been erroneously included in the data. Many IRT parameter estimation programs (e.g., LOGIST) effectively eliminate examinees who earn scores of 0 or all correct when estimating item parameters. Editing rules should be tailored to the particular testing program.

Less conservative rules might negatively affect equating. Suppose that in a random groups design a sizable number of examinees typically earn scores below "chance" (number of multiple-choice items divided by the number of alternatives per item) on a test, and that more examinees scored below chance on the more difficult of the two forms. Eliminating these below "chance" examinees from the equating process could destroy the random equivalence between the samples taking Form X and Form Y, and it would result in the loss of all data in the lower tail of the distributions. We recommend using conservative editing rules whenever possible.

Another consideration is whether to eliminate test centers or testing sessions that had administration problems. For example, in the random groups design, each of the forms to be equated would be expected to be administered to approximately the same number of examinees in each test session. Numbers that are grossly unequal suggest administrative problems. In this case, elimination of the data for a test center or session can be considered. Elimination of data from test centers or test sessions with significant irregularities, such as a power failure that disrupted testing, also can be considered.

Sample Size Requirements

Sample size has a direct effect on random equating error. Livingston (1993) provided an interesting study of equating when the sample size is small. Harris (1993) reviewed research on sample size in equating and suggested that larger samples lead to better equating. In this section, schemes for estimating sample size requirements are considered that are mainly based on considerations in estimating random error in equating.

Rules of Thumb Using Standard Deviation Units as a Criterion. In Chapter 7, procedures were provided for estimating the sample size required to achieve a given level of equating precision. For the random groups design under normality assumptions, the standard error of equating between z-scores of -2 and $+2$ was shown to be less than .1 raw score standard deviation unit when the sample size was 400 per form for linear equating and slightly over 1500 per form for equipercentile equating. In any given situation, however, the shapes of the distributions, the degree of equating precision required, and the effects of smoothing if equipercentile equating is used (see the sample size discussion in Chapter 3) can be taken into account when developing sample size requirements. In addition, if a passing score is to be used in the testing program, then the precision at that passing score might be of primary concern (see Brennan and Kolen, 1987b, pp. 285–286).

Our experience suggests that these figures are also useful rules of thumb for sample size requirements for linear and equipercentile equating in the common-item nonequivalent groups design. Sample size considerations under this design, however, are complicated in that the degree of relationship between the total score and common-item score (see Budescu, 1985), along with the distribution shapes, have a strong influence on the standard errors.

Comprehensive standard error of equating expressions have not been derived for IRT equating procedures. The procedure used to estimate item parameters will likely affect the sample sizes required. A rule of thumb that is loosely based on the literature surveyed by Harris (1993) would be to require the same number of examinees for the three-parameter model as for equipercentile equating (approximately 1500 per form) and to require the same number of examinees for the Rasch (one-parameter) model as for the linear methods (400 per form).

Rules of Thumb Based on Comparisons with the Identity Equating. The rules of thumb just developed for the traditional methods were based on using a conservative criterion (standard errors of equating being less than .1 raw score standard deviation unit). The sample size issue can be addressed by asking a different question: What is the smallest sample size that would be expected to reduce equating error as compared to identity equating?

If identity equating is used, the Form Y equivalent of a Form X score is set to equal to the Form X score. That is, the Form Y equivalent of a Form X score of x_i is x_i. If equipercentile equating is the most appropriate method, then the bias incorporated by using identity equating is $x_i - e_Y(x_i)$. As was indicated in Chapter 3, the sum of random equating error variance and squared bias equals mean squared error in equating. Based on this relationship, *the identity equating is preferable to equipercentile equating if the squared bias associated with the identity equating is less than the random equating error variance associated with using equipercentile equating.*

The following example illustrates the application of this principle. In developing the rules of thumb mentioned earlier, a sample size of approximately 1500 per form was found to be required for the standard error of equating at any score to be less than .1 raw score standard deviation unit over the z-score range of -2 to $+2$. *Assume* that the largest absolute difference in equivalents between identity equating and equipercentile equating, $|x_i - e_Y(x_i)|$, is .1 standard deviation unit over the z-score range of -2 to $+2$. Thus, over this range, the maximum absolute equating bias associated with identity equating is *assumed* to be .1 standard deviation unit. Because squared bias and squared standard errors contribute equally to mean squared error, the same maximum level of mean squared error will accrue over the z-score range of -2 to $+2$ through the use of identity equating or equipercentile equating with a sample size of approximately 1500. Thus, in this situation, a sample size over 1500 would be required for equipercentile equating to result in less mean squared error than identity equating.

What if the largest difference in equivalents between using identity equating and equipercentile equating was *assumed* to be .2 standard deviation unit over the z-score range -2 to $+2$? Using equation (7.19) with $u = .2$, the sample size per form is approximately 382. Assuming a maximum difference in equivalents of .2 standard deviation unit, a sample size of over 382 would be required for equipercentile equating to produce less mean squared error than identity equating.

As was just demonstrated, this scheme is very sensitive to the extent that the forms are assumed to differ. Assuming that the forms are similar enough to be equated, *the larger the anticipated difference between forms, the smaller the sample size needed for equating to be useful.* However, larger representative samples lead to less random error. This scheme depends on the distributions of the scores (normal distributions were assumed here). However, if reasonable approximations to the distribution shapes can be found, and if reasonable assumptions about the degree of difference between forms can be made, then this scheme can be used to decide whether identity equating is preferable to another equating method.

Choosing From Among the Statistical Procedures

Various statistical methods for equating have been presented. For any of these methods to be used appropriately, the test specifications, the data collected, and the standardization and quality control procedures should be adequate. Otherwise, not equating (or using identity equating) might be the preferred option. Although it might be possible to implement all of the methods that have been discussed in a particular testing program, practical circumstances often rule out implementing some methods and suggest ruling out others.

Equating Criteria in Research Studies

Considerable research has been conducted that can be consulted when deciding which procedures to use. The reviews of research that have been done (Brennan and Kolen, 1987b; Cook and Petersen, 1987; Harris, 1993; Harris and Crouse, 1993; Skaggs, 1990; Skaggs and Lissitz, 1986) should be consulted for more comprehensive summaries of research than is presented in this chapter. Conducting research on equating and using the results from research on similar tests can help in deciding which statistical procedures to implement when equating a test. Consider the following questions:

1. In what situations do each of the equating methods perform adequately, and in what situations do each of the equating methods perform inadequately?
2. In a particular testing program, which equating method(s) should be applied?

Equating criteria are useful for addressing these questions.

Harris and Crouse (1993) conducted an exhaustive survey of criteria for comparing equating methods and results that have been used to address these questions. Some criteria for addressing these questions have been discussed previously in this book. In this section, some criteria are considered that have been used to address Question 1. Refer to Harris and Crouse (1993) for a more complete discussion.

One of the criteria identified by Harris and Crouse (1993) that has been used in research studies is *equating in a circle*. To use this paradigm in a situation with three forms, Form X is equated to Form Y, Form Y is equated to Form Z, and Form Z is equated back to Form X. Following through this chain, Form X is equated to itself. In this paradigm, equating is adequate to the extent that a Form X raw score of 1 converts to a score of 1, a raw score of 2 to a score of 2, etc. This paradigm can be used if Forms X, Y, and Z are equated using a random groups design. This paradigm also can be used with the common-item nonequivalent groups design if there are items in common between Forms X and Y, between Forms Y and Z, and between Forms Z and X. Angoff (1987) considered this criterion to be useful because "it provides advance knowledge of what the errorless result should be ..." (p. 298).

Although equating in a circle appears sensible, Brennan and Kolen (1987a,b) pointed out concerns with this paradigm. First, they indicated that identity equating will always be preferable to equating when using this paradigm. They demonstrated that equating methods which estimate fewer parameters (e.g., linear equating) tend to perform better than methods that estimate more parameters (e.g., equipercentile equating). They also demonstrated that, under the common-item nonequivalent groups design, the results of the comparison depend on the form used to start the circle. That is, different results are found when Form X is equated to itself through Forms Y and Z than when Form Z is equated to itself through Forms X and Y. These problems suggest cautious use of the equating in a circle paradigm. However, this procedure could be useful in identifying methods that produce poor equating results, in that if a method does not work well when equating a form to itself, it might not work well when equating alternate forms.

Another criterion reported by Harris and Crouse (1993) involves using *simulated equating* by defining true equating using a psychometric model and then generating data to fit the model. In this way, the true equating is known and can be used as the criterion equating. They pointed out, however, that the particular model used to generate the data might tend to favor certain methods over others, and that the usefulness of this approach depends on how well the generated data mimic data from real testing programs.

Harris and Crouse (1993) also discussed *large sample criteria*, in which data from a large sample of examinees are used to represent the population relationship. Smaller samples are drawn and the results are compared to the large sample results. For example, Livingston *et al.* (1990) used SAT forms from a spiral administration in which approximately 250,000 examinees were tested. The equating relationship based on the random groups design was used as the "true" equating relationship. The forms also had common-item sections. Nonequivalent examinee groups were formed from the available random groups. Various procedures were used to equate the forms using the common-item nonequivalent groups design. The results for the different procedures were compared to the random groups equating based on all 250,000 examinees. Harris and Crouse (1993) pointed out that this criterion is limited in its usefulness, because large groups of examinees typically are not available. In addition, the results are meaningful only to the extent that the examinee groups are formed in a manner that is similar to how groups occur in practice.

Based on their review of these criteria and others, Harris and Crouse (1993) concluded that "... no definitive criterion for evaluating equating exists ..." (p. 230). They went on to say that

> Given the controversy regarding which criterion is best, whether certain criteria are useful, and whether a criterion is needed at all, much work needs to be done in the area of equating criteria. As long as equating is performed, equating criteria will be needed to evaluate the results ... The fact that equating results appear to be so situation specific demands that studies be replicated and that some method of comparing results across studies be developed. (p. 232)

This discussion of criteria suggests that research can provide information about which method to use. However, it is unlikely that such research will lead to an unambiguous choice of an equating method, in part because different criteria might lead to the choice of different methods (see Harris and Crouse, 1993, for examples and additional references).

Characteristics of Equating Situations

Table 8.5 presents a list of characteristics of equating situations for which each of the methods is most appropriate. Mean and linear equating are most useful to consider when the sample size is small, the test forms are not too dissimilar, and a great degree of accuracy is needed only at scores that are not too far from the mean. The conversions for these methods are easy to express (a linear equation, with rounding and truncation rules), the analyses are relatively easy to conduct (summary statistics such as means, variances, and covariances are all that are needed), and the methods are relatively easy to explain to individuals who do not routinely conduct equating. Many applied situations exist in which these methods are adequate.

For example, many certification testing programs are concerned only that equating be accurate near a single passing score. In some programs, the equating might be used only to ensure that the passing score indicates the same level of achievement from administration to administration. If the passing score is not too far from the mean, then linear equating could be the most complex equating method that should be considered.

As another example, small samples of examinees often are administered

Table 8.5. Testing Situations in Which Various Equating Methods Are Most Appropriate.

Method	Situation
Identity	Random Groups and Common-Item Nonequivalent Groups Designs 1. Poor quality control or standardization conditions. 2. Very small samples. 3. Similar test form difficulties. 4. Simplicity in conversion tables or equations, in conducting analyses, and in describing procedures to non-psychometricians is desirable. 5. Possibly inaccurate results can be tolerated. Common-Item Nonequivalent Groups Design 6. Assumptions used to disconfound group and form differences do not hold reasonably well. Likely causes of problems are common item sets that are not representative of the full length test or examinee groups that differ considerably in overall achievement level.
Mean	Random Groups and Common-Item Nonequivalent Groups Designs 1. Adequate quality control and standardization conditions. Alternate forms built to same specifications. 2. Very small samples. 3. Similar test form difficulties. 4. Simplicity in conversion tables or equations, in conducting analyses, and in describing procedures to non-psychometricians is desirable. 5. Accuracy of results is most important near the mean. Common-Item Nonequivalent Groups Design 6. Assumptions used to disconfound group and form differences hold reasonably well. For these assumptions to hold, common items need to be representative, and examinee groups cannot differ too much in overall achievement level.
Linear	Random Groups and Common-Item Nonequivalent Groups Designs 1. Adequate quality control or standardization conditions. Alternate forms built to same specifications. 2. Small samples. 3. Similar test form difficulties. 4. Simplicity in conversion tables or equations, in conducting analyses, and in describing procedures to non-psychometricians is desirable. 5. Accuracy of results is most important in area that is not very far from the mean. Common-Item Nonequivalent Groups Design 6. Assumptions used to disconfound group and form differences hold reasonably well. For these assumptions to hold, common items need to be representative, and examinee groups cannot differ too much in overall achievement level.

Table 8.5. (continued)

Method	Situation
Equi-percentile	Random Groups and Common-Item Nonequivalent Groups Designs 1. Adequate quality control and standardization conditions. Alternate forms built to same specifications. 2. Large samples. 3. Test forms can differ in difficulty level more than for a linear method. 4. Complexity in conversion tables or equations, in conducting analyses, and in describing procedures to non-psychometricians can be tolerated. 5. Accuracy of results is important all along the score scale. Common-Item Nonequivalent Groups Design 6. Assumptions used to disconfound group and form differences hold reasonably well. For these assumptions to hold, common items need to be representative, and examinee groups cannot differ too much in overall achievement level.
Rasch	Random Groups and Common-Item Nonequivalent Groups Designs 1. Adequate quality control and standardization conditions. Alternate forms built to same specifications. 2. Small samples. 3. Similar test form difficulties. 4. Complexity in conversion tables, in parameter estimation, in conducting analyses, and in describing procedures to non-psychometricians can be tolerated. 5. Accuracy of results is most important in area that is not very far from the mean. 6. IRT model assumptions hold reasonably well. Common-Item Nonequivalent Groups Design 7. Assumptions used to disconfound group and form differences hold reasonably well. For these assumptions to hold, common items need to be representative, and examinee groups cannot differ too much in overall achievement level.
Three-Parameter IRT	Random Groups and Common-Item Nonequivalent Groups Designs 1. Adequate quality control and standardization conditions. Alternate forms built to same specifications. 2. Large samples. 3. Test forms can differ in difficulty level more than for a linear method. 4. Complexity in conversion tables, in parameter estimation, in conducting analyses, and in describing procedures to non-psychometricians can be tolerated. 5. Can tolerate computationally intensive item parameter estimation procedure. This problem is mitigated if item parameter estimates are needed for other purposes, such as for test construction.

Table 8.5. (continued)

Method	Situation
	6. Accuracy of results is important all along the score scale.
	7. IRT model assumptions hold reasonably well.
	Common-Item Nonequivalent Groups Design
	8. Assumptions used to disconfound group and form differences hold reasonably well. For these assumptions to hold, common items need to be representative, and examinee groups cannot differ too much in overall achievement level.

tests on test dates in which equating is conducted. In these small sample situations, mean or linear equating might be the most complicated method that would be needed, especially if the interest is in accuracy near the mean.

Assuming that the equating relationship is not linear, nonlinear methods (equipercentile and three-parameter IRT) are most often required when the sample sizes are large and accuracy is required all along the score scale. For example, the ACT (1989) Assessment uses equipercentile equating with large sample sizes because decisions are made at points all along the score scale. The SAT (Donlon, 1984) uses equipercentile and three-parameter IRT methods, along with linear methods, for similar reasons.

For any equating design, the use of IRT methods requires making strong assumptions. Research should be conducted in the context of the testing program to make sure that the methods are robust to the violations of these assumptions which are likely to occur in practice. Because Rasch equating is an IRT method, it requires strong statistical assumptions. However, Rasch equating has considerably smaller sample size requirements than do the three-parameter model methods.

For any equating method, the assumptions required for the common-item nonequivalent groups design (or common-item equating to an IRT calibrated item pool) are very strong. These assumptions can be especially problematic when examinee groups differ substantially, when alternate forms differ substantially, or when the specifications of the common-item sets differ from the specifications for the total test. In these situations, perhaps none of the equating methods would work well. Because of the strong assumptions that are required, methods based on different assumptions can be implemented and the results compared to each other and to results from previous test dates.

Situations can arise in which none of the methods produces an adequate equating. Suppose that (a) high equating accuracy is required at all points along the score scale, (b) the forms are expected to differ more than a little in difficulty, and (c) the sample size is small. In this situation, the objective of high equating accuracy might not be achieved by any of the equating methods. Other similar situations sometimes arise in practice.

Choosing From Among Equating Results

When various equating methods are applied in a particular situation, a process should be developed to choose from among the results. The use of double linking increases the choices that should be considered. Various statistical indices, procedures, and criteria can be used for comparing results from different equatings. Harris (1993) and Harris and Crouse (1993) reviewed many of these indices, criteria, and procedures.

Equating Versus Not Equating

Assuming that the test specifications, design, data collection, and quality control procedures are adequate, it is still possible that using the identity function will lead to less equating error than using one of the other equating methods. Hanson (1992) developed an approach that can be used to help decide whether to equate or use the identity function when using the random groups design. This approach includes using a significance test with the null hypothesis that the distribution of alternate forms is the same. If the null hypothesis is rejected, it is concluded that the distributions differ in the population and that equating should be considered. If the null hypothesis is retained, identity equating is used. Only random error is considered in Hanson's (1992) approach. However, systematic error can be even more problematic than random error. (See Dorans and Lawrence, 1990 for a similar procedure that considers only the mean and standard deviation.)

In small sample situations it is recommended that Hanson's (1992) procedure be used to help decide whether identity equating is preferable to another equating. If the significance test suggests that the distributions are the same, then identity equating could be used. Otherwise, the procedure described previously in this chapter can be used to estimate whether equating would result in more or less error than identity equating. Only if equating is expected to add in less error than identity equating, should an equating other than identity equating be considered.

Use of Robustness Checks

Many procedures have been suggested for estimating the equating relationship for a population using data from a sample. In any equating situation, a relevant question is: How robust is the estimation to the choice of method or procedure? To address this question of *robustness*, various methods and procedures can be applied, and if all of the results are similar, then the results are said to be robust with respect to the choice of method. If the results differ, then the results are not robust with respect to the choice of method. In this case, the choice of method is crucial, although a clear-cut basis for making the choice typically is not available.

In addition, equating can be conducted for various subgroups of examinees (e.g., males versus females). To the extent that the equating is robust, the equating should be similar in the various subgroups. For a particular method, substantial differences in equating results for different subgroups are suggestive of problems with that method.

Choosing From Among Results in the Random Groups Design

A general scheme for choosing from among different equipercentile smoothing results was presented in Chapter 3. Identity equating, mean equating, and linear equating can be considered as more drastic smoothing, and can be compared with unsmoothed equipercentile equating and with each other. In the discussion of postsmoothing in Chapter 3, it was suggested that a method be chosen which results in a smooth relationship without departing more than necessary (based on standard error bands) from the unsmoothed relationship. A process for choosing from among the different degrees of smoothing was described. Statistical tests were incorporated in the choice of presmoothing method. The methods that were presented depend on judgment at various stages in the process.

Statistical procedures other than those described so far in this book have been suggested for choosing from among results. Budescu (1987) and Jaeger (1981) considered statistical indices that could help in choosing between linear and equipercentile equating. Zeng (1995) developed a computerized expert system that chooses between postsmoothing results in a manner intended to mimic the procedures used by psychometricians.

Thomasson *et al.* (1994) presented a detailed set of heuristics for choosing among different smoothed equatings in the ASVAB program. In these procedures, statistical summary indices between the smoothed and unsmoothed relationships for different degrees of smoothing are calculated. Heuristics lead to a single relationship being chosen, based on the similarity of smoothed equating with unsmoothed equating. Graphic inspection and other judgmental procedures are used to make sure that the relationship chosen results in an apparently reasonable conversion which is consistent with previous experience.

Heuristics should be developed within the context of the testing program. Also, heuristics should not be applied blindly or followed rigidly. New wrinkles constantly are occurring in equating in testing programs. Therefore, the procedures should be flexible.

When double linking is used, a method must be chosen for combining the results from the two links. The results might be combined by first conducting the equating separately for the two links. After each equating is conducted, the results could be combined using a weighted average, and properties of this weighted average studied. If problems are detected, different combinations of results from the two links can be tried. Again, procedures should be tailored to the specific testing program.

Choosing From Among Results in the Common-Item Nonequivalent Groups Design

The choice among results in the common-item nonequivalent groups design is complicated further because so many sets of assumptions can be used to disconfound group and form differences. For example, in linear equating, results based on Tucker and Levine observed score method assumptions could be compared. If nonlinear methods are to be considered, IRT observed score (Chapter 6) and frequency estimation (Chapter 5) results (with various smoothing degrees and smoothing methods) can enter into the decision process. In theory, the choice of synthetic population weights is also of some concern, as was indicated in Chapter 4.

Some of the assumptions required for methods can be assessed. For example, the linearity of the regression of X on V that is required for the Tucker method could be checked (Braun and Holland, 1982, p. 25). If the regression were found to be nonlinear, the Braun-Holland (see Chapter 5) method might be used. The disattenuated correlation between X and V could be estimated. A disattenuated correlation substantially less than 1 would suggest problems with assumptions for the Levine method. IRT assumptions could be tested (see Hambleton *et al.*, 1991).

A major problem with this design is that it is impossible to test some of the crucial assumptions. For example, no direct way exists to assess the Tucker method assumption that the regression of X on V in Population 2 is the same as the regression of X on V in Population 1. Similarly, no direct way exists to assess the Levine method assumption that the correlation between true scores for X and V equals 1 in Population 2.

The assumptions required for the methods might lead to a preference of one method over another. For example, Tucker and frequency estimation equipercentile equating might be preferred when groups are similar. When groups are very different, Levine observed score or three-parameter IRT might be preferred, if the assumptions for these methods hold well enough. Sample size might also affect which method would perform better in a situation. Only general guidelines can be given here: The choice among results should be made in the context of the testing program.

Use of Consistency Checks

When conducting equating, the consistency of current results with past results is often the most informative data for choosing a method. For example,

Table 8.6. Scale Score Means and Standard Deviations
for a Hypothetical Example.

Year	Number Tested	Mean	Standard Deviation
1	1005	33.8	5.4
2	1051	33.1	5.6
3	1161	33.0	5.7
4	1192	32.8	5.8
5 (Tucker)	1210	32.5	5.9
5 (Levine Obs. Score)	1210	33.4	5.7

consider the scale score means and standard deviations in Table 8.6 for
Years 1 through 4. Over the period from Year 1 to Year 4, the tested group
became larger, overall lower achieving, and more variable. Assume that we
are in Year 5. Equating has been conducted, and the scale score means and
standard deviations that resulted from applying Tucker and Levine observed score equating are shown in Table 8.6. Which method gives results
that appear more sensible assuming that the past results were accurate? In
this case, the sample size is increasing, which is consistent with the past 4
years. Scale scores using the Tucker method have a lower mean than the
previous year and a higher standard deviation that is consistent with trends
over the past 4 years. The mean and standard deviation for the Levine observed score method are not consistent with this trend. Thus, the Tucker
results are more consistent with past trends than are the Levine observed
score results. The greater consistency of the Tucker method might lead to
the choice of the Tucker method results in this situation, although the
method that actually produced the most accurate results would never be
known for sure.

The example in Table 8.6 is based on comparing means and standard
deviations. Examining the consistency of entire score distributions can be
useful, too, especially when accuracy is important all along the score scale.
Also, examining the consistency of pass rates or consistency at particular
important score points also can be helpful. Suppose that approximately
40% of the examinees have passed a test on previous test dates. In a current equating, 41% would pass using the Levine observed score method
and 32% would pass using the Tucker method. In this case, the Levine observed score results might be preferred for consistency reasons, especially if
the major uses of the test involve a passing score.

Large unexpected differences in consistency checks might suggest either
quality control problems or problems with the assumptions of a particular
method. When these differences are found, the implementation of the
equating should be checked including the functioning of the common items
(if appropriate), the execution of the equating design, and other quality
control issues. Problems might have existed with past equatings, suggesting

that they should be checked as well. These potential sources of problems should be examined before accepting the results from an equating.

Equating and Score Scales

As was indicated in Chapter 1, equating is part of a scaling and equating process. Petersen *et al.* (1989) discussed score scales in detail, and indicated that the score scale often is chosen to facilitate score interpretation. The choice of score scale is especially important for tests in which decisions are made along a range of scores. The particular score scale is much less important if a test is used only in making pass-fail decisions, where decision consistency is crucial.

The choice of score scale affects equating. For example, in Chapter 2, rounding scale scores to integers was shown to have a significant effect on the similarity, across forms, of the scale score means, standard deviations, and other moments. Also, Petersen *et al.* (1989) discussed the problems that can result when raw scores on a form are used as the score scale—in particular, raw scores become easily confused with scale scores.

Typically, rounded scale scores are reported to examinees. These rounded scores might have some properties that appear to be undesirable. For example, in ACT Assessment (ACT, 1989) equating, a conversion table might result in many number-correct scores converting to a single scale score. Also, gaps can occur in conversion tables, in which no raw score converts to a particular scale score. These occurrences can be viewed as problematic by examinees. If the scale score increment is 1 point, an examinee might justifiably question why earning 1 number-correct score less than someone else would result in a 2- or 3-point difference in scale scores. Under the assumption that gaps, and too many raw scores converting to a single scale score, would not occur except for sampling error, results for a method or degree of smoothing might be chosen that minimize these problems.

In testing programs such as the ACT Assessment (ACT, 1989) and the SAT (Donlon, 1984, pp. 19–20), for practical reasons a number-correct score of all correct is forced to convert to the highest possible scale score, even if the equating suggests that some other score would be more appropriate. This process is used with the SAT and the ACT to ensure that the highest possible scale score can be earned on any form. However, doing so makes it easier to earn a top score on some forms than on others. For this reason, other testing programs allow the top score to differ depending on the difficulty of the form for high-scoring examinees. The effects of adjustments to the score scale and choosing methods to avoid gaps in the conversion should be evaluated on a case-by-case basis. The effects on moments and on score distributions should be carefully monitored.

Kolen *et al.* (1992, 1994) suggested procedures that can be used to compare the conditional (on true score or latent ability) means and conditional standard errors of measurement of rounded scale scores across forms. These procedures can be used to evaluate, at least partially, the equity property of equating described in Chapter 1. The effects on the conditional means and conditional standard errors of measurement of the choice of equating results and adjustments to the conversions can be evaluated using this scheme. The conditional means and conditional standard errors of measurement should be comparable across forms if the forms are to be used interchangeably.

Comparability of Composite Scores

Composites of test scores often are reported to examinees. Composites are typically sums or averages of test scores. For example, the composite score in the ACT Assessment is calculated by the following procedure:

1. Find the arithmetic average of the four ACT Assessment scale scores.
2. Round the average to an integer, where xx.5 is rounded up (e.g., 19.5 is rounded to 20).

In the ACT Assessment, the four test scores are equated, and the composite is not equated. Because the composite is used for many decisions, it is important that composite scores on alternate forms be comparable. Does the equating of the four test scores ensure that the composite score is equated?

Consider the following simplified situation. Tests 1 and 2 on Form Y are both normally distributed with a mean of 0 and a standard deviation of 1. Assume that the correlation between scores is .7. Also, the composite score is the sum of scores on Test 1 and Test 2. In this case, the composite score has a mean of 0 and a standard deviation equal to $\sqrt{1^2 + 1^2 + 2(.70)} = 1.8439$. Assume that Form X is linearly equated to Form Y. After equating, Form X scores on Tests 1 and 2 are both normally distributed with a mean of 0 and a standard deviation of 1, and the correlation is .60. The Form X composite score has a mean of 0 and a standard deviation of $\sqrt{1^2 + 1^2 + 2(.60)} = 1.7889$. This simplified illustration demonstrates that even when test scores are equated, composite scores are not necessarily comparable.

With the ACT Assessment, the distributions of composite scores are checked after each equating. If the relationships among the equated tests are similar across forms, then distributions of the composite scores would be similar across forms, and not rescaling the composite would be acceptable. Rescaling might be needed, however, if the relationships among forms differed.

The following example uses data from the ASVAB program (Bloxom and McCully, 1992). In the ASVAB program, individual tests are equated. The tests are used to form various composites, which are weighted sums of various ASVAB tests, and the composites are not equated. After an operational equating, Bloxom and McCully (1992) found that the standard deviations for some of the composites were noticeably smaller on the new forms than on the old form. In examining the reasons for the differences, the correlations between some of the ASVAB tests were found to have become lower. For example, one of the correlations between the tests that contributed to the composites was .598 on the old forms and .412 on the new forms. Bloxom and McCully concluded that the differences in correlations were "... due to the fact that the new forms assess some types of knowledge more independently than is the case in the reference [old] form" (p. 14). In this example, an attempt was made to equate the individual tests, but the composite distributions differed. If scores on the composites were used without adjustment, then certain examinees would be expected to score higher on one form than on another. This example illustrates how changes in specifications of tests can influence the relationships among tests and lead to composites that are not comparable.

Importance of Standardization Conditions and Quality Control

For equating to be adequate, testing conditions should be standardized and quality control procedures should be followed. Otherwise, identity equating, rescaling, or scaling to achieve comparability might be the best options. Quality control procedures are vital to adequate equating, and they often take more effort than other aspects of the equating process.

Test Development

The following is a list of changes in how the test forms are developed that can cause problems for equating:

1. *Test specifications change.* (See Chapter 1 and previous portions of this chapter.)
2. *In a common-item nonequivalent groups design or an item preequating design, the context of the common-items changes.* For example, it could be problematic if common items appear in considerably different positions on the two forms, such as a common item appearing near the beginning of the old form and near the end of the new form (Cook and Petersen, 1987; Eignor, 1985; Kolen and Harris, 1990). Another example involves items associated with a common stimulus (such as a reading passage) that have interdependencies. If one item associated with the passage is removed from the test, other items associated with that passage might

be affected. To be safe, when items associated with a common stimulus are used as common items, the set of items associated with the common stimulus on the new form should be exactly the same set of items as the items that were associated with the common stimulus on the old form. For example, the context in which common items were administered resulted in a significant scaling problem for NAEP (Zwick, 1991), as was described in Chapter 1.

3. *In a common-item nonequivalent groups design or an item preequating design, the text of the common items changes.* The text should be *exactly* the same in the old and new forms. Otherwise, the items might function differently. Minor editorial changes and rearranging of answer choices (Cizek, 1994) in items should be avoided.

Test Administration and Standardization Conditions

The conditions under which a test is administered should be standardized in order for tests administered at different locations and at different times to be comparable to one another. Some issues related to standardization that could have significant effects on scaling and equating include the following:

1. *Changes in the number of items on the test.* (Harris, 1987, 1988; Linn and Hambleton, 1991; Way *et al.*, 1989)
2. *Changes in timing of the test.* Changes in timing can have a significant effect on the scores of examinees. For example, Hanson (1989) reported a study in which scores on a test were compared with scores on a lengthened version of the same test, with the testing time extended accordingly. The lengthening was accomplished by appending unscored items to the original (unlengthened) test. In this study, scores on the lengthened test (excluding the appended items) were substantially higher than scores on the unlengthened test. (Also see Brennan, 1992, for a discussion of this study).
3. *Changes in motivation conditions.* Studies in which a new version of a test is administered under different motivation conditions than the old version of the test. This problem occurred in the ASVAB scaling example described in Chapter 1 (see Maier, 1993).
4. *Security breaches.* Examinees are found to have had prior exposure to test forms or items that appear in the forms involved in the equating, which suggests that a security breach occurred.
5. *Changes in the answer sheet design.* These changes can affect test performance (Bloxom *et al.*, 1993; Burke *et al.*, 1989; Harris, 1986).
6. *Scrambling of test items for security purposes.* Sometimes, test items within forms are scrambled to discourage examinee copying. However, scrambling can affect score distributions (e.g., Harris, 1991b). Dorans

and Lawrence (1990) and Hanson (1992) developed procedures for testing whether score distributions on scrambled forms differ.

7. *Changes in the font used in printing the test or in the pagination used.* These changes can affect scores.

8. *Section preequating in which preequating and operational sections appear in different positions in different forms* (e.g., Brennan, 1992).

9. *Use of calculators.* If calculators are allowed in some administrations and not in others, then scores from administrations that allow calculators are not directly comparable to scores from administrations that do not allow calculators. In these cases, separate calculator and non-calculator norms and scales might be needed. Loyd (1991) and Morgan and Stevens (1991), for example, investigated the effects of calculators. Other, similar changes in standardization conditions that might affect scores include allowing students to use dictionaries or word processors.

10. *Administration under nonstandard conditions,* such as large type, Braille, or extra time for handicapping conditions (Tenopyr *et al.,* 1993) or translations of tests into other languages.

Variations in standardization conditions can affect scores. The research cited suggests that such variations might lead to scores that are not comparable. The effects of variations in standardization procedures, and how to deal with them, should be considered in the context of the testing program.

Quality Control

Quality control checks are vital to adequate equating. They can be quite elaborate and extraordinarily time-consuming. Some of the quality control checks that can be made are as follows:

1. *Check that the test administration conditions are followed properly.* Some examples of problematic circumstances include test administrators giving examinees extra time to take the test, examinees found to be copying from one another, test administrators not spiraling the tests properly in a random groups design, and noise in test centers.

2. *The answer keys are correctly specified.* The correct key should be applied when scoring examinee records. Correctly applying answer keys requires special care when more than one form is administered and when different versions of a form exist, such as when items are scrambled for security purposes.

3. *The items appear as intended.* The text of the items, and especially the common items, should be checked.

4. *The equating procedures that are specified are followed correctly.* Typically, equating involves many interrelated steps, often necessitating the involvement of many people and the use of multiple computer programs. Without careful checking, an important step can be forgotten.

5. *The score distributions and score statistics are consistent with those observed in the past.* These consistency checks sometimes can suggest problems in scoring or data processing.

6. *The correct conversion table or equation is used with the operational scoring.* In general, the result of equating is an equation or conversion table that is supplied to whomever is to do the operational scoring. Usually, a few steps occur between the choice of the conversion and the creation of the table to be supplied. In our experience, it is vitally important to check the table or equation that is supplied with the one that was developed when the conversion was chosen.

Reequating[1]

Consider a situation in which a form of a test has been administered and equated, and subsequently it is discovered that an item possesses some type of ambiguity that makes the keyed alternative technically incorrect, or that the keyed alternative is only one of two or more technically correct answers. After reconsidering such an item, suppose that content matter specialists decide that the originally keyed alternative (say, *a*) is indeed correct, but the other alternatives (say, *b*, *c*, and *d*) also can be defended as correct, based on an obscure fact or facts. Clearly, decisions must be made about whether to give all examinees credit for the item and whether to reequate the form with that item scored correct for all examinees. (For the sake of this discussion, assume that even examinees who omitted the item would be given credit for it.)

Suppose that a firm decision on these matters is postponed until the form is reequated with all examinees being given credit for the item. There are then four conceivable ways to arrive at examinee "equated" scores:

1. original key applied with original equating relationship;
2. original key applied with revised equating relationship;
3. revised key applied with original equating relationship; and
4. revised key applied with revised equating relationship.

Applying the first option produces the scores that were originally reported to examinees, and essentially means acting as if the item is *not* flawed. The examinee who discovered the flaw may well consider this option to be unfair, and, in all likelihood, the public will share the examinee's concern. However, an examinee who is insightful enough to recognize such a flaw is also often insightful enough to choose the alternative that was intended as the correct answer. If so, the first option does not really treat that particular examinee unfairly, although it would be unfair for some other unidentified examinee who chose one of the other alternatives *for a correct reason.*

---

[1] This section is largely from Brennan and Kolen (1987b, pp. 286–287).

The second option, using the original key with the revised equating relationship, is difficult to defend under any reasonable scenario.

The third option, using the revised key with the original equating relationship, may appear to be an option that is generous to examinees. In effect, all examinees who selected alternatives *b*, *c*, or *d* (or omitted the item) will receive a higher "equated" score, whatever the reason for selecting that alternative. However, those examinees who are given credit unjustifiably (e.g., those who had misinformation or no information about the item) will fare better than their equally achieving counterparts, especially in a quota-based decision process. Thus, while this option is generous for some examinees, that very generosity may create a potential disservice to other examinees. In evaluating the fairness or reasonableness of any of these options, it is necessary to consider the consequences for not only examinees who are directly affected, but also examinees who are indirectly affected by the decision.

The fourth option, using the revised key with the revised equating relationship, essentially avoids the problems mentioned above with the third option, and the fourth option has considerable face validity. Indeed, this appearance of face validity is almost always judged to be an overwhelming argument in favor of the fourth option.

However, under some circumstances it can be argued that the first option may well be preferable *psychometrically* to the fourth if the goal is to be as fair as possible to *all* examinees, not just those who voice a legitimate complaint. For example, when all examinees are given credit for an item, the effective test length is reduced by one item, which, on average, benefits lower achieving examinees and works to the disadvantage of higher achieving examinees. To put it another way, when all alternatives are keyed correct because an item possesses an obscure ambiguity, it is likely that many examinees will be given credit for the item who would not otherwise have answered the item correctly. This fact will cause these examinees to appear higher achieving than they actually are, and other examinees will appear lower achieving by comparison. Indeed, examinees who selected alternative *a* (the response originally keyed as correct) will receive a *lower* equated score under the fourth option than under the first option. Reequating cannot really eradicate these problems. Indeed, reequating can never completely remove a test development flaw; the best it can do is mitigate the impact of such a flaw.

The above points are not intended to be interpreted as arguments in favor of never rescoring or reequating when a flawed item is discovered. Even if the psychometric arguments were compelling, arguments from other perspectives could be even more compelling. Nor are these points to be interpreted as arguments about the differential utility of benefiting lower achieving examinees versus disadvantaging higher achieving examinees. When such judgments need to be made, they should be based on a much broader set of considerations than merely psychometrics. The point here is that the issues involved in rescoring and reequating are quite complex, and certain unintended negative consequences are easily overlooked. (These problems become even more complex when the flawed item is in a common-item equating section.)

If reequating is judged necessary and scores have already been reported to examinees, then questions arise about what the effects of the reequating will be on examinees' scores. Specifically, how many scores will increase, how many will decrease, and how many will stay the same? Other practical questions arise, such as should scores be reissued for examinees whose scores would decrease after reequating? In addition, what is the affect on the test specifications and on the technical properties of the test when an item is removed? Can the test with the item removed be considered to be equated? These questions often can be very difficult to answer. Brennan and Kolen (1987b) and Dorans (1986) have addressed some of these questions. Reequating also sometimes needs to be considered when a security breach occurs, in which examinees obtain answers or questions prior to a test administration. Brennan and Kolen (1987b) and Gilmer (1989) illustrated some of the consequences of security breaches on equating relationships and on examinee scores.

Conditions Conducive to Satisfactory Equating

Conditions that are conducive to a satisfactory equating can be distilled from the various practical issues in equating which have been considered in this chapter. A list of some of these conditions, which is a modified version of the list provided by Brennan and Kolen (1987b), is given in Table 8.7. This table lists many of the characteristics of testing programs that are conducive to a satisfactory equating. Satisfactory equating does not require that all of these conditions hold. However, it might be best not to equate when some of these do not hold. For example, equating could not be conducted if the tests were built to different content specifications.

Comparability Issues in Special Circumstances

Various special issues affect equating and how the results are used. In addition, situations arise that are similar to equating situations, but which can best be characterized as being associated with some other process. Some of these issues are discussed in this section.

As has been stressed, scores on alternate forms of a test can be used interchangeably only if the forms are built carefully to well-defined test specifications and adequate test equating procedures are used. The test development process is crucial to being able to use scores on test forms inter-

Table 8.7. Conditions Conducive to a Satisfactory Equating.[a]

A. General
    1. The goals of equating, such as equating accuracy and the extent to which
       scores are to be comparable over long time periods, are clearly specified.
    2. The design for data collection, the equating linkage plan, the statistical
       methods used, and the procedures for choosing among results, are appro-
       priate for achieving the goals in the particular practical context in which
       equating is conducted.
    3. Adequate quality control procedures are followed.

B. Test Development—All designs
    1. Test content and statistical specifications are well defined and stable over
       time.
    2. When the test form is constructed, statistics on all or most of the items are
       available from pretesting or previous use.
    3. The test is reasonably long (e.g., 30 items, and preferably longer).
    4. Scoring keys are stable when items or forms are used on multiple occasions.

C. Test Development—Common-Item Nonequivalent Groups Design
    1. Each common item set is representative of the total test in content and stat-
       istical characteristics.
    2. Each common-item set is of sufficient length (e.g., at least 20% of the test for
       tests of 40 items or more; at least 30 items for long tests).
    3. Each common item is in approximately the same position in the old and new
       forms. Common-item stems, alternatives, and stimulus materials (if appli-
       cable) are identical in the old and new forms. Other item level context effects
       are controlled.
    4. Double linking is used. One old form was administered during the same time
       of year as the form to be equated. One old form was administered within the
       last year or so.

D. Examinee Groups
    1. Examinee groups are representative of operationally tested examinees.
    2. Examinee groups are stable over time.
    3. Examinee groups are relatively large.
    4. In the common-item nonequivalent groups design, the groups taking the old
       and new forms are not extremely different.

E. Administration
    1. The test and test items are secure.
    2. The test is administered under carefully controlled standardized conditions
       that are the same each time the test is administered.

F. Field of Study/Training
    1. The curriculum, training materials, and/or field of study are stable.

[a] Adapted from Brennan and Kolen (1987b)

changeably. After equating, examinees are expected to earn the same scale score and be measured with the same precision, regardless of the form taken. In addition, accurate equating relationships are symmetric and approximately the same across subgroups of examinees.

Various other linking processes are used with educational tests that are not built to common specifications. These processes do not lead to score interchangeability. Mislevy (1992) and Linn (1993) reviewed many of these procedures, and a summary is provided here.

Calibration

Mislevy (1992) and Linn (1993) conceived of calibration as linking different measures of the same construct, where measures differ in precision. For example, sometimes short forms of a test are constructed so that they can be administered in less time than the regular form. Assuming that the short and long forms are built to the same test specifications, proportionally, then the main difference between the forms is that the short form is less reliable. In this case, scores on the short form could be calibrated to scores on the long form. If a true score equating method were used and a congeneric model held (see Chapter 4), then the property of weak equity (or tau-equivalence, Yen, 1983) would hold. That is, over repeated testings examinees would be expected to earn the same scale score whether the long or short form were taken. In this case, however, the short form scores would contain more measurement error than the long form scores.

In elementary achievement level batteries, the test level that examinees are administered is designed to have content that is appropriate for the grade level of the examinee. Different test levels are administered at different grades. Scores on the different test levels often are reported on a common score scale.

In their discussion of vertical scaling, Petersen et al. (1989) indicated that when administering these test batteries, examinees who have been exposed to material tend to perform better than examinees who have not been exposed to material. Petersen et al. (1989) reported that, for this reason, the results of the scaling depend very heavily on the design used for data collection.

The following example illustrates this dependency. Suppose Level 3 is designed to be appropriate for third graders and Level 4 is designed to be appropriate for fourth graders in a subject matter area that is highly curriculum dependent, such as mathematics problem solving. A Level 3 test and a Level 4 test are spiraled to third grade examinees and to fourth grade examinees. After scaling, the means on the two levels are more similar for fourth graders than for third graders, because the third graders have not been exposed to any of the fourth grade material, whereas the fourth graders have been exposed to all of the material.

Petersen *et al.* (1989) suggested the use of a scaling test to scale multi-level achievement batteries. A scaling test is a test that samples the content across all of the levels to be placed on the same scale, but is constructed to be administered in a reasonable amount of time. The scaling test is administered in all grades for which the multilevel battery is appropriate. The resulting data are used to construct the score scale across grades. The levels then are "calibrated" to the scaling test by administering the appropriate test level to each examine who is administered the scaling test. In this way, the scaling test is used to define the *developmental continuum* across grades. Additional details can be found in Petersen *et al.* (1989).

Prediction (Projection)

Often scores on tests are used to predict scores on other tests or scores on criterion variables. For example, scores on the ACT (1989) test often are used to predict college grades. Linn (1993) considered a situation in which essay test scores are predicted from multiple-choice test scores.

Recently, there has been interest in predicting distributions of scores. For example, Bloxom *et al.* (1995) attempted to predict the distribution of NAEP scores for examinees who had taken only the ASVAB. In developing the prediction, some examinees were administered both NAEP and ASVAB. These data were used to predict the distributions. Beaton and Gonzalez (1994), Pashley and Phillips (1993), and Pashley *et al.* (1994) reported on linking NAEP with international assessments. Kiplinger and Linn (1994) and Ercikan (1994) reported on linking statewide assessments to the NAEP. Again, the focus of these efforts was on estimating score distributions.

Prediction systems are heavily dependent on the group used to conduct the prediction. In addition, predictions are necessarily one directional, which illustrates that prediction is not equating because it violates the symmetry property of equating discussed in Chapter 1. The directionality can be illustrated by noting that the regression of X on Y is not the same as (or the inverse of) the regression of Y on X.

Moderation

Often scores on different tests are used to make the same decision. In these circumstances, a table that relates scores on one test to scores on another test is developed. The scores that result are comparable in some sense. One procedure used to obtain such "comparable" scores is referred to as *statistical moderation*. For example, Marco and Abdel-Fattah (1991) and Houston and Sawyer (1991) developed tables relating the ACT and SAT tests. The tables were intended to be used by colleges which required that students take either the ACT or the SAT, where the colleges needed to have comparable scores on the two instruments. Equipercentile methods were used to develop the comparable scores. These reports clearly show that different comparable scores are found for different examinee groups, which illustrates that moderation is not equating because it violates the group invariance property of equating that was described in Chapter 1.

After comparable scores are developed, some examinees still will be expected to perform better on one test and some on the other. That is, it is not a matter of indifference to an examinee which test is administered, which again illustrates that moderation is not equating because it violates Lord's equity property of equating described in Chapter 1. Also, some subgroups would prefer one test over the other.

Another example of statistical moderation is the score scales for the SAT Achievement tests (Donlon, 1984), which are constructed by using the SAT Verbal and Mathematical tests as moderators. The achievement tests are linked to the SAT Verbal and Mathematical scores. The mean achievement test score is related to the mean SAT Verbal and Mathematical scores of examinees who take the achievement tests. In this way, scores on achievement tests that are taken by examinees with high SAT scores will tend to be higher than those taken by examinees with low SAT scores.

As was suggested by the preceding discussion, scores produced using moderation cannot be used interchangeably. Moderation is highly situation specific, highly group dependent, and highly dependent on the particular procedures used.

Comparability Issues with Computerized Tests

The basic equating designs and methods that have been described previously can be used to equate test forms that are administered by computer in much the same way that they are applied in equating paper and pencil tests. For example, a random groups equating of two alternate forms could be accomplished by administering the first examinee Form X, the second examinee Form Y, the third examinee Form X, etc.

Additional issues arise when equating *computer adaptive tests* (Dorans, 1990b). For example, in adaptive testing with the ASVAB (Segall, 1993), two distinct item pools were developed. The pools are referred to as forms. The forms then were randomly assigned to examinees. Even though the item parameters were on the same ability scale, the resulting ability estimates were found to have different distributions, presumably because of differences in the items in the pools. In the ASVAB program, the differences in distributions were eliminated by using an equipercentile equating of the ability estimates on the forms. This finding illustrates that a need might exist for equating alternate adaptive test forms, even when the pools are on the same IRT scale.

Adaptive tests often are developed using IRT calibrated item pools. Over time, the items in the pools are updated for security reasons. Wainer and Mislevy (1990) consider a process of *on-line calibration* of new, uncalibrated items in an adaptive test, in which uncalibrated items are introduced into the pool by embedding them in operational adaptive tests. These uncalibrated items do not contribute to an examinee's score. Responses are tabulated over a sufficient number of examinees, and these responses are used to estimate item parameters. These new items then are added to the pool. As with other procedures for adding to IRT calibrated item pools, context effects should be considered in on-line calibration. An issue with adaptive testing is that, typically, examinees are administered items which are close to their ability level. In conducting on-line calibration, examinees might be administered items that are far from their ability level. These and a variety of other practical issues discussed by Wainer and Mislevy (1990) should be considered when conducting on-line calibration.

Significant issues arise when paper and pencil forms are to be used interchangeably with computerized forms. Administering a test on computer can be quite a different experience for examinees than taking a paper and pencil test. Some of these differences are (a) ease of reading passages; (b) ease of reviewing or changing answers to previous questions; (c) speed in taking the test, and the effects of time limits on test speededness; (d) clarity of figures and diagrams; and (e) responding on a keyboard versus responding on an answer sheet.

In their review of computerized and paper and pencil testing, Mazzeo and Harvey (1988) concluded that "... it is clear that test publishers need to perform separate equatings and/or norming studies when computer-administered versions of standardized tests are introduced" (p. 26), which is consistent with the standards presented in APA (1986). Some of the studies that they reviewed indicated that computer administration favors certain subgroups over others. Mazzeo et al. (1991) were able to eliminate mode effects by modifying administration conditions for one test but not for another test. Other reviews (Mead and Drasgow, 1993; Spray et al. 1989) have concluded that, in many instances computerization does not have a large effect on the construct being measured. Reese (1992) concluded that, although some mode effects were noticeable with the GRE, tests were behaving similarly in the two modes. Parshall and Kromrey (1993) concluded that there were mode effects on the GRE that had a differential effect on subgroups. Research should be conducted in the context of a particular testing program before testing under different modes can be assumed to be comparable. Mode effects of paper and pencil and computer administered tests appear to be very complex, and they are most likely dependent on the particular testing program. For this reason, Mazzeo et al. (1991) recommended that there is a "... need to determine empirically (rather than to just assume) the equivalence of computer and paper versions of an examination" (p. 1). Also, the relative effects of mode of administration on important subgroups of examinees should be examined.

Issues become even more complex when computer adaptive tests are used. Wainer (1993) presented a number of practical issues that should be considered when converting to adaptive tests. Wainer et al. (1990) indicated that the conditional errors of measurement can differ between adaptive and nonadaptive tests. In these situations, the scores cannot be made strictly interchangeable. In some testing programs, the adaptive test is constructed so that the conditional standard errors of measurement are the same for the adaptive and nonadaptive versions, as was done with the GRE (Mills et al. 1994). However, a further equipercentile transformation was needed with the GRE Analytical test. Eignor (1993) described issues and problems in deriving comparable scores for adaptive and paper and pencil SAT tests, and Eignor et al. (1994) described issues with an adaptive nursing licensure test. Eignor (1993) found that counterbalancing was not effective when the single group design with counterbalancing was used, and he argued that the random groups design should be used when establishing equivalent scores. In the ASVAB program (Segall, 1993), equipercentile procedures with a random groups design are used to convert scores on the adaptive test to scores on the paper and pencil test. In this way, the same score distributions can be expected to result for the overall group. Using these procedures, however, does not ensure comparable distributions for subgroups or for composites of test scores. See APA (1986) for some technical guidelines that should be followed in establishing comparable scores for adaptive and paper and pencil tests.

Comparability of Performance and Other Alternative Assessments

From the perspective of equating, at least three characteristics distinguish performance and other alternative assessments from multiple-choice assessments. First, the scoring of these assessments is subject to error by judges, whereas multiple-choice scoring is more straightforward. Training of raters is probably the best way to control differences in scoring among judges, although various ways for adjusting for judge leniency have been developed (e.g., Braun, 1988; Englehard, 1992, 1994; Linacre, 1988; Longford, 1994; Raymond and Viswesaran, 1993).

Second, often very few tasks are administered to examinees with performance and other alternative assessments, because of lengthy per-task administration times. Many authors (e.g., Baxter et al., 1992; Dunbar et al., 1991) have indicated that the use of a small number of tasks results in an inadequate sample of the domain of interest. If the domain is sampled inadequately, then it is likely to have been sampled differently on alternate forms. The result of the inadequate sampling is that scores on one form cannot be used interchangeably with scores on another form, even if equating is attempted. With inadequate domain specification, certain examinee subgroups would favor certain forms, and other examinee subgroups would favor other test forms (Ferrara, 1993). When the domain is sampled inadequately, it might be best not to equate. Alternatively, some type of score adjustment other than equating might be considered.

Third, it might be impossible to use any of the commonly used equating designs with performance and other alternative assessments. It is often difficult to implement a random groups design, because forms sometimes cannot be spiraled within test centers due to administration constraints. It may not be possible to implement a common-item nonequivalent groups design, either, if a content balanced common-item set cannot be developed because the tests contain too few items. If two forms cannot be administered to examinees, then even the single group design cannot be used.

Because of these complications, sometimes it might be best not to equate performance and other alternative assessments. In some situations, however, the domain is sampled adequately, and it may be possible to apply standard equating procedures. For example, if feasible, the random groups design could be applied by randomly assigning forms to examinees. If a content representative set of common items could be developed, then the common-item nonequivalent groups design also could be used. Linear or equipercentile methods could be applied in these situations (See Harris et al., 1994, and Huynh and Ferrara, 1994, for some examples).

IRT methods also might be applied, although models would need to be used that allow for polytomous item scoring (Linacre, 1988; Masters, 1982; Muraki, 1992; Samejima, 1969; Thissen and Steinberg, 1984, 1986). Note, however, that these models might require more responses per examinee for stable estimation than is feasible. Also, strategies should be developed for managing local item dependence (Ferrara, 1993; Yen, 1993) and assessing model fit. Harris et al. (1994) and Huynh and Ferrara (1994) compared traditional and IRT methods. Baker (1992b, 1993) and Cohen and Kim (1993) presented test characteristic curve methods for scale linking using some of these models.

A variety of practical problems are present when scaling and equating methods are applied to performance and other alternative assessments. For some such assessments, very few raw score points are available (Ferrara, 1993; Harris et al., 1994). In these situations, the coarseness of the scale might make it difficult to improve on not equating. In studies using a test with small numbers of score points, Harris and Welch (1993) and Harris et al. (1994) found few differences between the identity function, equipercentile methods, and Rasch methods.

For test security reasons, sometimes forms of performance and other alternative assessments cannot be administered in special equating administrations and forms cannot be reused. One approach that has been followed is to use a measure that is not constructed to be parallel to the assessment as an external anchor. For example, Hanson (1993) attempted to use multiple-choice items as an external common-item set and applied equating procedures. He found that the results were sensitive to the assumptions made about the relationship between the multiple-choice and performance assessments, and he concluded that the identity equating would be preferable to any of the other equatings in the situation studied. However, often the only measure that is available is one which is not constructed to be parallel to the performance assessment. More research is needed to address the question of when such measures can be used to adjust scores. The strength of the relationship between the available measure and the performance or other alternative assessment and the extent to which the groups are nonequivalent should be investigated regarding how they affect the adjustment procedures that are developed.

Currently, there is much activity in the development of performance and other alternative assessments (Baker et al., 1993). Many unresolved issues in equating and scaling such assessments (e.g., Ferrara, 1993; Fitzpatrick et al., 1994; Gordon et al., 1993; Harris et al., 1994; Loyd et al., 1993) and combining scores from performance and other alternative assessments and multiple-choice tests (Wainer and Thissen, 1993) exist, and there is still some question about the conditions under which such assessments can be equated. When equating cannot be conducted, other score adjustment methods could be investigated. Methods and procedures for addressing the comparability of alternate forms of performance and other alternative assessments continue to emerge.

Score Comparability with Optional Test Sections

On some tests, examinees can choose which sets of items they are going to take, as is the case for some College Board Achievement Tests (Donlon, 1984). For these tests, some of the items are taken by all of the examinees and the rest are in optional sections. Examinees choose which optional section to take, and the examinee groups that take the alternate forms typically differ in performance on the common portion. What if some optional sections are more difficult, in some sense, than other optional sections? A major issue in this situation is whether or not scores for examinees taking different optional sections can be equated.

If the optional sections measure different content, then the scores for examinees who take one optional section cannot be said to be equivalent to scores for examinees who take a different optional section, even after some score adjustment is attempted. The comparability problems are even more severe if examinee choice of optional sections is related to their overall level of skill or to their area of expertise. In general, it seems impossible in most practical situations for scores on optional sections to be treated interchangeably. Wainer and Thissen (1994) provided a discussion of these and related issues, and concluded that choice is inconsistent with the notion of standardized testing, "unless those aspects that characterize the choice are irrelevant to what is being tested" (p. 191).

Whether or not scores based on tests with optional sections can be adjusted might also be approached from the perspective that some sort of adjustment might make scores more equitable. Livingston (1988) presented a scheme that can be used to adjust scores by linking them to the common portion of the test. His approach appears to treat score adjustment as a problem in statistical moderation, with the goal being to make scores more comparable than if the adjustment were not made. Wainer *et al.* (1994) presented a study in which they attempted to use a unidimensional IRT model to produce equivalent scores, and they point out some serious problems in doing so. Wainer and Thissen (1994) also provided some interesting examples of problems with optional items. Also see Fitzpatrick and Yen (1993). Allen *et al.* (1994a,b) showed how adjustment procedures can vary depending on the assumptions made about the relationship between the common and the optional sections.

Conclusion

Equating is now an established part of the development of many tests. When conditions allow, scores from equated test forms can be used interchangeably. Equated scores for examinees can be compared even when the examinees are administered different test forms. Equating facilitates the charting of trends. Without equating, we might be unable to tell whether or not there have been trends in student achievement over time. Without equating, examinees could be advantaged by happening to be administered an easier form. Other examinees could be disadvantaged by happening to be administered a more difficult form. Effective equating results in tests being more useful for making many decisions and for making the process of testing more equitable.

As has been discussed in this chapter, equating requires that many practical issues be considered by the individual conducting the equating. How these issues are handled can have profound effects on the quality of the equating. The test construction process that is followed and how the equating study is designed are crucial to adequate equating. If problems exist with the test construction or with the data, then no amount of statistical manipulation can lead to adequate equating results. In this sense, the design of tests and the design of data collection are of central concern. In addition, thorough quality control procedures need to be implemented for the equating to be successful. Even though the ideal equating likely has never been conducted in practice, adequate equating requires that practical issues be effectively handled. Otherwise, it might be best not to even attempt to conduct equating. The diversity of practical issues, and deciding how to address them, is what makes the practice of equating so challenging.

As we have seen in Chapters 2–7, the statistical and psychometric techniques involved in equating are diverse and require considerable statistical sophistication to understand. These techniques have evolved considerably over the past 15 years, and likely will continue to do so. From a psychometric perspective, equating is a rich area because it draws from a wide variety of psychometric theories, such as congeneric test theory, strong true score theory, and IRT. Equating provides for an application of these theories to an important practical problem.

The field of testing currently is undergoing significant change. Many major testing programs are incorporating alternatives to the paper and pencil multiple choice tests that have dominated much of standardized testing for the past 50 years or so. One set of alternatives includes tests that require examinees to produce written and verbal responses to tasks. These responses often are scored by judges, although procedures also are being developed for machine scoring. In addition, many testing programs are implementing testing in which examinees can take a test at almost any time, rather than having to take the test on one of a few test dates. Often, this type of testing involves computer administration. Such on-demand testing creates new issues in test security, development, quality control, equating, and score comparability. All of these changes in testing are causing psychometricians to reevaluate the concepts of equating, comparability, and other forms of score adjustments, and to develop new procedures to handle these alternatives.

Exercises

8.1. Assume that scores on Forms X and Y are normally distributed and that the forms were administered using a random groups design. Also assume that the forms differed by .1 standard deviation unit at a z-score of .5.

    a. What sample size would be required for linear equating to be preferable to the identity equating at this z-score?

    b. What sample size would have been required for linear equating to be preferable to the identity equating at this z-score if the forms had differed by .2 standard deviation unit at a z-score of .5?

    c. Describe a practical situation where it would make sense to ask these questions.

8.2. The single link plans in Figure 8.1 each have a definable pattern that could be used to extend the pattern indefinitely. For example, consider Single Link Plan 1. For Form C and following, forms are linked to the form that was administered in the same time of the preceding year.

    a. Provide a verbal description for Single Link Plan 4 in Figure 8.1. (Hints: Different statements are needed for even-numbered and odd-numbered years. Begin the description with Form D.)

    b. Using this description, indicate to which form each of Forms K, L, M, and N would link.

8.3. Suppose that a psychometrician recommended Single Link Plan 4 in Figure 8.1 for equating in a testing program and subsequently found out that it was not possible to link to a form from the previous administration. In particular, suppose that in Single Link Plan 4, Form E could not link to Form D, and Form I could not link to Form H. The psychometrician developed two modified plans. In Modified Plan 1, Form E linked to Form B and Form I linked to Form F. In Modified Plan 2, Form E linked to Form C and Form I linked to Form G.

    a. Provide verbal descriptions for Modified Plan 1 and Modified Plan 2.

    b. Indicate to which forms Forms K, L, M, and N would link to in the two modified plans. (Try drawing a figure illustrating the plan.)

    c. Evaluate Modified Plans 1 and 2 with regard to the four rules for developing equating plans.

8.4. Consider the example using consistency checks in Table 8.6. Based on consistency checks the results for which method should be chosen if the number tested had been 1050 instead of 1210? Why?

8.5. A test has been previously administered in a paper and pencil mode. The test now is to be administered by computer. The computer version is built to exactly the same content specifications as the paper an pencil test. All items that were administered in the paper and pencil mode have item parameters that have been estimated using an IRT model. The computerized version is constructed using some items that had been previously administered in the paper and pencil mode and some items that are new. Suppose the paper and pencil and computerized versions are being equated.

    a. How could a random groups design be implemented in this situation?

    b. How could a common-item equating to an IRT calibrated item pool design be implemented?

    c. What are the limitations of each design?

    d. How might context effects influence the common-item equating to an IRT calibrated item pool design in this situation?

    e. Which design is preferable?

8.6. List as many causes as you can think of for common items to function differently on two testing occasions. Be sure to consider causes having to do with changes in the items themselves, changes in the examinees, and changes in the administration conditions.

8.7. Assume you are creating an equating design for a testing program. Some of the characteristics of the program are as follows:

    I. Form A was the first form of the test and was scaled previously. Form B is to be equated to Form A. For practical reasons, the equating must be conducted during an operational administration. Each examinee can take only one form.

    II. The test to be equated is a reading test. Each test form consists of three reading passages, with each passage being from a different content area (science, humanities, and social studies). There are 15 items associated with each passage. Testing time is 45 minutes.

III. It will be easy to get large numbers of examinees to participate in the study.

IV. Various different decisions are made using this test, so it is important that equating be accurate all along the score scale.

Which equating design should be used—single group with counterbalancing, random groups, or common-item nonequivalent groups? Why? Which equating method should be used—equipercentile, or linear? Why?

References

Advisory Panel on the Scholastic Aptitude Test Score Decline. (1977). *On further examination.* New York: College Entrance Examination Board.

Allen, N.S., Holland, P.W., & Thayer, D. (1994a). *A missing data approach to estimating distributions of scores for optional test sections* (Research Report 94-17). Princeton, NJ: Educational Testing Service.

Allen, N.S., Holland, P.W., & Thayer, D. (1994b). *Estimating scores for an optional section using information from a common section* (Research Report 94-18). Princeton, NJ: Educational Testing Service.

American College Testing Program (ACT) (1989). *Preliminary technical manual for the Enhanced ACT Assessment.* Iowa City, IA: Author.

American Educational Research Association, American Psychological Association, National Council on Measurement in Education (AERA, APA, NCME) (1985). *Standards for educational and psychological testing.* Washington, DC: Author.

American Psychological Association (APA) (1986). *Guidelines for computer-based tests and interpretations.* Washington, DC: Author.

Andrulis, R.S., Starr, L.M., & Furst, L.W. (1978). The effects of repeaters on test equating. *Educational and Psychological Measurement, 38,* 341–349.

Angoff, W.H. (1953). Test reliability and effective test length. *Psychometrika, 18,* 1–14.

Angoff, W.A. (1971). Scales, norms, and equivalent scores. In R.L. Thorndike (Ed.), *Educational measurement* (2nd ed., pp. 508–600). Washington, DC: American Council on Education. (Reprinted as W. A. Angoff, *Scales, norms, and equivalent scores.* Princeton, NJ: Educational Testing Service, 1984.)

Angoff, W.H. (1987). Technical and practical issues in equating: A discussion of four papers. *Applied Psychological Measurement, 11,* 291–300.

Angoff, W.H., & Cowell, W.R. (1986). An examination of the assumption that the equating of parallel forms is population-independent. *Journal of Educational Measurement, 23,* 327–345.

Baker, F.B. (1992a). *Item response theory parameter estimation techniques.* New York: Marcel Dekker.

Baker, F.B. (1992b). Equating tests under the graded response model. *Applied Psychological Measurement, 16,* 87–96.

Baker, F.B. (1993). Equate 2.0: A computer program for the characteristic curve method of IRT equating. *Applied Psychological Measurement, 17,* 20.

Baker, F.B., & Al-Karni, A. (1991). A comparison of two procedures for computing IRT equating coefficients. *Journal of Educational Measurement, 28,* 147–162.

Baker, E.L., O'Neil, H.F., & Linn, R.L. (1993). Policy and validity prospects for performance-based assessment. *American Psychologist, 48,* 1210–1218.

Baxter, G.P., Shavelson, R.J., Goldman, S.R., & Pine, J. (1992). Evaluation of procedure-based scoring for hands-on science assessment. *Journal of Educational Measurement, 29,* 1–17.

Beaton, A.E., & Gonzalez, E. (1994). *Comparing the NAEP trial state assessment results with the IAEP international results.* Paper presented at the annual meeting of the National Council on Measurement in Education, New Orleans.

Bishop, Y.M.M., Fienberg, S.E, & Holland, P.W. (1975). *Discrete multivariate analysis. Theory and practice.* Cambridge, MA: MIT Press.

Blommers, P.J., & Forsyth, R.A. (1977). *Elementary statistical methods in psychology and education* (2nd ed.). Boston: Houghton Mifflin.

Bloxom, B., & McCully, R. (1992). *Initial operational test and evaluation of forms 18 and 19 of the Armed Services Vocational Aptitude Battery.* Monterey, CA: Defense Manpower Data Center.

Bloxom, B., McCully, R., Branch, R., Waters, B.K., Barnes, J., & Gribben, M. (1993). *Operational calibration of the circular-response optical-mark-reader answer sheets for the Armed Services Vocational Aptitude Battery (ASVAB).* Monterey, CA: Defense Manpower Data Center.

Bloxom, B., Pashley, P.J., Nicewander, W.A., & Yan, D. (1995). Linking to a large scale assessment: An empirical evaluation. *Journal of Educational and Behavioral Statistics, 20,* 1–26.

Brandenburg, D.C., & Forsyth, R.A. (1974). Approximating standardized achievement test norms with a theoretical model. *Educational and Psychological Measurement, 34,* 3–9.

Braun, H.I. (1988). Understanding scoring reliability: Experiments in calibrating essay readers. *Journal of Educational Statistics, 13,* 1–18.

Braun, H.I., & Holland, P.W. (1982). Observed-score test equating: A mathematical analysis of some ETS equating procedures. In P. W. Holland and D. B. Rubin (Eds.), *Test equating* (pp. 9–49). New York: Academic.

Brennan, R.L. (Ed.) (1989). *Methodology used in scaling the ACT Assessment and P-ACT+.* Iowa City, IA: American College Testing.

Brennan, R.L. (1990). *Congeneric models and Levine's linear equating procedures* (ACT Research Report 90-12). Iowa City, IA: American College Testing.

Brennan, R.L. (1992). The context of context effects. *Applied Measurement in Education, 5,* 225–264.

Brennan, R.L., & Kolen, M.J. (1987a). A reply to Angoff. *Applied Psychological Measurement, 11,* 301–306.

Brennan, R.L., & Kolen, M.J. (1987b). Some practical issues in equating. *Applied Psychological Measurement, 11,* 279–290.

Budescu, D. (1985). Efficiency of linear equating as a function of the length of the anchor test. *Journal of Educational Measurement, 22,* 13–20.

Budescu, D. (1987). Selecting an equating method: Linear or equipercentile? *Journal of Educational Statistics, 12,* 33–43.

Burke, E.F., Hartke, D., & Shadow, L. (1989). *Print format effects on ASVAB test score performance: Literature review* (AFHRL Technical Paper 88-58). Brooks Air Force Base, TX: Air Force Human Resources Laboratory.

Carlin, J.B., & Rubin, D.B. (1991). Summarizing multiple-choice tests using three informative statistics. *Psychological Bulletin, 110,* 338–349.

Cizek, G.J. (1994). The effect of altering the position of options in a multiple-choice examination. *Educational and Psychological Measurement, 54,* 8–20.

Cohen, A.S., & Kim, S.H. (1993). *A comparison of equating methods under the graded response model.* Paper presented at the annual meeting of the National Council on Measurement in Education, Atlanta.

Congressional Budget Office. (1986). *Trends in educational achievement.* Washington, DC: Author.

Cook, L.L. (1994). *Recentering the SAT score scale: An overview and some policy considerations.* Paper presented at the annual meeting of the National Council on Measurement in Education, New Orleans.

Cook, L.L., & Eignor, D.R. (1991). An NCME instructional module on IRT equating methods. *Educational Measurement: Issues and Practice, 10,* 37–45.

Cook, L.L., & Petersen, N.S. (1987). Problems related to the use of conventional and item response theory equating methods in less than optimal circumstances. *Applied Psychological Measurement, 11,* 225–244.

Cope, R.T. (1986). *Use versus nonuse of repeater examinees in common item linear equating with nonequivalent populations* (ACT Technical Bulletin 51). Iowa City, IA: American College Testing.

Cope, R.T., & Kolen, M.J. (1990). *A study of methods for estimating distributions of test scores* (ACT Research Report 90-5). Iowa City, IA: American College Testing.

Crouse, J.D. (1991). *Comparing the equating accuracy from three data collection designs using bootstrap estimation methods.* Unpublished doctoral dissertation, The University of Iowa, Iowa City, IA.

Cureton, E.F., & Tukey, J.W. (1951). Smoothing frequency distributions, equating tests, and preparing norms. *American Psychologist, 6,* 404.

Darroch, J.N., & Ratcliff, D. (1972). Generalized iterative scaling for log-linear models. *Annals of Mathematical Statistics, 43,* 1470–1480.

de Boor, C. (1978). *A practical guide to splines* (Applied Mathematical Sciences, Volume 27). New York: Springer-Verlag.

Divgi, D.R. (1985). A minimum chi-square method for developing a common metric in item response theory. *Applied Psychological Measurement, 9,* 413–415.

Donlon, T. (Ed.) (1984). *The College Board technical handbook for the Scholastic Aptitude Test and Achievement Tests.* New York: College Entrance Examination Board.

Dorans, N.J. (1986). The impact of item deletion on equating conversions and reported score distributions. *Journal of Educational Measurement, 23,* 245–264.

Dorans, N.J. (1990a). Equating methods and sampling designs. *Applied Measurement in Education, 3,* 3–17.

Dorans, N.J. (1990b). Scaling and equating. In H. Wainer (Ed.), *Computerized adaptive testing: A primer* (pp. 137–160). Hillsdale, NJ: Erlbaum.

Dorans, N.J. (1994a). *Choosing and evaluating a scale transformation: Centering and realigning SAT score distributions.* Paper presented at the annual meeting of the National Council on Measurement in Education, New Orleans.

Dorans, N.J. (1994b). *Effects of scale choice on score distributions: Two views of subgroup performance on the SAT.* Paper presented at the annual meeting of the National Council on Measurement in Education, New Orleans.

Dorans, N.J., & Lawrence, I.M. (1990). Checking the statistical equivalence of nearly identical test editions. *Applied Measurement in Education, 3,* 245–254.

Draper, N.R., & Smith, H. (1981). *Applied regression analysis* (2nd ed.). New York: Wiley.

Dunbar, S.B., Koretz, D.M., & Hoover, H.D. (1991). Quality control in the development and use of performance assessments. *Applied Measurement in Education, 4,* 289–303.

Ebel, R.L., & Frisbie, D.A. (1991). *Essentials of educational measurement* (5th ed.). Englewood Cliffs, NJ: Prentice-Hall.

Efron, B. (1982). *The jackknife, the bootstrap, and other resampling plans.* Philadelphia, PA: Society for Industrial and Applied Mathematics.

Efron, B., & Tibshirani, R.J. (1993). *An introduction to the bootstrap* (Monographs on Statistics and Applied Probability 57). New York: Chapman & Hall.

Eignor, D.R. (1985). *An investigation of the feasibility and practical outcomes of pre-equating the SAT verbal and mathematical sections* (Research Report 85-10). Princeton, NJ: Educational Testing Service.

Eignor, D. (1993). *Deriving comparable scores for computer adaptive and conventional tests: An example using the SAT* (Research Report 93-55). Princeton, NJ: Educational Testing Service.

Eignor, D.R., & Stocking, M.L. (1986). *An investigation of the possible causes for the inadequacy of IRT preequating* (Research Report 86-14). Princeton, NJ: Educational Testing Service.

Eignor, D.R., Way, W.D., & Amoss, K.E. (1994). *Establishing the comparability of the NCLEX using CAT with traditional NCLEX examinations.* Paper presented at the annual meeting of the National Council on Measurement in Education, New Orleans.

Englehard, G. (1992). The measurement of writing ability with a many-faceted Rasch model. *Applied Measurement in Education, 5,* 171–191.

Englehard, G. (1994). Examining rater errors in the assessment of written composition with a many-faceted Rasch model. *Journal of Educational Measurement, 31,* 93–112.

Ercikan, K. (1994). *Linking state tests to NAEP.* Paper presented at the annual meeting of the National Council on Measurement in Education, New Orleans.

Fairbank, B.A. (1987). The use of presmoothing and postsmoothing to increase the precision of equipercentile equating. *Applied Psychological Measurement, 11,* 245–262.

Feldt, L.S., & Brennan, R.L. (1989). Reliability. In R.L. Linn (Ed.), *Educational measurement* (3rd ed., pp. 105–146). New York: Macmillan.

Ferrara, S. (1993). *Generalizability theory and scaling: Their roles in writing assessment and implications for performance assessments in other content areas.* Paper presented at the annual meeting of the National Council on Measurement in Education, Atlanta.

Fitzpatrick, A.R., & Yen, W.M. (1993). *The psychometric characteristics of choice items.* Paper presented at the annual meeting of the National Council on Measurement in Education, Atlanta.

Fitzpatrick, A.R., Ercikan, K., Yen, Y.M., & Ferrara, S. (1994). *The consistency between ratings collected in different test years.* Paper presented at the annual meeting of the National Council on Measurement in Education, New Orleans.

Gilmer, J.S. (1989). The effects of test disclosure on equated scores and pass rates. *Applied Psychological Measurement, 13,* 245–255.

Gordon, B., Englehard, G., Gabrielson, S., & Bernkopf, S. (1993). *Issues in equating performance assessments: Lessons from writing assessment.* Paper presented at the annual meeting of the American Educational Research Association, Atlanta.

Gulliksen, H. (1950). *Theory of mental tests.* New York: Wiley.

Haberman, S.J. (1974a). *The analysis of frequency data.* Chicago: University of Chicago.

Haberman, S.J. (1974b). Log-linear models for frequency tables with ordered classifications. *Biometrics, 30,* 589–600.

Haberman, S.J. (1978). *Analysis of qualitative data. Vol. 1. Introductory topics.* New York: Academic.

Haebara, T. (1980). Equating logistic ability scales by a weighted least squares method. *Japanese Psychological Research, 22,* 144–149.

Hambleton, R.K., & Swaminathan, H. (1985). *Item response theory. Principles and applications.* Boston: Kluwer.

Hambleton, R.K., Swaminathan, H., & Rogers, H.J. (1991). *Fundamentals of item response theory.* Newbury Park, CA: Sage.

Han, T. (1993). *Comparison of IRT observed-score equating with both IRT true-score and classical equipercentile equating.* Unpublished doctoral dissertation, Southern Illinois University, Carbondale, IL.

Hanson, B.A. (1989). Scaling the P-ACT+. In R.L. Brennan (Ed.), *Methodology used in scaling the ACT Assessment and P-ACT+* (pp. 57–73). Iowa City, IA: American College Testing.

Hanson, B.A. (1990). *An investigation of methods for improving estimation of test score distributions* (ACT Research Report 90–4). Iowa City, IA: American College Testing.

Hanson, B.A. (1991a). A note on Levine's formula for equating unequally reliable tests using data from the common item nonequivalent groups design. *Journal of Educational Statistics, 16,* 93–100.

Hanson, B.A. (1991b). *Method of moments estimates for the four-parameter beta compound binomial model and the calculation of classification consistency indexes* (ACT Research Report 91–5). Iowa City, IA: American College Testing.

Hanson, B.A. (1991c). A comparison of bivariate smoothing methods in common-item equipercentile equating. *Applied Psychological Measurement, 15,* 391–408.

Hanson, B.A. (1992). *Testing for differences in test score distributions using log-linear models.* Paper presented at the annual meeting of the American Educational Research Association, San Francisco.

Hanson, B.A. (1993). *A missing data approach to adjusting writing sample scores.* Paper presented at the annual meeting of the National Council on Measurement in Education, Atlanta.

Hanson, B.A., Zeng, L., & Kolen, M.J. (1993). Standard errors of Levine linear equating. *Applied Psychological Measurement, 17,* 225–237.

Hanson, B.A., Zeng, L., & Colton, D. (1994). *A comparison of presmoothing and postsmoothing methods in equipercentile equating* (ACT Research Report 94-4). Iowa City, IA: American College Testing.

Harnischfeger, A., & Wiley, D.E. (1975). *Achievement test score decline: Do we need to worry?* Chicago: CEMREL.

Harris, D.J. (1986). A comparison of two answer sheet formats. *Educational and Psychological Measurement, 46,* 475–478.

Harris, D.J. (1987). *Estimating examinee achievement using a customized test.* Paper presented at the annual meeting of the American Educational Research Association, Washington, DC.

Harris, D.J. (1988). *An examination of the effect of test length on customized testing using item response theory.* Paper presented at the annual meeting of the American Educational Research Association, New Orleans.

Harris, D.J. (1989). Comparison of 1-, 2-, and 3-parameter IRT models. *Educational Measurement: Issues and Practice, 8,* 35–41.

Harris, D.J. (1991a). *Equating with nonrepresentative common item sets and nonequivalent groups.* Paper presented at the annual meeting of the American Educational Research Association, Chicago.

Harris, D.J. (1991b). *Practical implications of the context effects resulting from the use of scrambled test forms.* Paper presented at the annual meeting of the American Educational Research Association, Chicago.

Harris, D.J. (1993). *Practical issues in equating.* Paper presented at the annual meeting of the American Educational Research Association, Atlanta.

Harris, D.J., & Crouse, J.D. (1993). A study of criteria used in equating. *Applied Measurement in Education, 6*, 195–240.

Harris, D.J., & Kolen, M.J. (1986). Effect of examinee group on equating relationships. *Applied Psychological Measurement, 10*, 35–43.

Harris, D.J., & Kolen, M.J. (1990). A comparison of two equipercentile equating methods for common item equating. *Educational and Psychological Measurement, 50*, 61–71.

Harris, D.J., & Welch, C.J. (1993). *Equating writing samples.* Paper presented at the annual meeting of the National Council on Measurement in Education, Atlanta.

Harris, D.J., Welch, C.J., & Wang, T. (1994). *Issues in equating performance assessments.* Paper presented at the annual meeting of the National Council on Measurement in Education, New Orleans.

Holland, P.W., & Rubin, D.B. (1982) *Test equating.* New York: Academic.

Holland, P.W., & Thayer, D.T. (1987). *Notes on the use of log-linear models for fitting discrete probability distributions* (Technical Report 87-79). Princeton, NJ: Educational Testing Service.

Holland, P.W., & Thayer, D.T. (1989). *The kernel method of equating score distributions* (Technical Report No. 89-84). Princeton, NJ: Educational Testing Service.

Holland, P.W., & Thayer, D.T. (1990). *Kernel equating and the counterbalanced design.* Paper presented at the annual meeting of the American Educational Research Association, Boston.

Holland, P.W., King, B.F., & Thayer, D.T. (1989). *The standard error of equating for the kernel method of equating score distributions* (Technical Report 89-83). Princeton, NJ: Educational Testing Service.

Houston, W., & Sawyer, R. (1991). Relating scores on the Enhanced ACT Assessment and the SAT test batteries. *College and University, 66*, 195–200.

Hung, P., Wu, Y., & Chen, Y. (1991). *IRT item parameter linking: Relevant issues for the purpose of item banking.* Paper presented at the International Academic Symposium on Psychological Measurement, Tainan, Taiwan.

Huynh, H., & Ferrara, S. (1994). A comparison of equal percentile and partial credit equatings for performance-based assessments composed of free-response items. *Journal of Educational Measurement, 31*, 125–141.

IMSL (1991). *Fortran subroutines for mathematical applications. Math/library.* Houston, TX: Author.

Jaeger, R.M. (1981). Some exploratory indices for selection of a test equating method. *Journal of Educational Measurement, 18*, 23–38.

Jarjoura, D., & Kolen, M.J. (1985). Standard errors of equipercentile equating for the common item nonequivalent populations design. *Journal of Educational Statistics, 10*, 143–160.

Keats, J.A., & Lord, F.M. (1962). A theoretical distribution for mental test scores. *Psychometrika, 27*, 59–72.

Kendall, M., & Stuart, A. (1977). *The advanced theory of statistics* (4th ed., Vol. 1). New York: Macmillan.

Kiplinger, V.L., & Linn, R.L. (1994). *Linking statewide tests to the National Assessment of Educational Progress: Stability of results.* Paper presented at the annual meeting of the National Council on Measurement in Education, New Orleans.

Klein L.W., & Jarjoura, D. (1985). The importance of content representation for common-item equating with nonrandom groups. *Journal of Educational Measurement, 22*, 197–206.

Kolen, M.J. (1981). Comparison of traditional and item response theory methods for equating tests. *Journal of Educational Measurement, 18*, 1–11.

Kolen, M.J. (1984). Effectiveness of analytic smoothing in equipercentile equating. *Journal of Educational Statistics, 9*, 25–44.

Kolen, M.J. (1985). Standard errors of Tucker equating. *Applied Psychological Measurement, 9,* 209–223.

Kolen, M.J. (1988). An NCME instructional module on traditional equating methodology. *Educational Measurement: Issues and Practice, 7,* 29–36.

Kolen, M.J. (1991). Smoothing methods for estimating test score distributions. *Journal of Educational Measurement, 28,* 257–282.

Kolen, M.J., & Brennan, R.L. (1987). Linear equating models for the common-item nonequivalent-populations design. *Applied Psychological Measurement, 11,* 263–277.

Kolen, M.J., & Harris, D.J. (1990). Comparison of item preequating and random groups equating using IRT and equipercentile methods. *Journal of Educational Measurement, 27,* 27–39.

Kolen, M.J., & Jarjoura, D. (1987). Analytic smoothing for equipercentile equating under the common item nonequivalent populations design. *Psychometrika, 52,* 43–59.

Kolen, M.J., Hanson, B.A., & Brennan, R.L. (1992). Conditional standard errors of measurement for scale scores. *Journal of Educational Measurement, 29,* 285–307.

Kolen, M.J., Zeng, L., & Hanson, B.A. (1994). *Conditional standard errors of measurement for scale scores using IRT.* Paper presented at the annual meeting of the National Council on Measurement in Education, New Orleans.

Lawrence, I.M., Dorans, N.J., Feigenbaum, M.D., Feryok, N.J., Schmitt, A.P., & Wright, N.K. (1994). *Technical issues related to the introduction of the new SAT and PSAT/NMSQT* (Research Memorandum 94-10). Princeton, NJ: Educational Testing Service.

Levine, R. (1955). *Equating the score scales of alternate forms administered to samples of different ability* (Research Bulletin 55-23). Princeton, NJ: Educational Testing Service.

Linacre, J.M. (1988). *Many-faceted Rasch measurement.* Chicago: MESA Press.

Linn, R.L. (1993). Linking results of distinct assessments. *Applied Measurement in Education, 6,* 83–102.

Linn, R.L., & Hambleton, R.K. (1991). Customized tests and customized norms. *Applied Measurement in Education, 4,* 185–207.

Livingston, S.A. (1988). *Adjusting scores on examinations offering a choice of essay questions* (Research Report 88-64). Princeton, NJ: Educational Testing Service.

Livingston, S.A. (1993). Small-sample equating with log-linear smoothing. *Journal of Educational Measurement, 30,* 23–39.

Livingston, S.A., & Feryok, N.J. (1987). *Univariate vs. bivariate smoothing in frequency estimation equating* (Research Report 87-36). Princeton, NJ: Educational Testing Service.

Livingston, S.A., Dorans, N.J., & Wright, N. K. (1990). What combination of sampling and equating methods works best? *Applied Measurement in Education, 3,* 73–95.

Longford, N.T. (1994). Reliability of essay rating and score adjustment. *Journal of Educational and Behavioral Statistics, 19,* 171–200.

Lord, F.M. (1950). *Notes on comparable scales for test scores* (Research Bulletin 50-48). Princeton, NJ: Educational Testing Service.

Lord, F.M. (1965). A strong true score theory with applications. *Psychometrika, 30,* 239–270.

Lord, F.M. (1969). Estimating true-score distributions in psychological testing. (An empirical Bayes estimation problem.) *Psychometrika, 34,* 259–299.

Lord, F.M. (1975). Automated hypothesis tests and standard errors for nonstandard problems. *The American Statistician, 29,* 56–59.

Lord, F.M. (1980). *Applications of item response theory to practical testing problems.* Hillsdale, NJ: Erlbaum.

Lord, F.M. (1982a). Item response theory and equating—A technical summary. In P.W. Holland and D.B. Rubin (Eds.), *Test equating* (pp. 141–148). New York: Academic.

Lord, F.M. (1982b). The standard error of equipercentile equating. *Journal of Educational Statistics, 7,* 165–174.

Lord, F.M. (1982c). Standard error of an equating by item response theory. *Applied Psychological Measurement, 6,* 463–472.

Lord, F.M., & Novick, M.R. (1968). *Statistical theories of mental test scores.* Reading, MA: Addison Wesley.

Lord, F.M., & Wingersky, M.S. (1984). Comparison of IRT true-score and equipercentile observed-score "equatings". *Applied Psychological Measurement, 8,* 452–461.

Loyd, B.H. (1991). Mathematics test performance: The effects of item type and calculator use. *Applied Measurement in Education, 4,* 11–22.

Loyd, B.H., & Hoover, H.D. (1980). Vertical equating using the Rasch Model. *Journal of Educational Measurement, 17,* 179–193.

Loyd, B., Englehard, G., & Crocker, L. (1993). *Equity, equivalence, and equating: Fundamental issues and proposed strategies for the National Board for Professional Teaching Standards.* Paper presented at the annual meeting of the National Council on Measurement in Education, Atlanta.

MacCann, R.G. (1990). Derivations of observed score equating methods that cater to populations differing in ability. *Journal of Educational Statistics, 15,* 146–170.

Maier, M.H. (1993). *Military aptitude testing: The past fifty years* (DMDC Technical Report 93-007). Monterey, CA: Defense Manpower Data Center.

Marco, G.L. (1977). Item characteristic curve solutions to three intractable testing problems. *Journal of Educational Measurement, 14,* 139–160.

Marco, G.L. (1981). Equating tests in the era of test disclosure. In B.F. Green (Ed.), *New directions for testing and measurement: Issues in testing—coaching, disclosure, and ethnic bias* (pp. 105–122). San Francisco: Jossey-Bass.

Marco, G.L., & Abdel-Fattah, A.A. (1991). Developing concordance tables for scores on the Enhanced ACT Assessment and the SAT. *College and University, 66,* 187–194.

Marco, G.L., Petersen, N.S., & Stewart, E.E. (1983). A test of the adequacy of curvilinear score equating models. In D. Weiss (Ed.), *New horizons in testing* (pp. 147–176). New York: Academic.

Masters, G.N. (1982). A Rasch model for partial credit scoring. *Psychometrika, 47,* 149–174.

Mazzeo, J., & Harvey, A.L. (1988). *The equivalence of scores from automated and conventional educational and psychological tests. A review of the literature* (College Board Report 88-8). New York: College Entrance Examination Board.

Mazzeo, J., Druesne, B., Raffeld, P.C., Checketts, K.T., & Muhlstein, A. (1991). *Comparability of computer and paper-and-pencil scores for two CLEP general examinations* (College Board Report 91-5). New York: College Entrance Examination Board.

McKinley, R.L., & Schaeffer, G.A. (1989). *Reducing test form overlap of the GRE subject test in mathematics using IRT triple-part equating* (Research Report 89-8). Princeton, NJ: Educational Testing Service.

Mead, A.D., & Drasgow, F. (1993). Equivalence of computerized and paper-and-pencil cognitive ability tests: A meta-analysis. *Psychological Bulletin, 114,* 449–458.

Millman, J., & Greene, J. (1989). The specification and development of tests of achievement and ability. In R.L. Linn (Ed.), *Educational measurement* (3rd ed., pp. 335–366). New York: Macmillan.

Mills, C., Durso, R., Golub-Smith, M., Schaeffer, G., & Steffen, M. (1994). *The in-*

troduction and comparability of the computer adaptive GRE general test. Paper presented at the annual meeting of the National Council on Measurement in Education, New Orleans.

Mislevy, R.J. (1992). *Linking educational assessments: Concepts, issues, methods, and prospects*. Princeton, NJ: ETS Policy Information Center.

Mislevy, R.J., & Bock, R.D. (1990). *BILOG 3. Item analysis and test scoring with binary logistic models* (2nd ed.). Mooresville, IN: Scientific Software.

Mislevy, R.J., & Stocking, M.L. (1989). A consumers guide to LOGIST and BILOG. *Applied Psychological Measurement, 13*, 57–75.

Morgan, R., & Stevens, J. (1991). *Experimental study of the effects of calculator use in the advanced placement calculus examinations* (Research Report 91-5). Princeton, NJ: Educational Testing Service.

Morris, C.N. (1982). On the foundations of test equating. In P.W. Holland and D.B. Rubin (Eds.), *Test equating* (pp. 169–191). New York: Academic.

Muraki, E. (1992). A generalized partial credit model: Application of an EM algorithm. *Applied Psychological Measurement, 16*, 159–176.

Parshall, C.G., & Kromrey, J.D. (1993). *Computer testing versus paper-and-pencil testing: An analysis of examinee characteristics associated with mode effect*. Paper presented at the annual meeting of the American Educational Research Association, Atlanta.

Pashley, P.J. & Phillips, G.W. (1993). *Toward world class standards. A research study linking international and national assessments*. Princeton, NJ: Educational Testing Service.

Pashley, P.J., Lewis, C., & Yan, D. (1994). *Statistical linking procedures for deriving point estimates and associated standard errors*. Paper presented at the annual meeting of the National Council on Measurement in Education, New Orleans.

Petersen, N.S., Cook, L.L., & Stocking, M.L. (1983). IRT versus conventional equating methods: A comparative study of scale stability. *Journal of Educational Statistics, 8*, 137–156.

Petersen, N.S., Kolen, M.J., & Hoover, H.D. (1989). Scaling, norming, and equating. In R.L. Linn (Ed.), *Educational measurement* (3rd ed., pp. 221–262). New York: Macmillan.

Press, W.H., Flannery, B.P., Teukolsky, S.A., & Vetterling, W.T. (1989). *Numerical recipes. The art of scientific computing (Fortran version)*. Cambridge, UK: Cambridge University Press.

Rasch, G. (1960). *Probabilistic models for some intelligence and attainment tests*. Copenhagen: Danish Institute for Educational Research.

Raymond, M.R., & Viswesvaran, C. (1993). Least squares models to correct for rater effects in performance assessment. *Journal of Educational Measurement, 30*, 253–268.

Reese, C. (1992). *Development of a computer-based test for the GRE general test*. Paper presented at the annual meeting of the National Council on Measurement in Education, San Francisco.

Reinsch, C.H. (1967). Smoothing by spline functions. *Numerische Mathematik, 10*, 177–183.

Rosenbaum, P.R., & Thayer, D. (1987). Smoothing the joint and marginal distributions of scored two-way contingency tables in test equating. *British Journal of Mathematical and Statistical Psychology, 40*, 43–49.

Samejima, F. (1969). *Estimation of latent ability using a response pattern of graded scores*. (Psychometrika Monograph No. 17) Richmond, VA Psychometrics Society.

Segall, D.O. (1993). *Score equating verification analyses of the CAT-ASVAB*. Briefing presented to the Defense Advisory Committee on Military Personnel Testing, Williamsburg, VA.

Skaggs, G. (1990). *Assessing the utility of item response theory models for test equat-*

ing. Paper presented at the annual meeting of the National Council on Measurement in Education, Boston.

Skaggs, G., & Lissitz, R.W. (1986). IRT test equating: Relevant issues and a review of recent research. *Review of Educational Research, 56*, 495–529.

Spray, J.A., Ackerman, T.A., Reckase, M.D., & Carlson, J.E. (1989). Effect of medium of item presentation on examinee performance and item characteristics. *Journal of Educational Measurement, 26*, 261–271.

Stocking, M.L., & Eignor, D.R. (1986). *The impact of different ability distributions on IRT preequating* (Research Report 86-49). Princeton, NJ: Educational Testing Service.

Stocking, M.L., & Lord, F.M. (1983). Developing a common metric in item response theory. *Applied Psychological Measurement, 7*, 201–210.

Tenopyr, M.L., Angoff, W.H., Butcher, J.N., Geisinger, K.F., & Reilly, R.R. (1993). Psychometric and assessment issues raised by the Americans with Disabilities Act (ADA). *The Score, 15*, 1–15.

Thissen, D., & Steinberg, L. (1984). A response model for multiple choice items. *Psychometrika, 49*, 501–519.

Thissen, D., & Steinberg, L. (1986). A taxonomy of item response models. *Psychometrika, 51*, 567–577.

Thomasson, G.L., Bloxom, B., & Wise, L. (1994). *Initial operational test and evaluation of forms 20, 21, and 22 of the Armed Services Vocational Aptitude Battery (ASVAB)* (DMDC Technical Report 94-001). Monterey, CA: Defense Manpower Data Center.

Wainer, H. (1993). Some practical considerations when converting a linearly administered test to an adaptive format. *Educational Measurement: Issues and Practice, 12*, 15–20.

Wainer, H., & Mislevy, R.J. (1990). Item response theory, item calibration, and proficiency estimation. In H. Wainer (Ed.), *Computerized adaptive testing: A primer* (pp. 65–102). Hillsdale, NJ: Erlbaum.

Wainer, H., & Thissen, D. (1993). Combining multiple-choice and constructed-response test scores: Toward a Marxist theory of test construction. *Applied Measurement in Education, 6*, 103–118.

Wainer, H., & Thissen, D. (1994). On examinee choice in educational testing. *Review of Educational Research, 64*, 159–195.

Wainer, H., Dorans, N.J., Green, B.F., Mislevy, R.J., Steinberg, L., & Thissen, D. (1990). Future challenges. In H. Wainer (Ed.), *Computerized adaptive testing: A primer* (pp. 233–286). Hillsdale, NJ: Erlbaum.

Wainer, H., Wang, X., & Thissen, D. (1994). How well can we compare scores on test forms that are constructed by examinees choice? *Journal of Educational Measurement, 31*, 183–199.

Wang, T., & Kolen, M.J. (1994). *A quadratic curve equating method to equate the first three moments in equipercentile equating* (ACT Research Report 94-2). Iowa City, IA: American College Testing.

Way, W.D., & Tang, K.L. (1991). *A comparison of four logistic model equating methods*. Paper presented at the annual meeting of the American Educational Research Association, Chicago.

Way, W.D., Forsyth, R.A., & Ansley, T.N. (1989). IRT ability estimates from customized achievement tests without representative content sampling. *Applied Measurement in Education, 2*, 15–35.

Wingersky, M.S., Barton, M.A., & Lord, F.M. (1982). *LOGIST users guide*. Princeton, NJ: Educational Testing Service.

Wingersky, M.S., Cook, L.L., & Eignor, D.R. (1987). *Specifying the characteristics of linking items used for item response theory item calibration* (Research Report 87–24). Princeton, NJ: Educational Testing Service.

Woodruff, D.J. (1986). Derivations of observed score linear equating methods based on test score models for the common item nonequivalent populations design. *Journal of Educational Statistics, 11,* 245–257.

Woodruff, D.J. (1989). A comparison of three linear equating methods for the common-item nonequivalent-populations design. *Applied Psychological Measurement, 13,* 257–261.

Wright, B.D., & Stone, M.H. (1979). *Best test design.* Chicago: MESA Press.

Yen, W.M. (1983). Tau-equivalence and equipercentile equating. *Psychometrika, 48,* 353–369.

Yen, W.M. (1993). Scaling performance assessments: Strategies for managing local item dependence. *Journal of Educational Measurement, 30,* 187–213.

Zeng, L. (1993). A numerical approach for computing standard errors of linear equating. *Applied Psychological Measurement, 17,* 177–186.

Zeng, L. (1995). The optimal degree of smoothing in equipercentile equating with postsmoothing. *Applied Psychological Measurement, 19,* 177–190.

Zeng, L., & Cope, R.T. (1995). Standard errors of linear equating for the counterbalanced design. *Journal of Educational and Behavioral Statistics, 4,* 337–348.

Zeng, L., & Kolen, M.J. (1994). IRT scale transformations using numerical integration. Paper presented at the Annual Meeting of the American Educational Research Association, New Orleans.

Zeng, L., Hanson, B.A. & Kolen, M.J. (1994). Standard errors of a chain of linear equatings. *Applied Psychological Measurement, 18,* 369–378.

Zwick, R. (1991). Effects of item order and context on estimation of NAEP Reading Proficiency. *Educational Measurement: Issues and Practice, 10,* 10–16.

Appendix A: Answers to Exercises

Chapter 1

1.1.a. Because the top 1% of the examinees on a particular test date will be the same regardless of whether or not an equating process is used, equating likely would not affect who was awarded a scholarship.

1.1.b. In order to identify the top 1% of the examinees during the whole year, it is necessary to consider examinees who were administered two forms as one group. If the forms on the two test dates were unequally difficult, then the use of equating could result in scholarships being awarded to different examinees as compared to just using the raw score on the form each examinee happened to be administered.

1.2. Because Form X_3 is easier than Form X_2, a raw score of 29 on Form X_3 indicates the same level of achievement as a raw score of 28 on Form X_2. From the table, a Form X_2 raw score of 28 corresponds to a scale score of 13. Thus, a raw score of 29 on Form X_3 also corresponds to a scale score of 13.

1.3. Because the test is to be secure, items that are going to be used as scored items in subsequent administrations cannot be released to examinees. Of the designs listed, the common-item nonequivalent groups design with external common items can be most easily implemented. On a particular administration, each examinee would receive a test form containing the scored items, a set of unscored items that had been administered along with a previous form, and possibly another set of unscored items to be used as a common-item section in subsequent equatings. Thus, all items that contribute to an examinee's score would be new items that would never be reused. The single group design with counterbalancing (assuming no differential order effects) and random groups design also could be implemented using examinees from other states. For example, using the random groups design, forms could be spiraled in another state which did not require that the test be released. The equated forms could be used subsequently in the state that required disclosure. The common-item nonequivalent groups design with internal common items may also be used in this way.

1.4. Random groups design. This design requires that only one form be administered to each examinee.

1.5. Only the common-item nonequivalent groups design can be used. Both the random groups and single group designs require the administration of more than one form on a given test date.

1.6. a. Group 2. b. Group 1. c. The content of the common items should be representative of the total test; otherwise, inaccurate equating might result.

1.7. Statement I is consistent with an observed score definition. Statement II is consistent with an equity definition.

1.8. Random. Systematic.

Chapter 2

2.1. $P(2.7) = 100\{.7 + [2.7 - (3 - .5)][.9 - .7]\} = 74$;
$P(.2) = 100\{0 + [.2 - (0 - .5)][.2 - 0]\} = 14$;
$P^{-1}(25) = (.25 - .2)/(.5 - .2) + (1 - .5) = .67$;
$P^{-1}(97) = (.97 - .90)/(1 - .90) + (4 - .5) = 4.2$.

2.2. $\mu(X) = 1.70$; $\sigma(X) = 1.2689$; $\mu(Y) = 2.30$; $\sigma(Y) = 1.2689$; $m(x) = x + .60$; $l(x) = x + .60$.

2.3. $\mu[e_Y(x)] = .2(.50) + .3(1.75) + .2(2.8333) + .2(3.50) + .1(4.25) = 2.3167$;
$\sigma[e_Y(x)]$
$$= \sqrt{[.2(.50^2) + .3(1.75^2) + .2(2.8333^2) + .2(3.50^2) + .1(4.25^2)] - 2.3167^2}$$
$$= 1.2098.$$

2.4. Note: $\mu(X) = 6.7500$; $\sigma(X) = 1.8131$; $\mu(Y) = 5.0500$; $\sigma(Y) = 1.7284$. See Tables A.1 and A.2.

2.5. The mean and linear methods will produce the same results. This can be seen by applying the formulas. Note that the equipercentile method will not produce the same results as the mean and linear methods under these conditions unless the higher order moments (skewness, kurtosis, etc.) are identical for the two forms.

2.6. $21.4793 + [(23.15 - 23)/(24 - 23)][22.2695 - 21.4793] = 21.5978$.

2.7. $1.1(.8x + 1.2) + 10 = .88x + 1.32 + 10 = .88x + 11.32$.

2.8. In general, the shapes will be the same under mean and linear equating. Under equipercentile equating, the shape will be the same only if the shape of the Form X and Form Y distributions are the same. Actually, the shape of the Form X scores converted to the Form Y scale will be approximately the same as the shape of the Form Y distribution.

Table A.1. Score Distributions for Exercise 2.4.

x	$f(x)$	$F(x)$	$P(x)$	y	$g(y)$	$G(y)$	$Q(y)$
0	.00	.00	.0	0	.00	.00	.0
1	.01	.01	.5	1	.02	.02	1.0
2	.02	.03	2.0	2	.05	.07	4.5
3	.03	.06	4.5	3	.10	.17	12.0
4	.04	.10	8.0	4	.20	.37	27.0
5	.10	.20	15.0	5	.25	.62	49.5
6	.20	.40	30.0	6	.20	.82	72.0
7	.25	.65	52.5	7	.10	.92	87.0
8	.20	.85	75.0	8	.05	.97	94.5
9	.10	.95	90.0	9	.02	.99	98.0
10	.05	1.00	97.5	10	.01	1.00	99.5

Table A.2. Equated Scores for Exercise 2.4.

x	$m_Y(x)$	$l_Y(x)$	$e_Y(x)$
0	−1.7000	−1.3846	.0000
1	−.7000	−.4314	.7500
2	.3000	.5219	1.5000
3	1.3000	1.4752	2.0000
4	2.3000	2.4285	2.6000
5	3.3000	3.3818	3.3000
6	4.3000	4.3350	4.1500
7	5.3000	5.2883	5.1200
8	6.3000	6.2416	6.1500
9	7.3000	7.1949	7.3000
10	8.3000	8.1482	8.7500

Chapter 3

3.1. Note: $e_Y(x_i) = 28.3$; $t_Y(x_i) = 29.1$; $\hat{e}_Y(x_i) = 31.1$; $\hat{t}_Y(x_i) = 31.3$.

a. $29.1 − 28.3 = .8$. b. $31.1 − 28.3 = 2.8$. c. $31.3 − 28.3 = 3.0$. d. We cannot tell from the information given—we would need to have an indication of the variability of sample values over many replications, rather than the one replication that is given. e. Unsmoothed at $x_i = 26$. f. We cannot tell from the information given—we would need to have an indication of the variability of sample values over many replications, rather than the one replication that is given.

3.2. Mean, standard deviation, and skewness.

3.3. For Form Y, $C = 7$ is the highest value of C with a nominally significant χ^2. So, of the models evaluated, those with $C \leq 7$ would be eliminated. The model with the smallest value of C that is not eliminated using a nominal significance level of .30 is $C = 8$. For Form X, $C \leq 5$ are eliminated. $C = 6$ is the smallest value of C that is not eliminated.

3.4. Using equation (3.11), $\hat{d}_Y(28.6) = 28.0321 + 1.0557(.6) - .0075(.6)^2 + .0003(.6)^3 = 28.6629$.

3.5. Conversions for $S = .20$ and $S = .30$. Conversions for $S = .75$ and $S = 1.00$. It would matter which was chosen if Form X was used later as the old form for equating a new form, because in this process the unrounded conversion for Form X would be used.

3.6. It appears that the relationships for all S-parameters examined would fall within the ± 2 standard error bands. The identity equating relationship would fall outside the bands from 4 to 20 (refer to the standard errors in Table 3.2 to help answer this question).

3.7. For $N = 100$ on the Science Reasoning test, the identity equating was better than any of the other equating methods. Even with $N = 250$ on the Science Reasoning test, the identity equating performed as well as or better than any of the equipercentile methods. One factor that could have led to the identity equating appearing to be relatively better with small samples for the Science Reasoning test than for the English test would be if the two Science Reasoning forms were more similar to one another than were the two English forms. In the extreme case, suppose that two Science Reasoning forms were actually identical. In this case, the identity equating always would be better than any of the other equating methods.

Chapter 4

4.1. Denote $\mu_1 \equiv \mu_1(X)$, $\sigma_1 \equiv \sigma_1(X)$, etc. We want to show that $\sigma_s^2 = w_1\sigma_1^2 + w_2\sigma_2^2 + w_1w_2(\mu_1 - \mu_2)^2$. By definition, $\sigma_s^2 = w_1 E_1(X - \mu_s)^2 + w_2 E_2(X - \mu_s)^2$. Noting that $\mu_s = w_1\mu_1 + w_2\mu_2$ and $w_1 + w_2 = 1$,

$$w_1 E_1(X - \mu_s)^2 = w_1 E_1(X - w_1\mu_1 - w_2\mu_2)^2$$
$$= w_1 E_1[(X - \mu_1) + w_2(\mu_1 - \mu_2)]^2$$
$$= w_1 E_1(X - \mu_1)^2 + w_1 w_2^2(\mu_1 - \mu_2)^2$$
$$= w_1\sigma_1^2 + w_1 w_2^2(\mu_1 - \mu_2)^2.$$

By similar reasoning,

$$w_2 E_2(X - \mu_s)^2 = w_2\sigma_2^2 + w_1^2 w_2(\mu_1 - \mu_2)^2.$$

Thus,

$$\sigma_s^2 = w_1 E_1(X - \mu_s)^2 + w_2 E_2(X - \mu_s)^2$$
$$= w_1\sigma_1^2 + w_1 w_2^2(\mu_1 - \mu_2)^2 + w_2\sigma_2^2 + w_1^2 w_2(\mu_1 - \mu_2)^2$$
$$= w_1\sigma_1^2 + w_2\sigma_2^2 + (w_1 + w_2)w_1 w_2(\mu_1 - \mu_2)^2$$
$$= w_1\sigma_1^2 + w_2\sigma_2^2 + w_1 w_2(\mu_1 - \mu_2)^2.$$

4.2. To prove that Angoff's $\mu_s(X)$ gives results identical to equation (4.17), note that $\mu_s(V) = w_1\mu_1(V) + w_2\mu_2(V)$, and recall that $w_1 + w_2 = 1$. Therefore, Angoff's $\mu_s(X)$ is

$$\mu_s(X) = \mu_1(X) + \alpha_1(X|V)[w_1\mu_1(V) + w_2\mu_2(V) - \mu_1(V)]$$
$$= \mu_1(X) + \alpha_1(X|V)[-w_2\mu_1(V) + w_2\mu_2(V)]$$
$$= \mu_1(X) - w_2\alpha_1(X|V)[\mu_1(V) - \mu_2(V)],$$

which is equation (4.17) since $\gamma_1 = \alpha_1(X|V)$.

To prove that Angoff's $\sigma_s^2(X)$ gives results identical to equation (4.19), note that

$$\sigma_s^2(V) = w_1\sigma_1^2(V) + w_2\sigma_2^2(V) + w_1w_2[\mu_1(V) - \mu_2(V)]^2.$$

(This result is analogous to the result proved in Exercise 4.1.) Therefore, Angoff's $\sigma_s^2(X)$ is

$$\sigma_s^2(X) = \sigma_1^2(X) + \alpha_1^2(X|V)\{w_1\sigma_1^2(V) + w_2\sigma_2^2(V)$$
$$+ w_1w_2[\mu_1(V) - \mu_2(V)]^2 - \sigma_1^2(V)]\}$$
$$= \sigma_1^2(X) + \alpha_1^2(X|V)[-w_2\sigma_1^2(V) + w_2\sigma_2^2(V)]$$
$$+ w_1w_2\alpha_1^2(X|V)[\mu_1(V) - \mu_2(V)]^2$$
$$= \sigma_1^2(X) - w_2\alpha_1^2(X|V)[\sigma_1^2(V) - \sigma_2^2(V)] + w_1w_2\alpha_1^2(X|V)[\mu_1(V) - \mu_2(V)]^2,$$

which is equation (4.19) since $\gamma_1 = \alpha_1(X|V)$. Similar proofs can be provided for $\mu_s(Y)$ and $\sigma_s^2(Y)$.

4.4. The Tucker results are the same as those provided in the third row of Table 4.4. For the Levine method, using equations (4.58) and (4.59), respectively,

$$\gamma_1 = \frac{6.5278^2 + 13.4088}{2.3760^2 + 13.4088} = 2.9401$$

$$\gamma_2 = \frac{6.8784^2 + 14.7603}{2.4515^2 + 14.7603} = 2.9886.$$

Note that

$$\mu_1(V) - \mu_2(V) = 5.1063 - 5.8626 = -.7563 \qquad \text{and}$$
$$\sigma_1^2(V) - \sigma_2^2(V) = 2.3760^2 - 2.4515^2 = -.3645.$$

Therefore, equations (4.17)–(4.20) give

$$\mu_s(X) = 15.8205 - .5(2.9401)(-.7563) = 16.9323$$
$$\mu_s(Y) = 18.6728 + .5(2.9886)(-.7563) = 17.5427$$
$$\sigma_s^2(X) = 6.5278^2 - .5(2.9401^2)(-.3645) + .25(2.9401^2)(-.7563^2) = 45.4237$$
$$\sigma_s^2(Y) = 6.8784^2 + .5(2.9886^2)(-.3645) + .25(2.9886^2)(-.7563^2) = 46.9618.$$

Using equation (4.1),

$$l_{Ys}(x) = \sqrt{46.9618/45.4237}\,(x - 16.9323) + 17.5427 = .33 + 1.02x.$$

4.5. Using the formula in Table 4.1,

$$\rho_1(X, X') = \frac{\gamma_1^2[\sigma_1(X, V) - \sigma_1^2(V)]}{(\gamma_1 - 1)\sigma_1^2(X)},$$

where $\gamma_1 = \sigma_1^2(X)/\sigma_1(X, V)$. For the illustrative example,

$$\gamma_1 = 6.5278^2/13.4088 = 3.1779 \qquad \text{and}$$

$$\rho_1(X, X') = \frac{3.1779^2(13.4088 - 2.3760^2)}{(3.1779 - 1)6.5278^2} = .845.$$

Similarly,

$$\rho_2(Y, Y') = \frac{\gamma_2^2[\sigma_2(Y, V) - \sigma_2^2(V)]}{(\gamma_2 - 1)\sigma_2^2(Y)},$$

where $\gamma_2 = \sigma_2^2(Y)/\sigma_2(Y, V)$. For the illustrative example,

$$\gamma_2 = 6.8784^2/14.7603 = 3.2054$$

$$\rho_2(Y, Y') = \frac{3.2054^2(14.7603 - 2.4515^2)}{(3.2054 - 1)6.8784^2} = .862.$$

4.6.a. From equation (4.38), the most general equation for γ_1 is $\gamma_1 = \sigma_1(T_X)/\sigma_1(T_V)$. It follows that

$$\gamma_1 = \frac{(K_X/K_V)\sigma_1(T_V)}{\sigma_1(T_V)} = \frac{K_X}{K_V}.$$

Similarly, $\gamma_2 = K_Y/K_V$.

4.6.b. Under the classical model, the γs are ratios of actual test lengths; whereas under the classical congeneric model, the γs are ratios of effective test lengths.

4.7.   All of it [see equation 4.82].

4.8.   No, it is not good practice from the perspective of equating alternate forms. All other things being equal, using more highly discriminating items will cause the variance for the new form to be larger than the variance for previous forms. Consequently, form differences likely will be a large percent of the observed differences in variances, and equating becomes more suspect as forms become more different in their statistical characteristics. These and related issues are discussed in more depth in Chapter 8.

4.9.   From equation (4.59),

$$\gamma_2 = \frac{\sigma_2^2(Y) + \sigma_2(Y, V)}{\sigma_2^2(V) + \sigma_2(Y, V)}.$$

Recall that, since γ_2 is for an external anchor, $\sigma_2(E_Y, E_V) = 0$. Replacing the quantities in equation (4.59) with the corresponding expressions in equation set (4.70) gives

$$\gamma_2 = \frac{[\lambda_Y^2 \sigma_2^2(T) + \lambda_Y \sigma_2^2(E)] + \lambda_Y \lambda_V \sigma_2^2(T)}{[\lambda_V^2 \sigma_2^2(T) + \lambda_V \sigma_2^2(E)] + \lambda_Y \lambda_V \sigma_2^2(T)}$$

$$= \frac{\lambda_Y[(\lambda_Y + \lambda_V)\sigma_2^2(T) + \sigma_2^2(E)]}{\lambda_V[(\lambda_V + \lambda_Y)\sigma_2^2(T) + \sigma_2^2(E)]}$$

$$= \lambda_Y/\lambda_V.$$

4.10.a. Since $X = A + V$,

$$\sigma_1(X, V) = \sigma_1(A + V, V) = \sigma_1^2(V) + \sigma_1(A, V).$$

The assumption that $\rho_1(X, V) > 0$ implies that $\sigma_1(X, V) > 0$. Since $\sigma_1^2(V) \geq 0$ by definition, the above equation leads to the conclusion that $\sigma_1(A, V) > 0$ and, therefore, $\sigma_1^2(V) < \sigma_1(X, V)$. Also,

$$\begin{aligned} \sigma_1^2(X) &= \sigma_1(A + V, A + V) = \sigma_1^2(A) + \sigma_1^2(V) + 2\sigma_1(A, V) \\ &= [\sigma_1^2(V) + \sigma_1(A, V)] + [\sigma_1^2(A) + \sigma_1(A, V)] \\ &= \sigma_1(X, V) + [\sigma_1^2(A) + \sigma_1(A, V)]. \end{aligned}$$

Since $\sigma_1^2(A) \geq 0$ by definition and it has been shown that $\sigma_1(A, V) > 0$, it necessarily follows that $\sigma_1(X, V) < \sigma_1^2(X)$. Consequently, $\sigma_1^2(V) < \sigma_1(X, V) < \sigma_1^2(X)$.

4.10.b. $\gamma_{1T} = \sigma_1(X, V)/\sigma_1^2(V)$, which must be greater than 1 because $\sigma_1(X, V) > \sigma_1^2(V)$. Now, $\gamma_{1L} = \sigma_1^2(X)/\sigma_1(X, V)$. To show that $\gamma_{1T} < \gamma_{1L}$, it must be shown that

$$\sigma_1(X, V)/\sigma_1^2(V) < \sigma_1^2(X)/\sigma_1(X, V) \qquad \text{or}$$

$$\sigma_1^2(X, V) < \sigma_1^2(X)\sigma_1^2(V) \qquad \text{or} \qquad \left[\frac{\sigma_1(X, V)}{\sigma_1(X)\sigma_1(V)}\right]^2 < 1,$$

which must be true because the term in brackets is $\rho_1(X, V)$, which is less than 1 by assumption.

4.10.c. Suppose that V and X measure the same construct and both satisfy the classical test theory model. If V is longer than X, then $\sigma^2(V) > \sigma^2(X)$. This, of course, cannot occur with an internal set of common items because V can be no longer than X.

Chapter 5

5.1. See Table A.3.

5.2. See Table A.4.

5.3. See Table A.5.

5.4. For the Tucker method, the means and standard deviations for the synthetic group for Form X are 2.5606 and 1.4331, and for Form Y they are 2.4288 and 1.4261. The linear equation for the Tucker method is $l(x) = .9951x - .1192$. For the Braun-Holland method, the means and standard deviations for the synthetic group for Form X are 2.5525 and 1.4482, and for Form Y they are 2.4400 and 1.4531. The linear equation for the Braun-Holland method is $l(x) = 1.0034x - .1211$.

5.5. For X, V in Population 1, linear *regression slope* $= .6058$, and linear *regression intercept* $= 1.6519$. The means of X given V for $v = 0, 1, 2, 3$ are 1.9, 2.125, 2.65, 3.7. The residual means for $v = 0, 1, 2, 3$ are .2481, $-.1327$, $-.2135$, and

Table A.3. Conditional Distributions of Form X
Given Common-Item Scores for Population 1 in
Exercise 5.1.

	v			
x	0	1	2	3
0	.20	.10	.10	.00
1	.20	.20	.10	.05
2	.30	.30	.25	.10
3	.15	.30	.25	.25
4	.10	.075	.20	.30
5	.05	.025	.10	.30
$h_1(v)$	.20	.40	.20	.20

Table A.4. Calculation of Distribution of Form X and Common-Item Scores for
Population 1 Using Frequency Estimation Assumptions in Exercise 5.2.

	v					
x	0	1	2	3	$f_2(x)$	$F_2(x)$
0	.20(.20) = .04	.10(.20) = .02	.10(.40) = .04	.00(.20) = .00	.10	.10
1	.20(.20) = .04	.20(.20) = .04	.10(.40) = .04	.05(.20) = .01	.13	.23
2	.30(.20) = .06	.30(.20) = .06	.25(.40) = .10	.10(.20) = .02	.24	.47
3	.15(.20) = .03	.30(.20) = .06	.25(.40) = .10	.25(.20) = .05	.24	.71
4	.10(.20) = .02	.075(.20) = .015	.20(.40) = .08	.30(.20) = .06	.175	.885
5	.05(.20) = .01	.025(.20) = .005	.10(.40) = .04	.30(.20) = .06	.115	1.00
$h_2(v)$	.20	.20	.40	.20		

Table A.5. Cumulative Distributions and Finding Equipercentile
Equivalents for $w_1 = .5$ in Exercise 5.3.

x	$F_s(x)$	$P_s(x)$	y	$G_s(y)$	$Q_s(y)$	x	$e_{Ys}(x)$
0	.1000	5.00	0	.0925	4.62	0	.04
1	.2400	17.00	1	.3000	19.62	1	.87
2	.4850	36.25	2	.5150	40.75	2	1.79
3	.7300	60.75	3	.7525	63.38	3	2.89
4	.8925	81.12	4	.9000	82.62	4	3.90
5	1.0000	94.62	5	1.0000	95.00	5	4.96

.2308. Because the residuals tend to be negative in the middle and positive at the ends, the regression of X on V for Population 1 appears to be nonlinear. Similarly, for Population 2, the mean residuals for the regression of Y on V are .2385, $-.1231$, $-.2346$, .3538, also suggesting nonlinear regression. This nonlinearity of regression would likely cause the Tucker and Braun-Holland methods to differ.

5.6. For $x = 1$; $P_1(x = 1) = 17.50$; 17.5th percentile for V in Population $1 = .375$; Percentile Rank of $v = .375$ in Population $2 = 17.5$; $Q_2^{-1}(17.5) = .975$. Thus, $x = 1$ is equivalent to $y = .975$ using chained equipercentile. For $x = 3$; $P_1(x = 3) = 62.50$; 62.5th percentile for V in Population $1 = 1.625$; Percentile Rank of $v = 1.625$ in Population $2 = 45$; $Q_2^{-1}(45) = 2.273$. Thus, $x = 3$ is equivalent to $y = 2.273$ using the chained equipercentile method.

Chapter 6

6.1. For the first item, using equation (6.1),

$$p_{ij} = .10 + (1 - .10)\frac{\exp[1.7(1.30)(.5 - -1.30)]}{1 + \exp[1.7(1.30)(.5 - -1.30)]} = .9835.$$

For the two other items, $p_{ij} = .7082$, and .3763.

6.2. For $\theta_I = .5$, $f(x = 0) = .0030$; $f(x = 1) = .1881$; $f(x = 2) = .5468$; $f(x = 3) = .2621$.

6.3.a. From equation (6.4), $b_{Jj} = Ab_{Ij} + B$ and $b_{Jj^*} = Ab_{Ij^*} + B$. Subtract the second equation from the first to get $b_{Jj} - b_{Jj^*} = A(b_{Jj} - b_{Jj^*})$, which implies that $A = (b_{Jj} - b_{Jj^*})/(b_{Ij} - b_{Ij^*})$.

6.3.b. From equation (6.3), $a_{Jj} = a_{Ij}/A$. Solving for A, $A = a_{Ij}/a_{Jj}$.

6.3.c. From equation (6.4), $b_{Jj} = Ab_{Ij} + B$. Taking the variance over items (j), $\sigma^2(b_J) = A^2\sigma^2(b_I)$. Solving for A and recognizing that variances must be positive, $A = \sigma(b_J)/\sigma(b_I)$.

6.3.d. From Exercise 6.3b., $A = a_{Ij}/a_{Jj}$. Taking the expectation, over items (j), $A = \mu(a_I)/\mu(a_J)$.

6.4. For $\theta_{Ii} = -2.00$, the value of the test characteristic curve is $.26 + .27 + .18 = .71$; at the other abilities, it is 2.07, 2.44, .71, and 2.44.

6.5. See Table A.6.

6.6. See Table A.7, which was constructed from Table 6.4.

6.7. See Table A.8.

6.8. Equating to a particular old form allows the use of traditional methods as a check. The traditional methods are based on different assumptions than the IRT methods, which allows for a comparison of how robust the equating is to the assumptions used. In addition, when equating to a particular old form,

Table A.6. IRT Observed Score Equating Answer to Exercise 6.5.

Probability of Correct Answers and True Scores

θ_i	Item					τ
	$j = 1$	$j = 2$	$j = 3$	$j = 4$	$j = 5$	
Form X						
-1.0000	.7370	.6000	.2836	.2531	.2133	2.0871
.0000	.8799	.9079	.4032	.2825	.2678	2.7414
1.0000	.9521	.9867	.6881	.4965	.4690	3.5925
Form Y						
-1.0000	.7156	.6757	.2791	.2686	.2074	2.1464
.0000	.8851	.8773	.6000	.3288	.2456	2.9368
1.0000	.9611	.9642	.9209	.5137	.4255	3.7855

Form X Distribution

x	$f(x\mid\theta = -1)$	$f(x\mid\theta = 0)$	$f(x\mid\theta = 1)$	$f(x)$	$F(x)$	$P(x)$
0	.0443	.0035	.0001	.0159	.0159	.7966
1	.2351	.0646	.0052	.1016	.1175	6.6734
2	.3925	.3383	.0989	.2766	.3941	25.5831
3	.2524	.3990	.3443	.3319	.7260	56.0064
4	.0690	.1704	.4009	.2134	.9394	83.2720
5	.0068	.0244	.1506	.0606	1.0000	96.9718

Form Y Distribution

y	$g(y\mid\theta = -1)$	$g(y\mid\theta = 0)$	$g(y\mid\theta = 1)$	$g(y)$	$G(y)$	$Q(y)$
0	.0385	.0029	.0000	.0138	.0138	.6905
1	.2165	.0490	.0020	.0892	.1030	5.8393
2	.3953	.2594	.0425	.2324	.3354	21.9178
3	.2670	.4235	.3100	.3335	.6688	50.2114
4	.0752	.2276	.4589	.2539	.9228	79.5807
5	.0075	.0376	.1866	.0772	1.0000	96.1384

Form Y Equivalents of Form X scores

x	$e_Y(x)$
0	.0772
1	1.0936
2	2.1577
3	3.1738
4	4.1454
5	5.1079

Table A.7. Answer to Exercise 6.6.

r	x	$f_r(x)$ for $r \leq 4$		Probability	
4	0	$f_4(0) = f_3(0)(1-p_4)$		$= .4430(1-.4)$	$= .2658$
	1	$f_4(1) = f_3(1)(1-p_4) + f_3(0)p_4$		$= .4167(1-.4) + .4430(.4)$	$= .4272$
	2	$f_4(2) = f_3(2)(1-p_4) + f_3(1)p_4$		$= .1277(1-.4) + .4167(.4)$	$= .2433$
	3	$f_4(3) = f_3(3)(1-p_4) + f_3(2)p_4$		$= .0126(1-.4) + .1277(.4)$	$= .0586$
	4	$f_4(4) =$	$f_3(3)p_4 =$		$.0126(.4) = .0050$

Table A.8. Estimated Probability of Correct Response
Given $\theta = 1$ for Exercise 6.7.

Item	Scale J	Mean/sigma	Mean/mean
1	.9040	.8526	.8522
2	.8366	.8076	.8055
3	.2390	.2233	.2222
sum	1.9796	1.8835	1.8799
$H diff$		.0037	.0039
$SL diff$		.0092	.0099

the common items provide direct evidence about how the new group compares to the old group for two groups of examinees that actually can be observed. In IRT equating to a calibrated pool, the only group of examinees
who takes all of the common items is the new group. Thus, when equating to
a pool, there is no old group with which to compare the new group on the
common items, unless we rely on the assumptions of the IRT model, which is
a much weaker comparison than can be made when we have two groups who
actually took the common items.

6.9. Step (a) is similar, except that, with IRT, a design might be selected that involves linking to an IRT calibrated item pool. Step (b) is the same, in that the
same construction, administration, and scoring procedures could be used for
either type of equating method. In Step (c), IRT equating involves estimating
item parameters and scaling the item parameter estimates. These steps are
not needed in the traditional methods. In both types of methods, the raw
scores are converted to scale scores by using statistical methods. However,
traditional methods differ from the IRT methods. Also, the IRT methods
might involve equating using an item pool. Steps (d), (e), and (f) are the same
for the two types of methods.

Chapter 7

7.1. Answers to 7.1.a, 7.1.b, and 7.1.c are given in Table A.9. Using equation (7.10)
for Exercise 7.1.d, the standard error at $x = 3$ is 1.3467. The standard error at
$x = 5$ is 1.4291.

Table A.9. Bootstrap Standard Errors for Exercise 7.1a, b, and c.

Statistic	Sample				$\widehat{se}_{boot}$
	1	2	3	4	
$\hat{\mu}(X)$	4.0000	2.7500	4.2500	3.2500	
$\hat{\mu}(Y)$	3.0000	4.6667	3.6667	2.0000	
$\hat{\sigma}(X)$	2.1213	2.0463	1.9203	2.2776	
$\hat{\sigma}(Y)$	1.4142	.4714	1.8856	1.4142	
$\hat{l}_Y(x = 3)$	2.3333	4.7243	2.4392	1.8448	1.2856
$\hat{l}_Y(x = 5)$	3.6667	5.1850	4.4031	3.0866	.9098
$sc[\hat{l}_Y(x = 3)]$	10.9333	11.8897	10.9757	10.7379	.5142
$sc[\hat{l}_Y(x = 5)]$	11.4667	12.0740	11.7613	11.2346	.3639
$sc_{int}[\hat{l}_Y(x = 3)]$	11	12	11	11	.5000
$sc_{int}[\hat{l}_Y(x = 5)]$	11	12	12	11	.5774

7.2. Using equation (7.12),

$$\hat{var}[\hat{e}_Y(x_i)] \cong \frac{1}{[.7418 - .7100]^2} \left\{ \frac{(72.68/100)(1 - 72.68/100)(4329 + 4152)}{4329(4152)} \right.$$

$$\left. - \frac{(.7418 - 72.68/100)(72.68/100 - .7100)}{4329(.7418 - .7100)} \right\} = .09084.$$

Estimated standard error equals $\sqrt{.09084} = .3014$. Using equation (7.13),

$$\hat{var}[\hat{e}_Y(x_i)] \cong 8.9393^2 \frac{(72.68/100)(1 - 72.68/100)}{.33^2} \left(\frac{1}{4329} + \frac{1}{4152} \right) = .0687.$$

Estimated standard error equals $\sqrt{.0687} = .2621$. The differences between the standard errors could be caused by the distributions not being normal. Also, equation (7.12) assumes discrete distributions, whereas equation (7.13) assumes continuous distributions. Differences also could result from error in estimating the standard errors.

7.3. a. 150 total (75 per form). b. 228 total (114 per form). c. If the relationship was truly linear, it would be best to use linear, because linear has less random error.

7.4. Using equation (7.11), with a sample size of 100 per form, the error variance for linear equating equals .03, and the error variance for equipercentile equals .0456. The squared bias for linear is $(1.3 - 1.2)^2 = .01$. Thus, the mean squared error for linear is $.03 + .01 = .04$. Assuming no bias for equipercentile, the mean squared error for equipercentile $= .0456$. Therefore, linear leads to less error than equipercentile. With a sample size of 1000 per form, the mean squared error for linear is .013 and the mean squared for equipercentile is .0046. With a sample size of 1000, equipercentile leads to less error than linear. Thus, it appears that linear equating requires smaller sample sizes than equipercentile equating.

7.5. a. .2629 and .4382. b. .1351 and .2683. c. .3264 and .6993. d. 96 per form and 267 per form.

7.6. The identity equating does not require any estimation. Thus, the standard error for the identity equating is 0. If the population equating is similar to the identity equating, then the identity equating might be best. Otherwise, the identity equating can contain substantial systematic error (which is not reflected in the standard error). Thus, the identity equating is most attractive when the sample size is small or when there is reason to believe that the alternate forms are very similar.

Chapter 8

8.1.a. From equation (7.18), a sample size of more than $N_{tot} = (2/.1^2)(2 + .5^2) = $ 450 total (225 per form) would be needed.

8.1.b. From equation (7.18), a sample size of more than $N_{tot} = (2/.2^2)(2 + .5^2) = $ 112.5 total (approx. 67 per form) would be needed.

8.1.c. In a situation where a single passing score is used, the passing score is at a z-score of .5, and the equating relationship is linear in the population.

8.2.a. For Forms D and following: In even-numbered years, the spring form links to the previous spring form and the fall form links to the previous spring form. In odd-numbered years, the spring form links to the previous fall and the fall form links to the previous fall.

8.2.b. Form K links to Form I. Form L links to Form I. Form M links to Form L. Form N links to Form L.

8.3.a. For Forms D and following in Modified Plan 1 (changes from Link Plan 4 shown in bold italics): In even-numbered years, the spring form links to the previous spring form and the fall form links to the previous spring form. In odd-numbered years, *the spring form links to the fall form from two years earlier* and the fall form links to the previous fall.

For Forms D and following in Modified Plan 2: In even-numbered years, the spring form links to the previous spring form and the fall form links to the previous spring form. In odd-numbered years, *the spring form links to the previous spring* and the fall form links to the previous fall.

8.3.b. In Modified Plan 1, K links to I, L links to I, *M links to J*, and N links to L. In Modified Plan 2, K links to I, L links to I, *M links to K*, and N links to L.

8.3.c. For Modified Plan 1, Rule 1 is violated (this plan results in equating strains), and Rules 2 through 4 are met as well with this plan as with Single Link Plan 4. For Modified Plan 2, Rule 1 is achieved much better than for Modified Plan 1, Rule 2 is met better than for Single Link Plan 4 or for Modified Plan 1, and Rules 3 and 4 are met as well as for Modified Plan 1 or Single Link Plan 4. Modified Plan 2 seems to be the best of the two modified plans.

8.4. In Table 8.6, for the first 4 years the decrease in mean and increase in standard deviation were accompanied by an increase in the sample size. However, now in year 5 there is a decrease in the sample size. The Levine method results are most similar to the results when the sample size was near 1050 in year 2. For this reason, the Levine method might be considered to be preferable. However, the choice between methods is much more difficult in this situation, because a sample size decrease never happened previously. In practice, many additional issues would need to be considered.

8.5.a. Randomly assign examinees to the two modes. Convert parameter estimates for the computerized version to the base IRT scale using the random groups design. Probably two different classrooms would be needed, one for paper and pencil and one for computer.

8.5.b. Use the items that are in common between the two modes as common items in the common-item equating to an item pool design.

8.5.c. Random groups requires large sample sizes and a way to randomly assign examinees to different modes of testing. Common-item equating to an item pool requires that the common items behave the same on computerized and paper and pencil versions. This requirement likely would not be met. This design also requires that the groups taking the computerized and paper and pencil versions be reasonably similar in achievement level.

8.5.d. It is unlikely that all items will behave the same when administered by computer as when administered using paper and pencil. Therefore, the results from using this design would be suspect. At a minimum, a study should be conducted to discover the extent to which context effects affect the performance of the items.

8.5.e. The random groups design is preferable. Even with this design, it would be necessary to study whether or not the construct being measured by the test changes from a paper and pencil to a computerized mode. For example, there is evidence that reading tests with long reading passages can be affected greatly when they are adapted for computer administration. Note that with the random groups design, the effects of computerization could be studied for those items that had been previously administered in the paper and pencil mode.

8.6. Some causes due to changes in items include changes in item position, changes in surrounding items, changes in font, changes in wording, and rearranging alternatives. Some causes due to changes in examinees include changes in a field of study and changes in the composition of the examinee groups. For example, changes in country names, changes in laws, and new scientific discoveries might lead to changes in the functioning of an item. As another example, a vocabulary word like "exorcist" might become much more familiar after the release of a movie of the same name. Some causes due to changes in administration conditions include changes in time given to take the test, security breaches, changes in mode of administration, changes in test content, changes in test length, changes in motivation conditions, changes in calculator usage, and changes in directions given to examinees.

8.7. To consider equating, the forms must be built to the same content and statistical specifications. Assuming that they are, the single group design is eliminated because it would require that two forms be administered to each examinee, which would be difficult during an operational administration. The common-item nonequivalent groups design is eliminated because having many items associated with each reading passage would make it impossible to construct a content representative set of common items. The random groups design could be used. This design requires large samples, which would not be a problem in this example. Also, the random groups design is not affected by context, fatigue, and practice effects, and the only statistical assumption that it requires is that the process used to randomly assign forms was effective. Therefore, the random groups design is best in this situation. Equipercentile equating would be preferred because it generally provides more accuracy along the score scale (assuming that the relationship is not truly linear). Equipercentile equating also requires large sample sizes, which is not a problem in the situation described.

Appendix B: Computer Programs and Data Sets

We will make available some of the data sets used in this book to interested persons. We will also make available Macintosh computer programs, and associated documentation, that can be used to conduct many of the analyses in the book, free of charge. The data sets and programs can be found at the web address http://www.uiowa.edu/~itp/ or obtained from Michael Kolen, Iowa Testing Programs, 224C Lindquist Center, University of Iowa, Iowa City, IA 52242. Email: Michael-Kolen@uiowa.edu.

Data Sets

The data sets that we will send are as follows:

1. Frequency distributions of Form X and Form Y scores for the random groups examples used in the illustrative examples in Chapters 2 and 3.
2. Scored (0/1) data for Form X and Form Y for the data set used in the illustrative examples in Chapters 4 through 6. Each record contains a string of 36 item (0/1) scores, the total score, and the score on the common items.

Computer Programs

The following Macintosh computer application programs, along with associated documentation, also will be included:

1. **RAGE** by L. Zeng, M.J. Kolen, and B.A. Hanson. This program conducts linear and equipercentile equating as described in Chapter 2. The

program implements the cubic spline smoothing method described in Chapter 3.

2. **RG Equate** by B.A. Hanson. This program conducts equipercentile equating with log-linear smoothing, as described in Chapter 3.

3. **Usmooth** by B.A. Hanson. This program smoothes test score distributions using the log-linear smoothing method, as described in Chapter 3.

4. **CIPE** by M.J. Kolen. This program conducts observed score equating under the common-item nonequivalent groups design as described in Chapters 4 and 5. Tucker linear (external or internal common items), Levine linear observed score (internal common items only), and frequency estimation equipercentile equating with cubic spline smoothing are implemented.

5. **ST** by L. Zeng and B.A. Hanson. This program conducts IRT scale transformations using the mean/mean, mean/sigma, Stocking and Lord, and Haebara methods described in Chapter 6.

6. **PIE** by B.A. Hanson and L. Zeng. This program conducts IRT true and observed score equating using the methods described in Chapter 6.

7. **Equating Error** by B.A. Hanson. This program estimates bootstrap standard errors of equipercentile equating for the random groups design. Standard errors for both the cubic spline postsmoothing and log-linear presmoothing methods can be calculated.

Although these programs have been tested and we believe them to be free of errors, we do not warrant, guarantee, or make any representations regarding the use or the results of this software in terms of their appropriateness, correctness, accuracy, reliability, or otherwise. The entire responsibility for the use of this software rests with the user.

Index

Springer Series in Statistics

(continued from p. ii)